Fundamental Theories of Physics

Volume 183

The international monograph series "Fundamental Theories of Physics" aims to stretch the boundaries of mainstream physics by clarifying and developing the theoretical and conceptual framework of physics and by applying it to a wide range of interdisciplinary scientific fields. Original contributions in well-established fields such as Quantum Physics, Relativity Theory, Cosmology, Quantum Field Theory, Statistical Mechanics and Nonlinear Dynamics are welcome. The series also provides a forum for non-conventional approaches to these fields. Publications should present new and promising ideas, with prospects for their further development, and carefully show how they connect to conventional views of the topic. Although the aim of this series is to go beyond established mainstream physics, a high profile and open-minded Editorial Board will evaluate all contributions carefully to ensure a high scientific standard.

More information about this series at http://www.springer.com/series/6001

Torsten Asselmeyer-Maluga
Editor

At the Frontier of Spacetime

Scalar-Tensor Theory, Bells Inequality,
Machs Principle, Exotic Smoothness

 Springer

Editor
Torsten Asselmeyer-Maluga
German Aerospace Center
Berlin
Germany

ISSN 0168-1222   ISSN 2365-6425   (electronic)
Fundamental Theories of Physics
ISBN 978-3-319-81008-9   ISBN 978-3-319-31299-6   (eBook)
DOI 10.1007/978-3-319-31299-6

This Springer imprint is published by Springer Nature
The registered company is Springer International Publishing AG Switzerland

To Carl

Preface

In 2015, we celebrated the 100th anniversary of the development of general relativity theory (GRT). Einstein presented his theory at the Prussian Academy of Science in Berlin on November 25th, 1915. In GRT, he replaced the absolute space and time of Newton in favor of a changing arena called "spacetime," in which gravity appeared as curvature. The equivalence principle linked every acceleration locally with gravitation. In principle, GRT poses the possibility of understanding all forces in the world using geometry. Galileo Galilei expressed this thought nearly 400 years ago when he pronounced: *He who understands geometry, understands anything in the world*. Therefore it was logical that Einstein continued this program even after completing his GRT, with the development of proposals for a unified field theory.

Carl H. Brans chose to investigate such theories for his undergraduate thesis at Loyola University in New Orleans. It was the beginning of a lifelong engagement with GRT. Even the mathematical beauty of GRT and the unified field theory attracted him. As a 10-year-old boy, he taught himself differential and integral calculus, and difficult books on mathematics and mathematical physics held a great appeal for him. His preference for GRT was a little bit unusual at the time. Since the 1920s, the GRT had lost its prominent role in theoretical physics. The advent of quantum mechanics and elementary particle physics, together with new work on quantum electrodynamics, inspired more interest among physicists in those days.

In the 1950s, the situation began to change. John A. Wheeler, at Princeton University, started to develop geometrodynamics as a new representation of GRT. At the same time, Wheeler established his by-now famous working group, which focused on problems in GRT and on the foundations of quantum mechanics. In parallel, Robert Dicke, Wheeler's colleague at Princeton, began to work on the experimental problems in GRT during his sabbatical year of 1954. He also became interested in Mach's principle, which Einstein had used as a guide during the development of GRT. Dicke considered Mach's principle to imply: *The gravitational constant, κ, should be a function of the mass distribution in the universe*. Paul A.M. Dirac had earlier conjectured that there is a relation between the coupling constant of gravity and the mass and radius of the universe (now known as Dirac's "large number conjecture"). For an expanding universe, one thereby obtains a variable gravitational coupling constant!

In 1957, Carl Brans arrived in Princeton to undertake his graduate study and Ph.D. thesis. In his contribution to this book, he writes extensively about that time. He heard some lectures and visited the seminar of Wheeler, who had established his famous group. Charles Misner, who had recently completed his Ph.D. thesis, introduced Carl to fiber bundle theory. Hence, Carl planned to write his Ph.D. thesis about the application of fiber bundles in physics. At that time, he also began to be interested in the mathematical structure of spacetime. But the time was not yet ripe for these ideas; fiber bundles would only become commonly used in theoretical physics in the 1970s. Instead, Misner recommended that Carl should contact Dicke, who was searching for a theoretical physicist. It was the beginning of a lifelong and fruitful collaboration.

Mach's principle and Dirac's large-numbers hypothesis formed the basis for the discussions between Dicke and Brans. They wondered if they could create a version of GRT with a variable gravitational coupling. Brans pursued the idea and developed it in his Ph.D. thesis in 1961. Today this renowned theory is known as the Brans-Dicke theory. They introduced a scalar field to represent the variable coupling. Pascual Jordan had described a similar theory in his 1955 book, *Schwerkraft und Weltall*, though Jordan's work was not well known at the time. Brans and Dicke's work quickly received much more attention within the physics community, helping to establish the importance of "scalar-tensor" theories of gravitation, as Carl describes further in his contribution to this volume.

In the Brans-Dicke theory, one arbitrary parameter (usually denoted by ω) quantifies the coupling between the scalar field and spacetime curvature. Dicke proposed to express ω in terms of other physical constants; failing that, most experimental tests of the theory concentrated on possible restrictions on ω. An outstanding experimenter, Dicke was strongly interested in the experimental verification of the Brans-Dicke theory. As an important side-effect of these efforts, many effects of GRT were tested with unprecedented precision. Among them included classic experiments like the Eötvös experiment to confirm the weak equivalence principle, as well as various NASA missions. Martin McHugh's contribution in this volume presents an overview of these experiments, as well as Dicke's endeavor to confirm the Brans-Dicke theory.

In 1961, Brans and Dicke's paper appeared in the *Physical Review*. Following its publication, the Brans-Dicke theory had wide repercussions. The meaning and importance of scalar fields in physics increased significantly, from their role in spontaneous symmetry breaking, as in the Higgs mechanism, to the dynamics of the very early universe, as in models of cosmic inflation. Other theories, which incorporated a scalar field to model variable cosmological effects, such as quintessence, used Brans-Dicke theory as a prototype. Interest in Brans-Dicke theory increased further during the 1980s and 1990s in the context of string theory. Finally, the discovery of the Higgs boson in 2012 marked the first experimental detection of a fundamental scalar field in nature.

The present volume includes a collection of invited papers by renowned colleagues. The contributions range over various aspects of scalar fields to Mach's principle, Bell's inequality, and spacetime structure. Together, the chapters illustrate how Carl's ideas have been developed even further over the years. The volume is organized into three parts, reflecting the scientific foci of Carl's career.

The first part concerns the scalar-tensor theory. In the decades since the development of Brans-Dicke theory, scalar fields have come to play a diverse set of roles in physics, from the inflaton that drove cosmic inflation, to the axion that breaks chiral symmetry in QCD, to the Higgs boson that generates mass for elementary particles and the dilaton field that breaks global scale invariance (Weyl symmetry). Chapters in this part focus on this diversity of scalar fields in the context of GRT.

David Kaiser (MIT, USA) describes the role of Brans-Dicke (or non-minimal) couplings between scalar fields and spacetime curvature in the context of inflationary model-building. As he discusses, recent observational data, such as collected by the *Planck* satellite, place strong constraints on models of early-universe inflation. Models with Brans-Dicke couplings provide a natural way of realizing inflation while matching all the latest observations. Yasunori Fujii (Waseda University, Japan) focuses on a possible relation between microscopic physics and the cosmological model of Brans and Dicke. According to Brans-Dicke theory, the mass of an electron would not be constant in an expanding universe. However, Fujii demonstrates, one may introduce a massive scalar field (akin to a dilaton) to address this feature, and further estimate the dilaton mass. Roman Jackiw (MIT, USA) and So-Young Pi (Boston University, USA) focus on a special version of Brans-Dicke theory which is independent of the underlying scale (Weyl symmetry), which should affect short-scale behavior.

The appearance of different scalar fields naturally leads to the question of how those fields might relate or interact with each other. Friedrich Hehl (University of Cologne, Germany, and University of Missouri-Columbia, USA) addresses such questions. First he shows that the dilaton and axion fields appear naturally in the context of Einstein−Cartan theory. Next he constructs the metric as well as the axion and dilaton fields directly from an electromagnetic model of the universe ("premetric electrodynamics").

Many researchers have implicitly assumed that Brans-Dicke theory would yield small deviations from the usual predictions of GRT. But what about more radical departures, such as contributions that are quadratic in the curvature? This question is discussed by Tirthabir Biswas (Loyola University New Orleans, USA) in collaboration with Alexey Koshelev (Universidade da Beira Interior, Portugal) and Anupam Mazumdar (Lancaster University, UK). They demonstrate the appearance of the Brans-Dicke model as a stable solution to physically well-motivated consistency conditions.

What is the influence of the scalar field on objects in the universe and on the universe as a whole? These fascinating questions are investigated by Eckehard W. Mielke (Universidad Autónoma Metropolitana Iztapalapa, Mexico) and Israel Quiros (Universidad de Guanajuato, Mexico). As shown by Mielke, the gravitational collapse of a boson cloud of scalar fields would lead to a boson star as a new type of a compact object. Moreover, as a coherent state (like the vortices of Bose–Einstein condensates), such collapse would allow for rotating solutions with quantized angular momentum. Quiros focuses on the cosmological impact of Brans-Dicke theory. Is the standard model of cosmology (the so-called ΛCDM model) a stable solution of Brans-Dicke theory? Assuming a Friedmann-Robertson-Walker metric in the Brans-Dicke theory, he demonstrates that the de Sitter solution of GRT is an attractor of the Jordan frame (dilatonic) Brans-Dicke theory only for special values of the coupling constant ω and for special scalar-field potentials. Only for these values does one obtain the ΛCDM model from Brans-Dicke theory.

The first part of the volume closes with the contribution by Martin McHugh (Loyola University New Orleans, USA) about the history of the Brans-Dicke theory and its experimental tests. Dicke became famous for this experimental work and was a popular contact to discuss unexplainable experimental results. At the end of 1965, he received a call from Arno Penzias and Robert W. Wilson at nearby Bell Laboratory, who had found a mysterious microwave signal. They had spent nearly a year searching for the cause of the signal in their antenna. Dicke immediately identified the signal as the long-sought cosmic microwave background (CMB), which he had dubbed the "ash of the Big Bang." In 1978, Penzias and Wilson received the Nobel Prize for their discovery.

The Brans-Dicke theory occupied Carl Brans for twenty years after its initial publication in 1961, and he continued to return to the topic after that. But Brans made contributions to several other topics as well. (Indeed, even beyond the research topics covered in this volume, Carl made additional, important contributions to the Petrov classification, numerical GRT, and complex GRT.) The second part of this volume includes contributions reflecting on Carl's work during the 1980s.

The original motivation for Brans-Dicke theory concerned Mach's principle, and the notion that the gravitational constant, κ, should be a function of the mass distribution of the universe. In his contribution for this volume, Bahram Mashoon (University of Missouri-Columbia, USA) describes the application of Mach's principle to particles' inertial property of spin. The inertia of intrinsic spin is studied via the coupling of intrinsic spin with rotation, a coupling which has recently been measured in neutron polarimetry. The implications of the inertia of intrinsic spin are critically examined in the light of the hypothesis that an electromagnetic wave cannot stand completely still with respect to an accelerated observer.

The second chapter in this part, by Michael J.W. Hall (Griffith University Brisbane, Australia), concerns Bell's inequality. Carl's colleague A.R. Marlow (Loyola University New Orleans, USA) notes that Carl developed an interest in quantum logic and interpretational problems in quantum mechanics. In particular, Carl became interested in Bell's theorem and the effort to decide whether any hidden variables determine the outcomes of measurements, or if the probabilistic framework of quantum mechanics is complete. In 1988, Carl published an article in which he noticed a circular argument in the derivation of Bell's theorem. Bell had to assume that an experimenter's selection of detector settings in an experimental test of quantum entanglement was entirely uncorrelated with any possible hidden variables that could affect the outcomes of those measurements—even though the events that determined the detector settings presumably shared an enormous causal past with any events that could have influenced the outcome of the measurements. Put another way, whatever hidden variables could have classically determined the outcomes of measurements could also have determined the experimenter's selection of detector settings. Hence, in order to derive strong no-go results like Bell's inequality, one must assume "measurement independence." Hall discusses the importance of such an assumption as well as means to relax it within the context of Bell's inequality. He further generalizes Brans's 1988 model to demonstrate that no more than $2 \log d$ bits of prior correlation between the hidden variables and the detector settings are required for a local deterministic model to reproduce the quantum-mechanical predictions for any d-dimensional system.

More recently, Carl's research has focused on the structure of spacetime, and in particular on exotic smoothness. These topics occupy the third part of the volume. As noted above, Charles Misner introduced Carl to such questions with his lecture on fiber bundle theory in 1957, and Norman Steenrod's book on *The Topology of Fiber Bundles* (1951) provided further inspiration. Exploiting similar methods, including cobordism theory, John Milnor made an unexpected discovery in 1956: there exist exotic 7-spheres.

To appreciate the importance of this result, one must dig deeply into manifold theory. The weak equivalence principle in GRT implies the usage of the manifold concept: every neighborhood of a point in spacetime must be locally flat, that is, it must be a subset of $\mathbb{R}^n$. Then spacetime is a smooth manifold, i.e. it is covered by smooth charts with smooth transition functions forming an atlas. A smooth atlas is a smoothness structure. Conventional wisdom had long held that every topological manifold could be smoothed (by smoothing the corners), so that there would only be one smoothness structure (given by the smoothness structure of the $\mathbb{R}^n$). But Milnor found seven 7-dimensional spheres S^7 which agreed topologically but differed in their smoothness structure, thereby providing the first counterexample to the higher-dimensional Poincaré conjecture. Milnor thus founded the new topic of differential topology and received the highest mathematical honor, the Fields medal, in 1962.

As Carl noticed, this revolution occurred only "some doors away from him" at Princeton university. From the physics point of view, the 7-sphere is not particularly interesting, except perhaps in string theory (in which Edward Witten used it to cancel the global gravitational anomalies in 1985). Moreover, exotic smoothness is difficult to visualize, because no exotic smoothness structure exists in dimension smaller than four. For dimension 5 and higher, there are only finitely many exotic smoothness structures, as shown by Kervaire and Milnor in 1963. But what about 4-manifolds as models of our spacetime?

The riddle was solved in the 1980s with the work of many mathematicians, including Michael Freedman, Simon Donaldson, Robert Gompf, and Clifford Taubes. Most compact 4-manifolds admit (countable) infinitely many different smoothness structures, whereas most non-compact 4-manifolds—including $\mathbb{R}^4$—admit (uncountable) infinitely different ones. Therefore, the physical dimension 4 is mathematically distinguished from any other dimension!

Carl attended a lecture by Ron Fintushel at Tulane University to hear about these results. It is typical for Carl that he immediately asked about their relevance for physics. In his first article in collaboration with the mathematician Duane Randall, Brans published the first deep results. It was the start of a long and fruitful collaboration between mathematicians and physicists on this topic. Indeed, many of Carl's questions remain open to this day. His questions helped to shape the direction for current research.

A driving force was the Brans conjecture from 1994. In an article from that year, Carl constructed an exotic $\mathbb{R}^4$ in which the exoticness is localized (now known as small exotic $\mathbb{R}^4$). The Brans conjecture is that this localized exoticness can act as a source for some externally regular field, just as matter or a wormhole can. This conjecture was partly proven by Jan Sładkowski and Torsten Asselmeyer-Maluga. In a 2002 paper by Brans and Asselmeyer-Maluga, this conjecture was extended:

"... *In summary, what we want to emphasize is that without changing the Einstein equations or introducing exotic, yet undiscovered forms of matter, or even without changing topology, there is a vast resource of possible explanations for recently observed surprising astrophysical data at the cosmological scale provided by differential topology. ...*"

Results in this area of research up through 2007 may be found in Brans and Asselmeyer-Maluga's book, *Exotic Smoothness and Physics* (World Scientific, 2007), which has become a standard reference for the topic. An introduction to the topic may also be found in Carl's contribution to the present volume. The third part of this book describes more recent developments.

Jan Sładkowski (University of Silesia Katowice, Poland) aims to describe spacetime structure from the physics point of view. He considers the algebra of all real functions over a manifold containing the information about the topology of the manifold. A generalization of these functions leads to Alain Connes's model of noncommutative geometry as a possible description of the standard model in elementary particle physics.

Jerzy Król (University of Silesia Katowice, Poland) studies model-theoretic aspects of exotic smoothness, uncovering unexpected relations to noncommutative spaces and quantum theory. Forcing, as a special extension of the axioms in set theory, is used to obtain the deformation of the algebra of usual complex functions to the noncommutative algebra of operators on a Hilbert space. The results in the context of the Epstein-Glaser renormalization in QFT are also discussed.

In the contribution by Duane Randall (Loyola University New Orleans, USA), a question of Milnor is answered: is there always an exotic n–sphere for $n > 6$ and $n \neq 12, 61$? In the next chapter, Torsten Asselmeyer-Maluga (German Aerospace Center Berlin, Germany) extensively discusses the following questions: Is it possible to construct a quantum gravity theory by using exotic smoothness? Is it possible to construct quantum gravity directly, i.e. without any quantization of a classical theory? In his chapter, the richness of exotic smoothness in dimension 4 is used to construct a quantum gravity theory directly. The use of this geometrical approach implies one problem: one has to construct a geometrical expression for a quantum state (the ψ–ontic interpretation as implied by current experiments). This construction, using wild embeddings (like Alexander's horned sphere), gives a fractal space. Moreover, quantum fluctuations arise from an unpredictable chaotic dynamics. The consequences for decoherence, the measurement problem, and cosmology are discussed.

The contributions in this volume are dedicated to Carl Brans on the occasion of his 80th birthday, and were written exclusively for this volume. The chapters were contributed by renowned colleagues who collaborated directly with Carl or who were inspired by his ideas. Though Carl never founded a formal school or group, his influence has been felt by many young scientists, across many countries and communities.

Throughout his career, colleagues and students have appreciated Carl's critical questions and his ambition to understand problems at a very deep level. Always approachable, Carl has inspired generations with his deep questions and important insights. Israel Quiros expressed it best in his dedication: "He is one of the greatest minds of the twentieth century." It is a great pleasure to honor Carl Brans with this collection. Happy Birthday, Carl!

Berlin   Torsten Asselmeyer-Maluga
January 2016

Acknowledgements

At first I want to thank my wife Andrea and daugther Lucia for the idea of this volume and overall support.

Special thanks go to Anna Brans for support and corrections as well looking for some photographs.

In particular I acknowledged the impressive work of all contributors in this volume. In my opinion, it is a great collection of papers honoring Carl H. Brans.

Furthermore, I want to thank Dave Kaiser for the help to include more contributors as well for all corrections in the preface and Paul Schultz for reading my own contribution carefully. Special thanks go to Friedrich Hehl for many advices during the preparation process. I want to thank Angela Lahee from Springer for the support and assistance.

Torsten Asselmeyer-Maluga

Contents

1   65 Years in and Around Relativity   1
Carl H. Brans

Part I   Scalar-Tensor Theories (Brans-Dicke Theory)

**2   Nonminimal Couplings in the Early Universe: Multifield
Models of Inflation and the Latest Observations**   41
David I. Kaiser

**3   A New Estimate of the Mass of the Gravitational Scalar
Field for Dark Energy**   59
Yasunori Fujii

**4   Axion and Dilaton + Metric Emerge Jointly from an
Electromagnetic Model Universe with Local and Linear
Response Behavior**   77
Friedrich W. Hehl

**5   Gravitational Theories with Stable (anti-)de
Sitter Backgrounds**   97
Tirthabir Biswas, Alexey S. Koshelev and Anupam Mazumdar

6   Rotating Boson Stars   115
Eckehard W. Mielke

**7   The Lambda-CDM Model Is Not an Universal Attractor
of the Brans-Dicke Cosmology**   133
Israel Quiros

8   New Setting for Spontaneous Gauge Symmetry Breaking?   159
Roman Jackiw and So-Young Pi

9   The Brans-Dicke Theory and Its Experimental Tests   163
Martin P. McHugh

Part II Mach's Principle and Bell's Inequality

10 Mach's Principle and the Origin of Inertia 177
     Bahram Mashhoon

**11 The Significance of Measurement Independence
     for Bell Inequalities and Locality** 189
     Michael J.W. Hall

Part III Exotic Smoothness and Space-Time Models

12 Exotic Smoothness, Physics and Related Topics 207
     Jan Sładkowski

**13 Model and Set-Theoretic Aspects of Exotic Smoothness
     Structures on $\mathbb{R}^4$** 217
     Jerzy Król

14 Exotic Smoothness on Spheres 241
     Duane Randall

**15 Smooth Quantum Gravity: Exotic Smoothness
     and Quantum Gravity** 247
     Torsten Asselmeyer-Maluga

Contributors

Torsten Asselmeyer-Maluga German Aerospace Center, Berlin, Germany

Tirthabir Biswas Loyola University, New Orleans, USA

Carl H. Brans Loyola University, New Orleans, USA

Yasunori Fujii Advanced Research Institute for Science and Engineering, Waseda University, Okubo, Shinjuku-ku, Tokyo, Japan

Michael J.W. Hall Centre for Quantum Computation and Communication Technology (Australian Research Council), Centre for Quantum Dynamics, Griffith University, Brisbane, QLD, Australia

Friedrich W. Hehl Institute for Theoretical Physics, University of Cologne, Cologne, Germany; Department of Physics and Astronomy, University of Missouri, Columbia, MO, USA

Roman Jackiw Department of Physics, Center for Theoretical Physics, Massachusetts Institute of Technology, Cambridge, MA, USA

David I. Kaiser Department of Physics, Center for Theoretical Physics, Massachusetts Institute of Technology, Cambridge, MA, USA

Alexey S. Koshelev Departamento de Física and Centro de Matemática e Aplicações, Universidade da Beira Interior, Covilhã, Portugal

Jerzy Król Institute of Physics, University of Silesia, Katowice, Poland

Bahram Mashhoon Department of Physics and Astronomy, University of Missouri, Columbia, MO, USA

Anupam Mazumdar Consortium for Fundamental Physics, Lancaster University, Lancaster, UK

Martin P. McHugh Loyola University, New Orleans, USA

Eckehard W. Mielke Departamento de Física, Universidad Autónoma Metropolitana Iztapalapa, Mexico, D.F., Mexico

So-Young Pi Department of Physics, Boston University, Boston, MA, USA

Israel Quiros Dpto. Ingeniería Civil, División de Ingeniería, Universidad de Guanajuato, Leon, Guanajuato, Mexico

Duane Randall Loyola University, New Orleans, USA

Jan Sładkowski Institute of Physics, University of Silesia, Katowice, Poland

About Carl H. Brans

Thoughts of Colleagues, Friends, Family

With this volume, the 80th birthday of Carl H. Brans will be celebrated. Instead of a single foreword, colleagues and friends will present her/his personal view on Carl H. Brans and his influence (Fig. 1).

Fig. 1 Anna and Carl Brans (December, 13th 2015), 80th birthday of Carl

Carl's Influence on the Movie Interstellar

The ideas that Newton's gravitational constant G might change from place to place and time to time, and might be controlled by some sort of nongravitational field, were hot topics in the Princeton University physics department when I was a PhD student there in the early 1960s. These ideas had been proposed by Princeton'Ăs Professor Robert H. Dicke and his graduate student Carl Brans in connection with their "Brans-Dicke theory of gravity", an interesting alternative to Einstein's general relativity. The Brans-Dicke theory has motivated a number of experiments that searched for varying G, but no convincing variations were ever found. These ideas and experiments motivated my interpretation of some of *Interstellar*'s gravitational anomalies and how to control them: bulk fields control the strength of G and make it vary. The Professor's equation, as used in this movie and shown on a blackboard in one sequence, builds on these ideas.

Kip Thorne, Caltech
(see **The Science of Interstellar** by Kip Thorne, Norton & Company 2014)

Carl as Colleague

I have known Carl since 1964 when I came back to Loyola having graduated from there in 1952 a few years prior to Carl. I consider my career as somewhat unusual in having it sandwiched between a brief acquaintance with Charles Misner at Notre Dame and a long friendship with Carl, two extraordinary physicist. Unfortunately brilliance doesn't osmose and Carl has had to be very patient with my dumb questions and ever ready to discuss my mathematical and theoretical questions. He has over the years conducted numerous weekly seminars in relativity for the benefit of interested undergraduates and a few faculty and is even currently doing so with discussions of such mysterious topics as the Unruh effect and Bell's inequalities.

One thing that has puzzled me about Carl is why he returned and stayed at Loyola ever since completing his degree at Princeton. I have never asked him about this but my conjecture is that Carl is such and independent, innovative thinker and, importantly, disciplined hard worker that he doesn't need the intellectual stimulation produced by a larger department. Call it the *Keiffer conjecture*. Happy birthday Carl.

David Keiffer, Loyola University New Orleans

Unsurprisingly, I had worked on several different versions of Brans-Dicke theory before I actually met Carl during my job interview at Loyola. Thankfully, I didn't know that I was actually meeting Carl Brans (somehow I missed his profile on the Loyola physics faculty listings) because that would have completely overwhelmed me. It was only halfway through the interview that I realized that I was talking to someone who knew a lot more gravity than I did. Since then, we have become very

Fig. 2 Carl in Princeton (1959) "This picture has me "studying"(?) while sitting on the heater in student apartment. I was always cold!"

good friends, his good nature, his humility, and his commitment to rigor is something that I cherish and I am inspired by. So, here is to Carl for showing the path that many others like me could follow (Fig. 2).

Tirthabir Biswas, Loyola University New Orleans

Supporting Young Scientists

I first met Carl Brans about twenty years ago, in the mid-1990s, when I was a graduate student. Carl invited me to visit him at Loyola University in New Orleans, and he and his wife Anna kindly hosted me in their beautiful home. Our first meeting has always stood out in my mind: Carl picked me up at the airport, drove me straight to his office, and handed me a piece of chalk. I was to give him a lecture, right there at the blackboard, about cosmic inflation. I launched in, as best I could, and after a fun discussion Carl announced that it was time to pause and get some seafood gumbo; after all, we were in New Orleans. Ever since my first visit, I have found it terrifically inspiring to talk with Carl and to try to sharpen my own ideas in the face of his excellent questions, which he has always delivered in a gentle and encouraging way.

David I. Kaiser, MIT

In 1983, as a 12-years old boy living in the GDR, the socialist part of Germany, I came across the name Carl H. Brans while reading the book "The View of Modern Physics" ("Das Bild der modernen Physik", Urania Verlag). At the end of the chapter "Relativity?, I found a discussion about a variable coupling constant in General Relativity together with this footnote: "This extension of General Relativity was essentially developed by P. Jordan, C. Brans and R. Dicke in 1961.". I was deeply impressed and hoped to one day work with these scientists. But I lived on the wrong side of the iron curtain and was forbidden to have contacts with people in the USA. Eventually the political situation began to evolve as I started my study of Physics in September 1990, one month before the unification of East and West Germany. In the second semester I joined a lecture group on algebraic topology which included a discussion of the existence of exotic $\mathbb{R}^4$ and the exceptional role of exotic smoothness in dimension four. I was very excited about the possible relationship between this new mathematics and the classical dimension of space-time. So I enthusiastically began to learn everything I could about Exotic Smoothness. In 1992, I came across the paper written by Brans and Randall with many intriguing ideas. A professor at my university helped me to get an invitation to the summer school on Gravity and Torsion in Erice (Italy) where I met Friedrich Hehl who had been a colleague of Brans as a visitor to the Princeton Physics department in 1973. In September 1995, Hehl was organizing a Heraeus seminar "Relativity and Scientific Computing—Computer Algebra, Numerics, Visualisation" and invited Brans to talk about his work with exotic smoothness and physics. It was at this conference that I met Carl and we immediately recognized our common enthusiastic interest in modern differential topology. This then led to some intensive email exchanges.

In 1997, Carl offered me a scholarship (LaSpace Grant) to visit him in New Orleans the next year. It was then that we began to write our book "Exotic smoothness and physics". Carl was very impressed by the story above about a 12-years-old boy behind the "wall" who had a dream to someday meet him. Of course, in New Orleans there is always life after work and I still remember the warm welcome (with a seafood gumbo) by Carl's wife Anna. That was the beginning of our wonderful friendship and successful collaboration. I feel so privileged to have had the opportunity to work with him on numerous occasions (Fig. 3).

Torsten Asselmeyer-Maluga, German Aerospace Center

His Family:

Carl Henry Brans was born at St. Paul's Hospital in Dallas, Texas on Friday, December 13, 1935. For his parents, Carl Brans, who worked in the maintenance

Fig. 3 Carl in the student apartment (1957)

department at Sears Roebuck, and Delia Elizabeth Murrah, a housewife and talented pianist, Friday the thirteenth represented good luck from then on. Carlie was an only child, but had the support of a large extended family in Dallas.

Both of Carl's paternal grandparents had come to America as children, his grandfather from Anholt, Germany and his grandmother from Austria. The men in

Fig. 4 Anna and Carl in Mississippi City (1957) "It is a "selfie" taken on our honeymoon."

Fig. 5 1958: Carl with the first son Tommy in front of the student apartment (*left*) and Anna, Carl and Tommy in front of the church in Princeton when the son Tommy was baptized (*right*)

the family were brick masons. His maternal grandmother, Mary Mc Namara was born in Dallas to Irish immigrants, whereas his maternal grandfather, who was a blacksmith, came from a family with deep American roots, which led back to England.

On May 5, 1956, I met Carl on a blind date for the Physics Department's annual end of year crawfish boil. It was the end of his junior year and the end of my sophomore year as a major in Sociology. This was the beginning of a beautiful relationship. On February 9, 1957 we were married and went to Princeton together in September of that year. During the 3 long lean years of living on NSF fellowships, our oldest son was born.

After 58 years of marriage, our little dynasty numbers nineteen descendants: Four sons, Thomas Joseph, Henry Robert, Patrick David and John Edward. A daughter, Mary Elizabeth, died in infancy of hydrocephalus.

Our later years were enriched as we saw our progeny flourish. We now have eight grandchildren who have given us seven great-grandchildren (so far) (Figs. 4 and 5).

Anna Dora Monteiro Brans
September, 2015

Chapter 1
65 Years in and Around Relativity

Carl H. Brans

At the very beginning I must thank all of the contributors to this book for taking their valuable time to add to it. Of course, my special thanks goes to Torsten Asselmeyer-Maluga, not only for organizing this work, but also for his friendship as well as his expert advice and tutelage in the mathematics of differential topology over the last 20 (or more) years we have known each other. As always, "Much thanks to you Torsten!"

1.1 Undergraduate Days, 1953–1957

I was not exposed to anyone with expert knowledge about current research in mathematical physics until I got to Princeton at about 20 years of age. Until then I mostly learned on my own, as many of us probably did. Later experience confirmed to me that indeed there is a good bit of the truth in the old saying: "A self taught person has a fool for a teacher." Also, I think this background made it difficult for me to work well in groups or to collaborate, except recently with Torsten. More importantly when I arrived at Princeton University I had a lot of holes in my knowledge of mathematics and physics, and felt quite intimidated by both faculty and other graduate students who had more solid foundations in mathematics and physics. In spite of this, I managed to muddle through and passed both oral and written qualifying exams at the end of my first year.

For me perhaps the most fascinating and motivating aspect of modern physics, primarily relativity and quantum theory, is the marvelously counter-intuitive models for reality beyond that which we can directly observe with more or less "everyday" experience. Of course much of the apparent mystery is a result of trying to describe these "worlds" beyond our everyday experience in natural language which has developed

C.H. Brans (✉)
Loyola University, 6363 St. Charles Avenue, Box 92, New Orleans 70118, USA
e-mail: brans@loyno.edu

© Springer International Publishing Switzerland 2016
T. Asselmeyer-Maluga (ed.), *At the Frontier of Spacetime*,
Fundamental Theories of Physics 183, DOI 10.1007/978-3-319-31299-6_1

exclusively from such direct, human-scaled, experiences. When natural language is inadequate, it has been replaced by mathematics, often using formalisms previously thought to be purely abstract, and unrelated to "common sense." Fortunately for me, I had developed a taste (if not a real skill) in pure mathematics. One of the most interesting things for me was finding that much of the mathematics needed by relativity and quantum theory had been developed before these needs were known. In turn, these physical fields have inspired the further development of mathematical structures. I was very interested in the often asked question: "What is this curious interaction between physics and mathematics?"

Also, I was re-discovering the fact that questioning old assumptions is fun. I enjoy turning to books such as *Counterexamples in Topology* [1]. To paraphrase a remark that I heard from John Wheeler, "Aren't we so very lucky to be paid to do something that we otherwise would do as a hobby."

As an undergraduate at Loyola, I delved into as much mathematics and mathematical physics as I could. So from our library I got a taste of topology from the books *Introduction to Topology* and *Algebraic topology,* by Lefschetz and a bit of measure theory from the book of Halmos. For differential geometry and relativity I read mostly from Sokolnikoff's *Tensor Analysis* and especially Einstein's *Meaning of Relativity*, 3rd edition. I was most fascinated by the Appendix on a unified field theory constructed by generalizing the standard symmetric metric to an asymmetric one. Although this did not turn out to be a widely explored direction in the quest for the classical unified field theory, I found it fascinating and wrote a primitive review of the subject as my senior thesis for my Physics degree from Loyola in 1957 [2] (unpublished). My original readings on quantum theory were probably from the book Lande, my introduction to Quantum Field theory the book by Wentzel. Of course I also tried to read some of the available articles in the Physical Review (only one issue then) and the Annals of Mathematics.

I should also mention that I was very fortunate to have the strong support of Frank Benedetto, S.J., then chairman of the department who did his PhD work on cosmic rays under the Nobel prize winner, Victor Hess. At Loyola Benedetto continued observational work with cosmic rays involving an array of geiger counters with requisite very high voltage. In spite of my obvious difficulties with anything practical, Benedetto provided me with some much needed financial support by employing me to check this array occasionally. I know that at some point I was reaching across the apparatus, with exposed high voltage terminals, and then some undetermined time later I awoke on the concrete floor on my back, sore and somewhat confused. Probably I had received a high voltage shock. I suppose it must have rewired the synaptic patterns in my brain, but I will never know since I was too embarrassed to tell anyone about it at the time. Since then I have even more studiously avoided experimental equipment as much as possible. On a more positive note, during my senior year Benedetto encouraged me to apply to the National Science Foundation for one of their relatively new Research Fellowships. At his suggestion I also applied to Princeton, Einstein's last home. Somehow both applications were successful.

However, during this time the most important thing in my life was my marriage to Anna Monteiro who somehow has managed to raise our family and stay with me for almost 60 years now.

1.2   Arrival in Princeton 1957–1960: Misner, Dicke et al.

In the late fifties as a result of research related to WWII, times were very good for U.S. federal support of research in physics and mathematics in general and especially for people in my age bracket. The fairly new National Science Foundation was providing very generous research support for graduate student fellowships. This enabled me to concentrate entirely on research and fortunately no lab work or teaching was required.

When I arrived in Princeton, I was overwhelmed to be around people whose names I knew as world-class leaders in mathematics and physics. Most of them had even known Einstein personally. Unfortunately this did not include me since he died a couple of years before I arrived. It is now very clear to me that Misner and Dicke were most influential to me in these 3 years. Misner's relativity course opened my eyes to some of mathematical formalisms which were providing new (at least to me) tools and incentives for my nascent explorations into mathematical physics. I discovered Steenrod's fiber bundle book and was inspired by the potential tools presented by bundle structures. In fact, much of the foundation for work that would fascinate me the rest of my career was being done next door in Fine Hall, physically connected to Palmer Lab. Of course, the chief director and contributor to theoretical relativity at Princeton at that time was John Wheeler. Probably as a result of my early development in a somewhat isolated environment for mathematical physics, I have always found it hard to work in a group. So I didn't really "join" Wheeler's very productive group, although I did attend many of its seminars. As I matured later, I realize that I missed a golden opportunity to profit from working more closely with Wheeler himself. I do recall an unforgettable afternoon during this period when someone mentioned what might be an interesting seminar at some place about an hour's drive from Princeton. However, no one seemed to have a car except me so Wheeler asked me to give him and Dirac (who was visiting then) a ride to the seminar. Of course, I couldn't say no, but my ancient car had obviously seen much better days. Actually it was rather "beat up" (both mechanically and cosmetically) to use the common phrase. In spite of this I was looking forward to the prospect of spending a long ride with two such pre-eminent physicists when I was only beginning my graduate school experience. From what I recall, Dirac and Wheeler sat in the back and a long, mostly one sided, conversation ensued between the very taciturn Dirac and the very outgoing and loquacious Wheeler. As I recall, during this period much of Dirac's work concerned quantization of the first order metric field using then current quantum field theory techniques. On the other hand, I knew that Wheeler thought the special nature of geometry meant that any quantization would require an entirely different approach. The "conversation" consisted mostly of rather long, but always polite, questions from Wheeler, and usually only one syllable answers from Dirac. Looking back on this from almost 60 years later, I believe that had I recorded the conversation, it would not now seem to be entirely out of date. Fortunately, considering the prominence of the passengers, I was very happy that my poor car managed to hold up for the round trip.

Also, although I did not look carefully into it, but about that time, late 1950s, workers in quantum field theory and particle physics were beginning to explore symmetries and Lagrangian formulations in which the notions of bundle theory could provide the underpinning for the exploration of quantum force fields as connections on principal bundles. Actually this was presaged as early as 1918 by the work of Weyl on gauge transformations of the electromagnetic potential as being related to conformal transformation of the metric, thus providing some sort of geometric interpretation of the electromagnetic field. This particular approach to a unified field theory was not further studied for many reasons. But it was important as an introduction of what was later to be expressed as gauge symmetry in QFT. These gauge transformations were associated with spacetime scalars with perhaps some internal (bundle group) structure. This ultimately led to what is now referred to as the Higgs scalar and related work. So even before introduction of a scalar field into general relativity, scalars were blossoming in the QFT context.

Although I did not realize it at the time, much of the later surprising discoveries of exotic, i.e., non-standard, structures on topologically simple manifolds such as $\mathbb{R}^4$, $\mathbb{R} \times S^3$, were being presaged by the work of one of the most talented young mathematicians next door in Fine Hall, John Milnor. I refer especially to Milnor's exotic spheres, Σ^7, which are smooth manifolds that are homeomorphic to S^7 but not diffeomorphic to it in its standard structure inherited from $\mathbb{R}^8$. Milnor started with an S^3 bundle over S^4 but with non standard coordinate patch identification of the upper and lower hemispheres of the resulting Σ^7. Later I recognized that since Yang-Mills original non-commutative gauge group is equivalent to an S^3 bundle over $\mathbb{R}^4$, connections on Milnor's exotic spheres could be thought of as Yang-Mills models over compactified spacetime. These topics are in a branch of mathematics known as differential topology, which has held my interest for the last 20 years or so.

In addition to Misner's lectures I sat in other courses. One of these was Quantum Mechanics/Field theory, by Marvin Goldberger. I remember Goldberger's class especially since I found none other than Eugene Wigner sitting next to me and very politely asking questions and making comments. That was certainly a notable experience for a young student such as me. As is well known, Wigner was very kind and extremely polite. It was said that you could never follow Wigner through a door, he would always bow and open it for you.

At that time there was no class or grade requirement at Princeton just a comprehensive exam which I managed to pass at the end of the academic year 1957–1958. It was then time for me to start work on a thesis. Misner's course had exposed me to the mathematics I found most fascinating, so I asked his opinion on my idea to write my thesis on the application of bundle theory to physics. Instead, he suggested that a more realistic path would be to talk to Robert Dicke. He said that Dicke was looking for a theoretical student to develop some of his ideas. It is hard to believe now but at that time I had never heard of Bob and his group, even though he had already become one of the outstanding experimental physicist of his time. Dicke was very approachable and patient in explaining to me his thoughts on Mach's principle, the "large number coincidences," and finally Bob's ideas that

$$\frac{GM}{R} \approx 1 \tag{1.1}$$

for M and R as the mass and radius of the universe as known then might lead to a relativistic generalization of

$$G^{-1} \approx M/R \implies \nabla^2 G^{-1} \approx \rho_{mass} \tag{1.2}$$

Or on an expected local level

$$G^{-1} = G_0^{-1} + \Sigma \frac{m}{r} \tag{1.3}$$

Bob referred to (1.1) as the "large number coincidence," expressed in atomic units as

$$\frac{10^{-40} 10^{80}}{10^{40}} \approx 1 \tag{1.4}$$

as pointed out by Dirac [3] and others. Of course modern observational cosmology has caused drastic revisions to these ideas which seem simplistic to the modern reader, but they were more or less mainstream in terms of observations up to the 1950s. Bob was also very interested in Mach's ideas in terms of a possible causal relation of the state of motion of inertial reference frames relative to the fixed stars. Exactly what is meant by "Mach's Principle" has been debated probably for a century by now [4]. From what I recall Bob was most concerned about the fact that in almost all expositions, "inertial" forces are not granted the same status as "real" ones. Since both inertial and gravitational forces are proportional to the mass they act on, Bob thought that inertial forces should be as real as gravitational ones. In other words Bob wanted to understand inertial forces, such as centrifugal, in terms of gravitational ones due to acceleration relative to the fixed stars. He suggested that this might result in having inertial mass depend on the mass distribution of the universe.[1] To avoid dependence on choice of units, he suggested that this could be expressed in terms of a variable G. This part of the argument is somewhat involved. I think that Bob was rather disappointed that our formulation of a scalar-tensor theory does not seem to provide an explicit derivation of Newton's bucket, etc.

---

[1] And I believe so did Einstein. See the later discussion of this issue to which a good bit of my thesis was addressed.

As a start, Bob referred me to a paper by Sciama [5] which presented a toy model of vector gravity analogous to electromagentism. This turns out to produce gravitational forces to be seen by an observer accelerating relative to the shell. In the simplest model a mass m accelerating relative to a spherical shell of mass M and radius R would be subject to a gravitational force

$$\mathbf{F}^{\text{Sciama}} = \frac{GM}{R} m\mathbf{A} \qquad (1.5)$$

when it has an acceleration $\mathbf{A}$ relative to the shell which represents the rest of the universe. This would be exactly the observed inertial force if $\frac{GM}{R} = 1$. Of course it was known that such a purely vector form could not replace Einstein's for general relativistic gravitation. Sciama presented it only as a toy model to understand inertial forces as gravitational ones. Dicke was fascinated by Sciama's work. Unfortunately for Bob, as far as I know neither scalar-tensor nor standard Einstein theory can reproduce anything as explicit as (1.5). Einstein had devoted a few pages of *Meaning of Relativity* to remark that he had incorporated Mach's Principle in his standard equations of general relativity by having the inertial mass depend on certain metric components and thus the mass in the entire universe. It seemed obvious that this must be purely a coordinate effect, and I later published a paper on it [6].

Another aspect of inertial and gravitational forces could be expressed in terms of the equivalence principle(s). Roughly, in modern usage the **weak principle** asserts that all point masses have same gravitational acceleration. The fact that point masses have a limit of zero self energy later turns out to be critical. The **strong principle** can be summarized as asserting that all gravitational effects are due to the metric alone which requires assumptions not necessarily made and observed in the weak principle.

For many years the work of Eötvös around 1900 was the standard test of the weak principle. However, on closer examination Dicke claimed that the experimental errors inherent in the experiment as done by Eötvös were much greater than believed. In fact, I recall his saying that that the accuracy of the Eötvös apparatus would be affected by human movement only a few meters away. So Bob and others set about re-designing the experiment with modern tools and the results were published in [7]. This seemed to be a very satisfactory confirmation of the **weak** principle of equivalence, but Bob was still intrigued by the fact that Einstein theory makes the implicit assumptions contained in the yet-to-be-tested strong principle. In the 1950s and early 1960s Bob was unaware of the implications of the later discovery by Ken Nordtvedt that a variable G (violation of the strong principle) would in fact result in a violation of even the weak principle for massive bodies. This will be discussed below.

In any event during the 1950s, Bob was highly motivated by Mach's principle, and still unaware of a connection between strong and weak principles. After a few visits, Bob suggested to me that I develop a rigorous formalism consistent with the weak but breaking the strong principle of equivalence incorporating a general relativistic formulation of (1.3) with variable G^2. At first, consider allowing G to be a variable in an action principle based on

$$S_0 = \int \sqrt{-g} d^4x (R + 8\pi G L_{matter})$$

(1.6)

Then clearly, the translational invariance of the action would lead to $(G T^{\mu\nu}_{matter})_{;\nu} = 0$ so matter conservation would be violated if G is not constant. However starting from (1.6) multiply S_0 by the (still constant) G^{-1}, thus decoupling the possibly non-constant G from the matter Lagrangian, so matter conservation could still hold, $(T^{\mu\nu}_{matter})_{;\nu} = 0$, as in Einstein theory. Using ϕ as a scalar field variable associated with variable G we would still need to add a Lagrangian for ϕ,

$$S_{ST} = \int \sqrt{-g} d^4x (\phi R + 8\pi L_{matter} + L_\phi)$$

(1.7)

If $\phi \approx G^{-1}$ then the dimensionally consistent and simplest choice for L_ϕ would be functionally of the form

$$L_\phi \sim \frac{\phi_\mu \phi^\mu}{\phi}$$

(1.8)

At this point I realized that the $\sim$ in (1.8) must be replaced by $=$. I noticed that this would still allow the introduction of a dimensionless constant, the now infamous ω, so the study would begin with something like

$$S_{ST} = \int \sqrt{-g} d^4x (\phi R + 8\pi L_{matter} + \omega \frac{\phi_\mu \phi^\mu}{\phi})$$

(1.9)

The usual variational principle then leads to

$$2\frac{\omega}{\phi} \Box \phi - \frac{\omega}{\phi^2} \phi^{,\alpha} \phi_{,\alpha} + R = 0$$

(1.10)

for ϕ, and an equation that can be written as

$$R_{\mu\nu} - \frac{1}{2} g_{\mu\nu} R = \phi^{-1} [8\pi T^{matter}_{\mu\nu} + \omega(\phi_{,\mu}\phi_\nu - g_{\mu\nu}\phi^{,\alpha}\phi_{,\alpha})/\phi + (\phi_{,\mu;\nu} - g_{\mu\nu}\Box\phi)]$$

(1.11)

for the metric. It is interesting to note that the gravitational field equations (1.11) have the usual Einstein tensor on the left side of the equation and ϕ^{-1} times a generalized source tensor that involves not only "matter" but also terms involving derivatives of ϕ. So, in this form the ϕ part of (1.11) seems to act like ordinary matter in producing the geometry. However, from a purely formal point of view (1.11) could be written

---

[2]This ultimately led to the oxymoronic phrase in my thesis title: "...variable...constant.".

$$\phi(R_{\mu\nu} - \frac{1}{2}g_{\mu\nu}R) - \omega(\phi_{,\mu}\phi_{\nu} - g_{\mu\nu}\phi^{,\alpha}\phi_{,\alpha})/\phi - (\phi_{,\mu;\nu} - g_{\mu\nu}\Box\phi) = 8\pi T_{\mu\nu}^{matter}$$
$$(1.12)$$

Thus, in (1.12) matter contributes to differential equations for **both** the metric and ϕ, leading more directly to the idea of a scalar-tensor theory, *scalar* combined with metric *tensor*, in which both quantities have matter as source. But as written, (1.10) and either (1.11) or (1.12) do not directly show matter as the source for ϕ, which was one of the original motivations. To get this simply contract (1.11) to get another equation for R in terms of T and ϕ. Eliminating R from these two equations then results in the important.

$$\Box\phi = 8\pi T/(2\omega + 3) \qquad\qquad (1.13)$$

It is clear that in the weak field approximation (1.13) will result in some relativistic approximation of (1.2). When I brought this equation to Bob, he was at first pleased but then concerned that we had replaced the determination of a constant, G, with an undetermined ω. Bob naturally found this somewhat unsatisfying.[3] He suggested that I look for something giving the value of ω as a function of the fundamental mathematical constants such as e, π, or the physical fine structure constant, etc. But I could not get anywhere with this, so he suggested I proceed with the mathematical analysis of the equations resulting from an action of the form in (1.9) and associated equation (1.13) which I did, ultimately leaving ω as a dimensionless number which could only be determined by experiment. Of course, from my side, I was somewhat disappointed that none of this would involve any mathematics beyond basic algebra and calculus. I was very concerned that the thesis itself was too trivial with little interesting mathematics so that it might not be sufficient to get a Doctorate from Princeton. But I was told it could lead to a PhD. Little did I know that from this unassuming and not very challenging thesis, Bob Dicke and others would develop a remarkably significant re-examination of General Relativty both experimentally and theoretically.

When I was stuck with the problem of trying to find an exact solution analogous to that of Schwarzschild for standard General Relativity, Misner suggested I try the metric form in isotropic coordinates. In this form it was possible to give exact solutions in terms of elementary functions, but the expressions fell naturally into four separate functional forms depending on the ranges of ω and other constants. Some of these solutions involve negative ω and exponential behavior for the metric and are all presented in my thesis. Of course, (1.13) caused Bob to consider only the one form for which $2\omega + 3 > 0$ in all of his work. In my thesis I neglected any interpretation of the solution for other ranges of ω, although I sometimes think of looking at them again. In my thesis and later publications, I presented fairly thorough arguments suggesting that $1/\phi$ does indeed act as G to some approximation in $1/\omega$ as well as the question of how this scalar-tensor model might be consistent with some form of Mach's principle, still rather amorphous to me. I also made attempts

---

[3]He was also concerned that since electromagnetic radiation has $T = 0$, an electromagnetic radiation field would not contribute as a source for ϕ. This question was shoved aside and not further investigated, I believe.

to find exact cosmological solutions analogous to the familiar FRW ones of standard General Relativity, but the best I could do was to give some information in terms of Taylor series approximation. All of this is presented in my thesis and [8].

I was well along with this work when someone (I don't know who) told me to check on Jordan's work, probably even the book *Schwerkraft und Weltall* [9]. Of course I was disappointed when I found Jordan's expositions, and thought that I would need to begin again with a new thesis topic. As a very young researcher, I was not fully aware of the phenomena associated with the not uncommon "re-discovery" of results.[4] But Bob assured me that we could proceed with the subject since since we had started the work independently of Jordan et al., and that we could point out that the difference in motivation was Mach's Principle and some of the large number coincidences, etc. I have been most asked: "What did Dicke know of Jordan's work and when?" All I can say is that I do not recall Bob's having mentioned Jordan's work when I was starting my thesis. Later when I visited Princeton as a visiting professor at the University, and later as a visitor to the Institute for Advanced Studies, I do not recall his telling me that he knew of Jordan's work before I told him about it around 1958 or so. However, It is now clear that indeed Bob was in correspondence with Jordan and even mentioned his work before 1958. This will be examined a little more closely in the discussion in the following.

Most all of this work and consultation with Dicke et al. was done in the late 1950s, but publications did not come until the early 1960s.

1.3   1960s: Scalars, Tensors, Invariants, Invariance, Petrov, etc.

The original paper on this formalism appeared in 1961 [10], written mostly by Dicke using formalisms and results from my thesis, which, despite my doubts was accepted as sufficient for me to get my doctorate. In addition I followed this up with a couple of papers, [6, 8] extracted from my thesis to give my arguments on how this formalism did indeed give a variable G, but not a strong Mach's principle in the sense that Sciama's toy model did. The thesis also described in detail the class of Schwarzshild-like solutions I found. Also I gave the explicit argument against the physical reality of Einstein's claim that inertial mass is a function of the mass distribution in the rest of the universe. A brief summary of the topics covered in my thesis is in the Appendix below.

It was clear that the theory gave predictions observably different from Einstein's by amounts that decreased with decreasing $1/\omega$. In other words, the scalar-tensor formulation differed from the original Einstein theory significantly in the empirical sense only if ω is not too large. How large? Since 1961 many observations have narrowed the error bars sufficiently to make the scalar-tensor formulation effectively irrelevant in the solar system scale. The book by Poisson and Will [11] is a detailed

---

[4]Since then I have done work which turned out to be on both sides of this phenomenon.

source for the current status of ST and other modifications of standard Einstein theory. Bob himself led much of this work. As planetary orbital data pushed ω higher and higher, Bob proposed a solar oblateness of sufficient size as to allow ST theory with reasonable value of ω. Ironically observations of the sun by Henry Hill, a former student of Dicke and highly regarded by him, indicated that the solar oblateness did not exist to the degree that Dicke was proposing. My collaborator and fellow contributor to this volume, Martin McHugh, is actively at work on a scientific biography of Dicke that should help unify the history of this topic.

As mentioned earlier, it turns out that a variable G might even break the weak equivalence principle. A simplified version of this later described to me by Dicke based his conversations with Nordtvedt along the following lines. Consider a mass, M, which contains a significant gravitational binding energy. If G is variable, the total amount of this energy will vary from point to point. Ken Nordtvedt proposed a thought experiment to explore the implications of this. Using a constant gravitational field for simplicity suppose that originally M is at height h. It is then broken in half overcoming the binding energy by an amount by an amount $G(h)f(M)$ at the top. Here $f(M)$ is some function depending on the structure of M. During the fall, each $M/2$ piece falls with acceleration, g_1 gaining energy $2(M/2)g_1$. They are then reassembled at the bottom, gaining energy proportional to $G(0)$. So the net energy gained from M at top to M at the bottom is

$$E_1 = Mg_1 + G(0)f(M) - G(h)f(M) \tag{1.14}$$

Or, we could reach the final state letting M itself fall with acceleration g_2, gaining energy

$$E_2 = Mg_2 \tag{1.15}$$

Obviously the beginning and ending conditions are the same in both cases, so we should have $E_1 = E_2$, or

$$0 = E_1 - E_2 = M(g_1 - g_2) + (G(0) - G(h))f(M) \tag{1.16}$$

So energy conservation expressed in (1.16) results in

$$g_1 - g_2 = (G(h) - G(0))f(M)/M \tag{1.17}$$

and the weak principle of equivalence cannot be extended to bodies with significant gravitational binding, $|f(M)/M| \gg 0$.

Nordtvedt brought this argument to Dicke around 1967 who agreed that indeed a variable G would result in objects falling with different accelerations dependent upon their fraction of gravitational binding energy. This crude thought experiment was turned into rigorous predictions that would be most easily and precisely observed in the earth-moon orbital motion. Of course this difference would not arise in Einstein's purely geometric model of gravitation, and again the difference would depend on $1/\omega$. In 1969 astronauts placed a laser reflector on the moon which would allow for very precise measurement of the earth-moon relative motion. The result again put tighter limits on $1/\omega$ and also confirmed that a theory with variable G would not satisfy even the weak principle of equivalence for bodies with significant gravitational binding energies. For more details, see [11]. Thus, it is somewhat ironic that Dicke should suggest the non-constant G in scalar-tensor theories while he was working on a repetition of the Eötvös experiment to verify the weak principle for non-pont masses to much higher accuracy.

I now believe that when I proposed a formulation generated by a Lagrangian such as in [12] Bob regarded his version of Mach's principle and other motivations as sufficiently different from Jordan's as to justify our use of this formalism. I would also like to point out that it was Dicke (not me) who first used the phrase "Brans-Dicke" theory [13].

Of course later scalar fields have re-emerged in various forms in string theory and modern cosmology. There are many sources an interested reader can find for more thorough look at the development of scalar-tensor theories and their off shoots, In addition to my early papers [6, 8, 10], there is of course a wealth of papers by Dicke, many of which are summarized in his book [14]. Much later, at the suggestion of others, I worked on some online surveys [15, 16]. For a first hand historical review of the work of Jordan and others, see Schucking [17].

An early incident contributed to my discontinuing interest in ST theories. At one meeting shortly after our 1961 paper, while I was still a very young recent PhD, I was attacked by some of general relativity's elder statesmen for destroying the purity of Einstein's formulation by adding the scalar field. This was very intimidating and together with the lack of challenging mathematics caused me to pretty much lose interest in ST theory and its empirical status.

However, I became interested in the fact that, as far as I knew, no scalar field had survived the introduction of special and general relativity. Of course looking back on the times I am well aware of studies of the significance of conformal transformations of the metric by Weyl and others. I now realize that in the context of the times many people were investigating what is now generally referred to as the Higgs Boson(scalar) in the quantum field theory realm but I was regarding gravity purely as a classical field. More precisely, in the classical realm, I was not aware of fields (at the potential level) which were spacetime scalars. Thus, the electromagnetic field is a vector field while the gravitational field explicitly involves the metric, a two-tensor field. However the ϕ in scalar-tensor theories provides a physically invariant quantity which cannot be transformed away over any open set by a particular choice of coordinates. In part, this led me to the question of finding the invariant trail left by the metric tensor in standard Einstein theory, with or without some scalar attached. The principle of general relativity allows the use of any coordinate system and lacking even a single scalar field such as ϕ, I wanted to try to understand the explicit technology for extracting **coordinate independent, invariant** information describing gravity.

The problem of obtaining invariant information about a geometry is, of course, a very old one and goes back to the definitions of curvature for a line, one dimensional, and then Gaussian curvature for two dimensional surfaces. For higher dimensions the

Riemann tensor appeared, which, as a tensor, is usually expressed in terms of coordinate dependent components. Extracting invariant information from those tensor components leads naturally to two important issues

1. **Invariants** that is, finding functions whose values are coordinate independent, and
2. **Metric determination** exploring to what degree these invariants determine the actual geometry.

Actually this could be described more directly by saying that in Einstein theory the field is a cross section, $\mathfrak{g}$, the metric, on the principal bundle of tangent vector frames over the base space, M, with bundle group reduced to the Lorentz group. However, to ultimately observe and compare this field in the operational sense, we need to know how the real time measuring devices are related to the local coordinates chosen for M, and a specific form for $\mathfrak{g}$. In more familiar form, the invariant content of Einstein theory presents the observer with proper distance metric,

$$ds^2 = \eta_{\alpha\beta}\omega^\alpha(x^\mu)\omega^\beta(x^\nu) \tag{1.18}$$

and we must know how both the coordinates, x^μ, on M and the Lorentz orthonormal forms, ω^α, as functions of them are related to local observations. A large literature grew out of these problems. I was especially attracted to the approach of Petrov, at least for a class of metrics which are Ricci flat. Petrov [18–20] has provided a very nice description of one approach to extracting all invariant information from a coordinate expression of the Riemann tensor. In four dimensions using the Ricci flat, Lorentz signature metric, this four index tensor can be first reduced to a symmetric two by two matrix whose four components are described by a pair of three by three matrices. Using the following notation,

$$M_{ab} = R_{0a0b}, \quad N_{ab} = R_{0abc}, \quad (a,b,c) = \text{cyclic}(1,2,3) \tag{1.19}$$

Here M and N are symmetric and M has trace zero. As an intermediate step, write

$$\Re_{AB} = \begin{pmatrix} M & N \\ N & -M \end{pmatrix} \tag{1.20}$$

where $A, B = 1...6$. Next, use the fact that the Hodge star is a complex structure on fiber Λ^2,

$$\star : \Lambda^2 \to \Lambda^2, \quad \star\star = -1. \tag{1.21}$$

Thus Let (a, b, c) be a cyclic permutation of $(1, 2, 3)$ and

$$u^a \equiv \omega^0 \wedge \omega^a, \quad v^a \equiv -\star u^a = \omega^b \wedge \omega^c \tag{1.22}$$

which then satisfy

$$(u^a, u^b) = (v^a, v^b) = 0 \quad (u^a, v^b) = \delta^{ab} \tag{1.23}$$

This then leads naturally to the combining of M and N into the single three by three complex, symmetric matrix

$$\mathcal{R} \equiv M + iN \tag{1.24}$$

and the definition of complex 2-forms basis for $\Lambda^2(\mathbb{C})$,

$$f^a \equiv u^a + iv^a \tag{1.25}$$

It is then easy to see that (1.25) and (1.23) are equivalent to

$$f^a \wedge f^b = i\delta^{ab} \star 1 \tag{1.26}$$

and

$$f^a \wedge \bar{f}^b = 0 \tag{1.27}$$

It is then easy to complexify the connection forms,

$$X^a = \omega^b_{\ c} - i\omega^0_{\ a} \tag{1.28}$$

and arrive at the structure equations,

$$df^a = f^b \wedge X^c - f^c \wedge X^b, \quad (a, b, c) = \text{cyclic } (1, 2, 3) \tag{1.29}$$

and finally the curvature equation

$$dX^a - X^b \wedge X^c = \mathcal{R}^a_n f^n + \mathcal{Q}^a_n \bar{f}^n, \quad \text{sum } n = 1, 2, 3 \tag{1.30}$$

The vacuum Einstein equations, Ricci flatness, can then be expressed by requiring **only** f^a, **and not** $\bar{f}^a$ on the right hand side of (1.30), or $\mathcal{Q} = 0$. More simply, the vacuum Einstein equations are equivalent to (1.26), (1.27), and (1.31),

$$dX^a - X^b \wedge X^c = \mathcal{R}^a_n f^n \tag{1.31}$$

As in similar algebraic matrix problems, the final forms resulted in possible degeneracies and natural division into distinct types. Petrov presents in his book [19] a complete and extensive approach to his results. Certainly we know the relationship of the Lorentz group, $SO(1, 3, \mathbb{R})$ to $SL(2, \mathbb{C})$. However, this formalism seems to correspond to the group $SO(3, \mathbb{C})$. In terms of the Lie algebras, $\mathfrak{so}(1, 3, \mathbb{R})$ is clearly 6 real dimensions, $\mathfrak{sl}(2, \mathbb{C})$ consists of all traceless $\mathbb{C}$ two by two so $4 - 1 = 3$ complex $= 6$ real dimensions, and $\mathfrak{so}(3, \mathbb{C})$ consists of all traceless, antisymmetric three by three $\mathbb{C}$ matrices, again 3 complex $= 6$ real dimensions.

In the case of algebraic degeneracies, successive covariant derivatives of the Riemann tensor, or its complex form, may be required. Obviously the necessary algebra involved in this process is clearly very complicated. At that time, I was beginning to become slightly computer literate, primarily thru FORTRAN. I was never interested in the numerical calculations expedited by such languages, but the logical branching it provided led me to investigate the possibility to do purely formal, symbolic, algebra and differential manipulations in a computer program designed to represent the Riemann tensor and its covariant derivatives by indexed objects and then perform algebraic operations needed to extract invariant information from them as generalized eigenvalues/eigenfunctions. I knew that others were doing symbolic manipulations with computer programs, such as the developing system Macsyma and currently packages such as Maple and Mathematica. Nevertheless since this was relatively novel at the time, I managed to publish my first and only computer paper [21]. As clumsy as it was, my FORTRAN package did help in post-Petrov work [22].

Note that Petrov's formalism involves finding standard form for complex three by three matrix and results in preferred vectors and scalars. It occurred to me that this invariant information might be used to (at least partly) determine local tangent vector frames as well as coordinates to the extent allowed by possible degeneracies. Note that the problem of finding canonical forms for complex matrices operating on $\mathbb{C}^3$ differs from that for real matrices on $\mathbb{R}^3$ in that non-zero complex vectors may have null magnitude using the complex Euclidean pseudo norm, replacing $\wedge$ product in (1.26) with that provided by the Euclidean three metric. The form of these equations seduced me into a (so far useless) search for a possible relationship between the complex two-form basis, f^a, and an arbitrary holomorphic basis for the **one** forms over $\mathbb{C}^3$, but with the Hermitean metric replaced by the pseudo-Eulidean in (1.26), (1.27). The temptation was instigated by the fact that all solutions to the two dimensional vacuum electrostatic field equations, $\Box_2 V = 0$ can be explicitly stated simply as the real or imaginary parts of any arbitrary holomorphic function of one complex variable. So, could we simply replace the **two-forms,** f in the preceding equations by holomorphic complex one-forms, $\sigma^a(z^b)$ of three complex variables, replacing the $\wedge$ in (1.26), (1.27) by the euclidean inner product (not a norm over $\mathbb{C}$). Thus, the Einstein equations as expressed in (1.26), (1.27), (1.29) and (1.31), bear a striking resemblance to the complex three dimensional geometry of $\mathbb{C}^3$ with the only requirement being that the basis for two-forms, σ^a, be holomorphic. Needless to say, so far such a correspondence has eluded me.

Before ending this discussion of invariant studies of Ricci flat metric geometries, I must refer to the result of Schmidt [23] noting that the completeness of the invariants found using these techniques **for a Lorentz signature** metric, is **not** guaranteed by the classical arguments involving the positive definite Riemannian geometry of $SO(4)$. In particular, in this paper he noted that the calculation of **all** invariants from the Riemann tensor and all of its derivatives can be zero, without having the space be flat, for certain explicitly stated gravitational wave equations. Thus, for any

$$ds^2 = 2dudv + a^2(u)dw^2 + b^2(u)dz^2 \tag{1.32}$$

such that

$$a\frac{d^2b}{du^2} + b\frac{d^2a}{du^2} = 0 \qquad (1.33)$$

the Ricci tensor is zero, but not all components of the Riemann tensor are zero if at least one of the second derivatives in (1.33) is not zero. However, all invariants formed from the Riemann tensor must be zero. Schmidt also pointed out that the reason for this discrepancy lies in the fact that $SO(1, 3)$ is not compact, whereas $SO(4)$ is.

When I developed and published my work on this subject, first [21], and later, in the 1970s [22, 24–26], I was unaware of the complete analysis of these tools by Thorpe [27] and others years earlier. My work was certainly not the earliest on the subject, bringing back my "re-discovery" of the work of Jordan and others on ST theories in beginning my thesis.

During the 1960s I also briefly looked at some of the problems associated with Komar's claim of an explicit presentation of some form of Wheeler's "gravitational geon," some sort of object held together by its own gravitational energy, at least for some time [28, 29].

1.4   1970s Alternative Geometric Formalisms and a Bit on Quantum Logic

During the early 1970s I continued my work on the geometric invariants problems [25, 26, 30, 31], but had to completely abandon any hopes that a holomorphic one-form structure over Euclidean $\mathbb{C}^3$ would lead to any results on the actual real, four-dimensional Einstein equations in (1.26), (1.27), (1.29) and (1.31). There probably is no way to relate frames of complex one-forms $\Lambda^1(\mathbb{C})$ over $\mathbb{C}^3$ to $\Lambda^2 \otimes \mathbb{C}$, even though the formalisms are deceptively similar. Another intuition that did not pan out.

During 1973–1974 academic year I had the opportunity of again spending time in the Princeton physics department. It was there that I first met Dave Wilkinson, Friedrich Hehl, Bahram Mashhoon, Eckehardt Mielke, and others. Of course I again resumed conversations with Bob Dicke, but by then the experimental evidence was strongly suggesting that our proposed form of a scalar-tensor theory was effectively, i.e., for reasonable values of ω, no more valid, at least in the solar system scales, than standard Einstein theory. Dicke had tried to resurrect the notion of an oblateness in the sun's mass distribution sufficient to adjust predictions of orbital data to conform to observations even with a reasonably small value of ω. However, Henry Hill, one of Dicke's students then at Arizona had made observations sufficiently accurate that the sun was not sufficiently oblate to provide the Dicke's proposed adjustmenta of ω back down to reasonable values. So it soon became clear to me that Bob had resigned himself to the situation that observations had driven the value of ω so high as to make the theory irrelevantly different from standard Einstein theory. Jim Peebles and Dave Wilkinson were close workers with Dicke at that time, and Bob seemed to have reluctantly given up interest is scalar-tensor theory[5] and taken up more general cosmological problems, observational and theoretical. Penzias and Wilson had approached Bob and Jim to discuss background universal cosmological radiation they had found. Bob and Jim suggested that this might be the black body remnants of the big bang cooled off to a present temperature in the four degree range. Of course, it was Penzias and Wilson who received the Noble prize for their observation of this phenomenon. The two groups decided to publish sequential articles in the Astrophysical Journal [32, 33].

Bob Dicke and Dave Wilkinson had begun construction of models of antennae to provide confirmation of the 4° black body hypothesis. I seem to recall that they were driven to improve the tools to detect sufficient frequency vs intensity data to confirm the black body characteristics of the radiation. Specifically, I heard a lot of talk about the importance of getting points surrounding the maximum in the Planck curve to confirm black body radiation at this temperature. Dave was also particularly interested in observing possible anisotropy of the radiation and designing detectors to do this. I believe the inclusion of Dave's name in the widely used WMAP term is in honor of his work. I spent some time listening to these exchanges, but the details of equipment design were essentially incomprehensible to me. I now realize that I was just an occasional observer of the work of people who were laying the groundwork for much of modern cosmological studies. However, after learning more from Dave Wilkinson, I published a paper [34] pointing out that background gravitational anisotropies could scramble polarization properties of the big bang radiation as now observed.

In spite of Bob's deep disappointment that our theory was not really relevant, at least in the solar system context, it is clear to me now that the wide promulgation of a theory seriously competing with Einstein's general relativistic gravitational one served as a very important motivation to re-activate experimental interest in gravity. The time was coincidentally very propitious in that it was presented when satellite explorations of the solar system were beginning and such activity now was provided with the additional impetus of being important in testing theories of (classical) gravity. However, it seems to me that before our 1961 paper [10], the lack of a widely known alternative to Einstein's equations seemed to have put a damper on motivations to experimentally investigate its validity after the early confirmation of these equations in the well known three standard tests: frequency shift and deflection of light, and the rotation of Mercury's perihelion. Largely because of Dicke's high standing in experimental physics this paper seems to have reinvigorated interest in experimental studies of gravity. Perhaps this was the most important aspect of our form of a scalar-tensor theory.

While Dicke, Wilkinson et al., pursued their very well known and important experimental work, my interests continued to be focussed on the mathematical and theoretical aspects of our understanding of classical, i.e., non-quantum, spacetime.

---

[5]In fact some 20 years later inflationary cosmology was to lead to renewed interest in the addition of a classical scalar field to the metric.

The question of the validity of Einstein's argument [35] that in general relativity the inertial mass of a point particle is affected by the gravitational field of all other masses continued to be claimed in various papers. I wrote another brief reply [36] showing that Einstein had really only defined a purely coordinate effect and thus not invariantly observable.

On returning to Loyola, I began to learn a bit about quantum logic from a new colleague, A.R. Marlow, specifically quantum logic [37] developed from the notion of state as a projection operator in Hilbert space. This presents many interesting comparisons with the Boolean logic associated with the classical description of state as an element of a point set. Marlow managed to invite a wide variety of workers in the field to several meetings at Loyola [38]. Although I was mainly learning about these formalisms, the Einstein-Bohr debate was obviously at the basis of the argument that quantum theory could not be expressed using Boolean algebras as inclusion in sets produce for classical or "hidden variable" models. This later led me look more carefully at Bell's theorem, especially in the next decade.

In the meantime, returning to classical physics and the mixture of topology with relativity theory, there is an interesting variation on the returning twin paradox in flat space that I looked into with Stewart. In teaching and discussing the twins and their spacetime world lines with beginning students I usually referred to the common explanation that at least one of the twins might be distinguished from the other by having to accelerate to return for local clock comparisons. However, I had also been introducing the possibility of non-trivial topology. So I thought about another example of the paradox in only one space and one time, (x, t), coordinates by simply topologically identifying two points, say $x = 0, t = t_0$ with $x = 1, t = t_0$ thus describing a two-dimensional cylinder. Stewart and I wrote a brief article [39] on this explaining that one observer **is** distinguished by remaining at rest in a coordinate systems where the spatial identifications were made simultaneously.

1.5 1980s Bell's Theorem Comments and Further Studies of Scalar-Tensor Consequences

In the late 1970s I began to look into quantum logic. Marlow organized a meeting on these topics with the proceedings [38]. While these ideas are certainly very interesting I did not dig deeply in this area as such. However, the topics led me to look again into the status of the very early Einstein-Bohr debate about what is now generally called "hidden variables." These variables are classically deterministic, but currently beyond our direct perception but determine the result of each quantum measurement. The observed quantum probability spreads are then simply a result of our ignorance of the details of the hidden variables distribution. However, Bell [40] proved a very important theorem that the use of classically deterministic hidden variables for the statistics of a class of experiments in which detector settings were made by "random" settings of separated detectors in spacelike (thus not causal) events could not reproduce the values predicted by quantum theory. It is important to note the use of apparently contradictory descriptions: "random" and "deterministic."

In more detail, consider a pair of particles created with spin zero moving in opposite directions and two detectors are set to measure the spin of each particle in directions **a** and **b** respectively. For simplicity record each measurement as $A(\mathbf{a}) = \pm 1$, and $B(\mathbf{b}) = \pm 1$, respectively. Let σ_1, σ_2 be the spin operators for the two particles. Then it is easy to show that the average of the products of these measurements predicted by standard quantum theory (again, with spin normalized to ± 1)

$$\langle \sigma_1 \cdot \mathbf{a} \, \sigma_2 \cdot \mathbf{b} \rangle = -\mathbf{a} \cdot \mathbf{b} \qquad (1.34)$$

Following Bell's notation, let λ represent the hidden variables which completely determine the outcome of each individual measurement so the values of each measurement, A and B are determined by the value of λ, which can be statistically distributed arbitrarily with measure $\rho(\lambda)d\lambda$. Then, the overall average of the individual measurements will result in

$$\langle \sigma_1 \cdot \mathbf{a} \, \sigma_2 \cdot \mathbf{b} \rangle = \int \rho(\lambda) A(\mathbf{a}, \lambda) B(\mathbf{b}, \lambda d\lambda \qquad (1.35)$$

Bell is then able to show rigorously that the result in (1.35) cannot duplicate that of (1.34), nor even come arbitrarily close to it **if the choices of settings a, b are made "randomly" and at events that cannot be subluminally related**.

It seemed to me that there is some circularity in this argument, since it assumes that choices for the direction **a, b** are truly random and not themselves subject to some hidden varibles. That is, true randomness (non-causal) in the settings must be assumed to prove that quantum measurements are not predictable in the classical sense. In fact, clearly the two events, choices of the settings (**a, b**) have intersecting past light cones, so, in principle, some past values for λ will determine not only the result of each measurement but even the choice of setting for each measurement. In other words the result $A(\mathbf{a}, \lambda)$ must be replaced by one in which the vector **a** is also a function of λ so we should write $A(\mathbf{a}(\lambda), \lambda)$, $B(\mathbf{b}(\lambda), \lambda)$ in (1.35) With this assumption everything is determined and (1.34) can be exactly reproduced. I explained this is more detail in [41].

Of course many others have been involved in this discussion. Mermin [42] is a nice review as of 1993. For another important example, see the work of deRaedt et al. [43] for an actual realist model for the Einstein-Podolsky-Rosen-Bohm experiment.

During this time I have kept up with a small bit of the literature related to ST theories. Of course variations on our original formalism [10] were obviously to be expected. A couple of papers in this direction caught my eye and resulted in comments by me. Dicke and I had been motivated by conservation of matter in our formalism so that $(T^{\mu\nu}_{matter})$; $\nu = 0$. Note that originally we had tried to be keep separate the tensors on the right hand side of the field equation into those for matter, $T^{\mu\nu}_{matter}$ and for ϕ, $T^{\mu\nu}_{\phi}$. However, the exploration of alternatives to this formalism sometimes disregarded this separations.

One of these was was a model for a "Self-Creation Cosmology" [44, 45] involving the addition of a scalar field in such a way that the the stress energy tensor for matter does not satisfy the conservation equation, so that matter was created, It turns out that this formalism was in fact not self consistent.

Probably the most popular path was to explore conformal metric changes to replace the ϕ formalism by non-linear terms in the curvature scalar, R, in the Lagrangian. One example is [46, 47]. Actually Dicke [13] as early as 1962 considered such changes, but using the phrase "transformation of units," rather than "conformal changes of the metric." Perhaps the use of this different terminology prevented the recognition of Dicke's early work by later workers. I published a paper [48] discussing this matter [41], pointing out that a conformal change of the metric does indeed change the physics, detectable for example by changes in non-null geodesics. Thus, a conformal metric transformation of the metric resulting from one set of field equations would be detectable in observation of planetary orbits, etc.

Another question of non-standard terminology is the expression that a conformal metric transformation can take a ST theory into Einstein's field equations describing this process as using an "Einstein frame" rather than a "Jordan frame." I believe this terminology can be misleading, because, what is happening is a change in metric. Also, as mentioned above, a conformal metric transformation results in physically detectable changes in geodesics. However, in terms of arriving at some form of an ST action, the conformal transformation

$$g_{ij} \to \bar{g}_{ij} = e^{2\psi} g_{ij} \tag{1.36}$$

results in a **vacuum** action

$$S = \int \sqrt{-g}R \to \bar{S} = \int \sqrt{-\bar{g}}[\bar{R} - \frac{3\nabla\lambda \cdot \nabla\lambda}{2\lambda^2}] \tag{1.37}$$

Actually, Dicke included matter and the ST field ϕ and their conformally changed expressions ($\bar{\phi} = \phi/\lambda$) in his paper.

The vacuum form in (1.37) is presented here to show that a form of ST theory can be obtained from standard Einstein theory by a conformal transformation of **both the metric and matter fields**.

1.6   1990-Today: More ST History

This was the period in which a good number of people expressed interest in my thoughts and recollections of the history of so-called Brans-Dicke theories. Consequently I wrote a few papers, some for existing web pages, some based on talks [15, 16, 49, 50]. In most of these, I included at least some of my recollections of my interaction with Dicke as well as brief, incomplete, comments on the history of the influence of the introduction of scalar-tensor formalism on more recent cosmology and even string theory. I believe one of my talks was titled something like "The rise and fall and resurrection (?) of scalar fields in gravity." In addition, in 2003 Fujii and Maeda [51] published an excellent review book on the topic of scalar-tensor theories.

Many of the questions centered around when Dicke first knew about Jordan's scalar-tensor theory. As I recalled earlier in this paper, I first learned of it when I was essentially finished with establishing a formalism in my thesis that I thought would fit Bob's specifications. At that time, he told me simply to provide sufficient reference but that what would come to be known as the Brans-Dicke theory was independent of Jordan's by virtue of its being motivated by Mach's principle. In the 1990s I had the opportunity to pursue the Jordan-Dicke connection more definitively.

During my stay as a visitor to the Institute for Advanced Studies in Princeton in 1993 I took advantage of an opportunity to record a brief conversation with Bob and specifically mentioned that I first became aware of Jordan's ST theory sometime around 1958 or 1959 and brought up the subject with him for advice on the effect of this on my thesis. I then asked Bob when he learned of Jordan's work and he replied "...I think it was about that time for me also..." Of course in 1991 Bob's illness had taken a severe toll on him, so this obvious lapse (recall [52]) was entirely understandable. I also took advantage of a visit to NYU around this time. Peter Bergmann also agreed to allow me to tape a conversation with him in which I also referred to the question of ST and his and Einstein's involvement with some form of a ST theory. To clarify things, Peter sent me a letter in 1991 with more specific information on his papers, "Unified Field Theory with Fifteen Field Variables," [53] in 1948, and "Comments on the Scalar-Tensor Theory," [54] in 1968. The letter also mentioned Einstein's thoughts on the implications of Kaluza's theory sometime in the late 1930s. He also pointed out that Jordan was able to smuggle proofs of his own paper on ST out of then occupied Germany to Pauli, then at Princeton. According to Bergmann, Pauli then turned this paper over to him, describing it as "nonsense," but asking for his (Peter's) opinion of it. I believe that I was very fortunate indeed to have been able to communicate with both Bob Dicke and Peter Bergmann on these and other matters. This was described more fully in Bergmann's 1948 paper cited above.

During the 1990s Fred Hehl was especially helpful to me for various reasons. First was his introduction of me to Torsten Asselmeyer-Maluga in Bad Honeff, 1995. I had just begun to publish a little on the subject of exotic smoothness and Torsten was also interested in such things. So Hehl was instrumental in connecting me to my most important collaborator for the next 20 years. Later Hehl had generously hosted my visit to his University in Cologne in 1998. He also enabled me to visit other groups around Germany.

1.7 1990-Today: Exotica

My serious interest, which certainly continues today, in the strange happenings in the world of differential topology began during this period. Admittedly all of this work is on **classical**, that is, non-quantum spacetime models. However such models are still widely used much of theoretical physics. Sometime, probably in the 1980s, I heard Ron Fintushel, then at Tulane, give a mathematics seminar in which he talked about new, non-standard smooth (C^∞ differentiable) structures on $\mathbb{R}^4$. It took a while for the impact of this statement to sink in, but during the question period I really deluged him with very basic questions about what this might mean. The counter-intuitive impact of these ideas makes them quite surprising and I may even have said that this idea does not make sense at all. But fortunately Ron is a very nice person and patient with mathematical dilettantes like me. He assured me that this statement has to be analyzed carefully and its meaning is not at all obvious but does make sense and is very important. Perhaps the rather cryptic statement:

Differences: *There is a difference between different and non-diffeomorphic smoothness structures on a given topological manifold*

was difficult for me to understand, but fortunately many colleagues have helped me along the way, especially Torsten Asselmeyer-Maluga and Duane Randall. Many thanks to them. For the reader I would recommend especially three books on the subject [55–57].

In fact, it had been proven that for every $n \neq 4$ there was only one smoothness structure, up to global diffeomorphisms, on $\mathbb{R}^n$. If $n \neq 4$ that structure is generated by global topological coordinates, $(x^1, \ldots, x^n)$ and by defining a function, $f : \mathbb{R}^n \to \mathbb{R}$, as differentiable if derivatives to all orders are well-defined when f is expressed as $f(x^1, \ldots, x^n)$ in the usual sense of real analysis. In the 1980s the uniqueness of the standard smooth structure on $\mathbb{R}^n$ for $n < 4$ had been settled using more or less ad hoc methods, not practical for $n = 4$, whereas for $n > 4$ the methods of cobordism and handlebodies had succeeded in establishing uniqueness.

However these techniques could not be applied to dimension less than five. A popularized description of the problem was presented by Freedman [58]. By the mid 1980s the work of Donaldson, Freedman et al. led to an existence proof that there must exist at least one smoothness structure on spaces which are homeomorphic to $\mathbb{R}^4$ but not diffeomorphic to the standard one (i.e., identified with some global topological coordinates). It turns out that the same is true for the existence of exotic smoothness on spaces which are homeomorphic to $\mathbb{R} \times S^3$, $\mathbb{R}^2 \times S^2$ and perhaps other relatively topologically simple manifolds. From this time on I have been mesmerized by the mathematics that dealt with such questions, "differential topology".

How/why would this be important to physics? In fact diffeomorphisms are global extensions of the idea embodied in Einstein's principle of General Relativity requiring that physics can be done in any (implicitly assumed to be smooth) coordinate system. Different coordinate systems must lead to different **expressions**, but not **physical content**, of any physical theory. Of course almost all theories involve doing calculus on certain functions, for example simply expressing a differential equation means that we know how to and are able to take derivatives. Such apparently trivial operations may now actually involve hidden assumptions about how do this even on a topologically simple space such as $\mathbb{R}^4$. The discovery of exotic $\mathbb{R}^4$'s means that this assumption is not trivial and involves the choice of a particular global smoothness.

There are several facts that suggested to me that these results ought to be investigated as possibly physically significant:

- **Critical dimension four** The arena of much of the mathematical research on exotic smoothness involves the physically significant dimension four.
- **Yang-Mills type moduli spaces** The manifold of cross sections of $SU(2)$ bundles, modulo gauge transformations, was found to contain one of the first examples of an exotic $\mathbb{R}^4_\Theta$ [59, 60].
- **Cosmology** The existence of exotic $\mathbb{R} \times S^3$ opens the door to possible exotic cosmological models [61].
- **Black holes** The existence of exotic $\mathbb{R}^2 \times S^2$ which is the topology of the Kruskal description of the non-singular region of the Schwarzschild solution [62].
- **Milnor spheres** Skipping to dimension seven, as early as 1956, Milnor [63] constructed spaces, Σ^7, which are topological S^7 from Yang-Mills type $SU(2) = S^3$ bundles over S^4. These Σ's could not be diffeomorphic to the standard S^7.

It is important to note that factorization of the topological products above need not be extended to the smooth realm. Thus exotic $\mathbb{R}^4$ is homeomorphic, but not diffeomorphic to $\mathbb{R} \times \mathbb{R}^3$. Actually, these alternatives to the standard approach can even be extended to the topological level. For example Whitehead continua are topological manifolds which are homeomorphic to $\mathbb{R}^4$ and to $\mathbb{R} \times W$, where W is a three manifold **not** homeomorphic to $\mathbb{R}^3$.

My immediate attachment to this field is due to the fact that I have always been interested in all examples of the cross fertilization between physics and (otherwise) pure mathematics. Examples and anecdotes abound. The development of geometry (and its generalizations) as mathematics and as physics is especially noteworthy. In the 1700s mathematicians began in earnest the questioning of the minimal structure of Euclid's axioms, in particular, the necessity or not of including the postulate pertaining to parallel lines. Max Jammer [64] summarizes this history very well. Underlying the mathematical discussion is the additional question of whether or not the axioms are physical or mathematical in nature. Of course, today, we are quite comfortable with the separation of pure mathematics from physics, but this has not always been so clear. Thus, for example, as described in [64], Gauss actually performed a physical experiment with surveying equipment to determine if the sum of the angles in a triangle is indeed π, as it should be in flat, Euclidean, geometry. He bounced light off of three mirrors constituting the three vertices of a triangle. Of course, with the technology available to him at the time, such an experiment could be done with only crude accuracy, but it presaged a whole set of experiments on the behavior of light rays undertaken over the last 30 years or so within the solar system.

Such work by Gauss, Riemann, Lobachevski and others on the apparently very abstract and non-physical subject of "non-Euclidean" geometry was precisely what was needed to provide the foundation for Einstein's theory of General Relativity, in which gravity is described in terms of the geometric properties of spacetime. The path by which Einstein was led to consider what must have appeared to him to be very abstruse and abstract mathematics as a possible tool for physics has recently been reviewed in the various volumes celebrating the centennial of his birth.

Later investigations of Einstein's theory led to the natural introduction of non-trivial topology in addition to geometry. In the meantime, the parallel development of quantum theory and quantum field theory has led to the introduction into physics of branches of mathematics such as function theory, Hilbert spaces, bundle theory, moduli space structures, etc. In fact, the second half of the twentieth century has seen a virtual explosion of applications of various branches of mathematics, many of which were considered to be of only abstract interest, to physics. Conversely, in many cases the direction of "applicability" has been reversed, some of which will be touched on later. Questions of interest in theoretical physics have turned out to have value in the pursuit of "pure" mathematics.

In summary, the rich interplay between physics and mathematics is obvious to contemporary workers. Certainly, there is no theorem that says "Good mathematics makes good physics," but certainly there is strong anecdotal evidence that this has been true in many important situations. So I was strongly motivated to look into whether or not these new smoothness structures could have physical significance.

My initial puzzlement in approaching differential topology was due to confusion of several terms:

- **Different smoothness** Given a topological manifold, M, $p \in M$, and a choice of two coordinate patch atlases, locally $x^i(p)$ and $y^i(p)$, which are to be **smooth** then either

  1. **Coordinate tranformation** Around each p there is a neighborhood in which the $y's$ are smooth and smoothly invertible functions of the $x's$. The two coordinate systems clearly represent the same physics in the sense of general relativity, or
  2. **Homeomorphism** (1) is not true, but there is a **homeomorphism** of $M \to M$, i.e., rearranging points $p \to p' = h(p)$ so that the combination $y^i(h(p))$ is a smooth, smoothly invertible function of the $x^j(p)$. Thus, without tearing holes in space, making wormholes, etc., we can replace one coordinate system by another. An example of this is the transition from Schwarzschild to Kruskal coordinates, so there is no change in the physics but functions that are smooth in one system may not be in the other system. For example, the metric components in Schwarzschild are apparently singular at $r = 2M$, **are** smooth in Kruskal.

  In either (1) or (2) the choices of local coordinates are **diffeomorphic** to each other. or,

- **Non diffeomorphic** Neither (1) nor (2) is true. This means that without tearing or cutting or any other topological changes, we can have truly different physics, not just the same physics in different coordinate system. That this happens in a manifold as topologically trivial as $\mathbb{R}^4$ is the great surprise coming from the work of Freedman, Donaldson, et al.

Consider a simple 1-dimensional example. Let $p \in \mathbb{R}$ be a real number. Choose the global smooth coordinate system, $x = p$ numerically. With this choice the class of smooth functions, $\mathcal{F}$ is given by the usual definition of differentiable functions, $f(x)$. However, suppose we had first performed a homeomorphism of $\mathbb{R}$ onto itself, replacing p by p^3. The topology of the manifold clearly has been unchanged. Consider the choice of coordinate, y,

$$y(p) = p^3, \tag{1.38}$$

Now the class of smooth functions, $\mathcal{F}'$, is defined to be those functions $f'(p)$ such that $f_c'(y) \equiv f'(y^{1/3})$ is C^∞ in the usual sense. Clearly, $\mathcal{F} \neq \mathcal{F}'$. The identity map is an element of $\mathcal{F}$, but, since $y^{1/3}$ is not differentiable at the origin, it is not an element of $\mathcal{F}'$. Thus, we have a simple example of one point set, with two *different* smoothness structures, $\mathcal{S} \neq \mathcal{S}'$, provided by the homeomorphism,

$$h(p) = p^{1/3}, \tag{1.39}$$

of $\mathbb{R}$ onto itself which is actually also a **diffeomorphism**. Its coordinate expression is simply the identity map,

$$h_c(x) \equiv y(h(p(x))) = x. \tag{1.40}$$

Thus, from the viewpoint of differential topology, these two **different** smoothness structures on topological $\mathbb{R}$ are actually **diffeomorphic**. Diagrammatically,

$$
\begin{array}{ccc}
\mathbb{R}^1_{TOP} & \xrightarrow{\ x\ } & \mathbb{R}^1 \\
h\downarrow & & \mathbb{I}\downarrow \\
\mathbb{R}^1_{TOP} & \xrightarrow{\ y\ } & \mathbb{R}^1
\end{array}
\tag{1.41}
$$

is commutative. The two horizontal maps are coordinate maps, defining $\mathcal{S}$ and $\mathcal{S}'$, respectively, the left downward map, h, in (1.39) is a homeomorphism, and the combined map expressed in the two coordinate systems, $\mathbb{I}$, is the identity diffeomorphism. The existence of such a diagram provides the fundamental definition of the equivalence, mathematical **and** physical, of two different smoothness structures.

Consider also the Schwarzschild singularity at $r = 2GM$. It was soon discovered that rather than an essential singularity, the anomalous behavior of the metric components at $r = 2GM$ was actually a result of extending the spherical coordinate too far toward the origin. In fact, an alternative extension uses other coordinates, such as proposed by Kruskal et al. The differential geometry depends on the "global" topological question of whether or not $r = 2GM$ is a static sphere, $S^2 \times \mathbb{R}$, if we choose standard Schwarzschild as global (a differential topological choice!), or the set $S^2 \times \{(u, v) | u^2 = v^2\}$, if we choose Kruskal coordinates. Perhaps even more notable is the way in which the $r = 0$ singularity is interpreted. In standard Schwarzschild coordinates it is simply a point times time, $\{pt\} \times \mathbb{R}$, while in Kruskal coordinates it is a hyperboloid, $S^2 \times \{(u, v) | u^2 - v^2 = -1\}$. In fact, we now know that there is an exotic version of this topology.

For most of the twentieth century it has been assumed by physicists that the choice of which coordinate systems are to be **smooth** is trivially determined by the topological coordinates, since most topological models are based on subsets of topologically Euclidean spaces. However, one of the main points of my studies is to review the physical implications of the mathematical discovery that the choice of smoothness is not necessarily uniquely determined by the topology, even for relatively simple topological spacetime models such as $\mathbb{R}^4$, or the closed cosmology, $\mathbb{R} \times S^3$, etc. So, we ask: What are the differential geometric consequences of exotic smoothness?

The short answer to this is that the consequences are global, and refer to the extendibility of local coordinates and the way in which they are patched together. Unfortunately with the present state of the mathematical technology, we can't give a full answer to the question, but can only state some general facts and conjecture for exotic smoothness which will be discussed later.

> **Differential topology:** *the choice of the way in which local coordinates are patched together, influences the physical properties of the spacetime model supporting differential geometry.*

1.8   The Physics of Exotic Smoothness

Can differential topology really have anything to do with physical theories? Clearly the answer to this question must be "Yes" because of the principle of general relativity. In light of this principle, the physical content of theories must be invariant under changes of local coordinate patches, **provided that the new smoothness structure is diffeomorphic to the original one**. This is in fact the prototype of "gauge" theory. However, the discovery of exotic smoothness structures shows that there are **many**, often an infinity, of non-diffeomorphic and thus physically inequivalent smoothness structures on many topological spaces of interest to physics. Because of these discoveries, we must face the fact that there is no **a priori** basis for preferring one such structure to another, or to the "standard" one just as we have no **a priori** reason to prefer flat to curved spacetime models. We note that these exotic structures are by definition all **locally** equivalent, so the local expression of physical laws is unchanged. This leads to the apparently paradoxical fact that the implications of exotic smoothness are global, but not in the topological sense!

Unfortunately, the technical difficulties encountered in applying these new results have resulted in only qualitative results for physical applications so far. However the rich interplay between physics and mathematics is obvious to contemporary workers.

Fortunately there are a class of manageable exotic structures available in the **smooth** category. These were discovered in the late 1950s by Milnor [63] and known as Milnor spheres. The simplest one is an exotic S^7. This space can be realized naturally as the bundle space of an $SU(2) \approx S^3$ bundle over S^4 (which is compactified $\mathbb{R}^4$) using a construction of Hopf. From the physics viewpoint, a Yang-Mills field with appropriate asymptotic behavior is a cross section of such a principal bundle.

Such fields satisfying Yang-Mills field equations are called **instantons**[6] and turn out to be important later in the story of exotica. For now, however, consider the construction of S^7 as the subset of quaternion 2-space, $\{(q_1, q_2) : |q_1|^2 + |q_2|^2 = 1\}$. There is a natural projection of this space into projective quaternion space, $(q_1 : q_2)$. This space, however, turns out to be nothing more than S^4. The kernel of this map is the set of unit quaternions, $S^3 \approx SU(2)$. Equivalently, S^7 can be defined by two copies of $(\mathbf{H} - 0) \times S^3$, with identification

$$(q, u) \sim (q/|q|^2, qu/|q|)$$

Milnor was able to generalize this to produce a manifold, Σ^7 by means of the identification

$$(q, u) \sim (q/|q|^2, q^j u q^k /|q|)$$

Milnor then showed that if $j + k = 1$ the space Σ^7 is *topologically* identical (homeomorphic) to S^7. However, if $(j - k)^2$ is not equal to 1 mod 7, then Σ^7 is **exotic**, that is not diffeomorphic to standard S^7.

However, when we return to exotic $\mathbb{R}^4_\Theta$, the extensive number of mathematical tools to investigate this subject precludes any significant summary to the non-specialist. The existence of $\mathbb{R}^4_\Theta$ has resulted in much interest and many developments in the mathematics of differential topology. This is certainly good for mathematics, but not so good for their use in physics. Not many physicists are well versed in subjects such as Morse theory, characteristic classes, handlebody construction, cobordism, etc. so it is difficult to do more than skim over some of what is known.

After the initial discovery of one $\mathbb{R}^4_\Theta$, rapid progress was made in discovering (more in the sense of existence rather than construction) various $\mathbb{R}^4_\Theta$'s, and classifying them. For example, Gompf has a paper entitled "An Exotic Menagerie," [65], showing the existence of an uncountable number of non-diffeomorphic $\mathbb{R}^4_\Theta$'s. Gompf's construction makes extensive use of handlebody chains, which apparently must be infinite. Freedman and Taylor [66] show the existence of a universal $\mathbb{R}^4_\Theta$ in which all others can be smoothly embedded. Also, as a note for use below, it turns out that some $\mathbb{R}^4_\Theta$'s can be smoothly embedded in standard $\mathbb{R}^4$, and others cannot.

Recently, field equations suggested by Seiberg and Witten [67] show great promise for simplifications of the study of moduli spaces. However, to date, it is unfortunately true that

- **Fact** No finite effective coordinate patch presentation exists of any exotic $\mathbb{R}^4_\Theta$.

Nevertheless, even in the absence of a manageable coordinate patch presentation, certain features can be explored. Some are summarized in results from previous papers.

---

[6]There can be no global non-zero vector field on S^4 for topological reasons, and thus no Minkowski signature metric.

1.9   Some Geometry and Physics on $\mathbb{R}^4_\Theta$

Let us begin here by simply stating a strikingly counter intuitive fact as well established mathematically:

Fact: *There exist global smoothness structures on topological $\mathbb{R}^4$ which are not diffeomorphic to the standard one. We label such manifolds $\mathbb{R}^4_\Theta$.*

For our purposes a remarkable feature of these exotic $\mathbb{R}^4_\Theta$'s is that each of them contains a compact set that cannot be contained in the interior of any smoothly embedded S^3. See, for example, the discussion in pages 366ff of Gompf and Stipsicz [56].

For some $\mathbb{R}^4_\Theta$, there exist global topological coordinates (t, x, y, z) and numbers $R_1 < R_2$ such that the spheres, S_{R_0}, defined by $t^2 + x^2 + y^2 + z^2 = R_0^2$ are smooth for $R_0 < R_1$, but are not smoothly embedded for any $R_0 \geq R_2$. Choose one such, say M, for our spacetime model.

We can thus state that for M

No single covering patch: *We can choose two sets, a and b in M such that both cannot be included in one smooth coordinate patch in **any** diffeomorphic presentation of M*

So, in general the attempts to interpret information received in a from b involve the a priori assumption that there is a single smooth coordinate patch including both of them, which is not necessarily true.

Thus the null geodesics from b could still be smooth and well behaved throughout their length, and the Einstein equations satisfied with normal matter, but it might be incorrect to assume that we can extrapolate from these incoming geodesics in a information about b **because we do not know the non-trivial transition function between the smooth coordinate patches linking the two sets.**

- **Cosmology**

  The coordinates (t, r): *Observational cosmology uses these, but now we know that they may not be smoothly extendible indefinitely into the past.*

If this is the case then the standard extrapolation of earth based observations to distant phenomena may not be justified.

More specifically, in observational astronomy it is generally assumed that the metric can be written in the FRW form

$$ds^2 = -dt^2 + a(t)^2 d\sigma_3^2, \qquad (1.42)$$

where the spatial three metric is usually expressed in spherical coordinates in a form depending on assumptions of isotropy and homogeneity. The associated topology is thus $K_\Theta = \mathbb{R}^1 \times M^3$ for some three-manifold, M^3. In the standard models the three metric is one of the three constant curvature ones, each containing a "radial" coordinate,[7] r. Because of isotropy, the incoming geodesics are described globally (modulo the proviso in the footnote) by differential equations involving r, t only. However, if K_Θ is exotically smooth, these coordinates cannot be globally smooth. Hence the actual metric would have to be expressed in terms of more than one r, t coordinate region, and information extracted from the coordinate overlaps. Unfortunately, because the present mathematical technology does not provide us with an effective coordinate patch structure, more explicit statements than this cannot now be made. Nevertheless, the assumption that we can extrapolate information coming from incoming light rays back in time and out in space as if these geodesics would act as a radial type of coordinate system when indefinitely extended into their past is not valid if K_Θ is used as a spacetime model. We should also note that although we have discussed only the $\mathbb{R}^4_\Theta$ (which is actually $\mathbb{R}^1 \times_\Theta \mathbb{R}^3$) we could equally have chosen an exotic $\mathbb{R}^1 \times_\Theta S^3$.

A simple analogy is provided by gravitational lensing phenomena. Here we see two incoming null geodesics arriving at earth from different directions. However, the possibility that in some reasonable situations they cannot be extrapolated backward as "good" radial coordinates because they have been focused by the gravitational lens effect of an intervening massive object has been widely discussed and generally accepted as viable. Thus the extrapolation of the different angle data for the two incoming geodesics to different sources is incorrect.

- **Gravitational lensing analogy** Null geodesics arriving from different angles may intersect in the past because of gravitational curvature caused by intervening mass and thus may not be extrapolated back as good radial coordinate lines.

What we are proposing here is more radical, of course, but just as viable in the sense that we know of no physical principles to exclude it, and it could lead to an understanding of apparent anomalous distant time behavior without introducing exotic theories or matter, just exotic smoothness of the spacetime manifold model.

- **Exotic structures** Null geodesics arriving from distant sources may not be extrapolated back as good radial coordinate lines because of intervening coordinate patch transformations caused by global exotic smoothness.

In summary, what we want to emphasize is that without changing the Einstein equations or introducing exotic, yet undiscovered forms of matter, or even without changing topology, there is a vast resource of possible explanations for recently observed surprising astrophysical data at the cosmological scale provided by differential topology.

While it is true that at this stage of development of the mathematical technology it is not possible to give explicitly the coordinate patch overlap functions, research along these lines is being actively pursued. Furthermore, Sładkowski [68], has shown

---

[7]Of course in the spherical case the "radial" coordinate is not indefinitely continuable because it is essentially an angular one. However, this is not the sort of coordinate anomaly we are addressing here and can certainly be accommodated in standard models.

that it is possible to relate isometry groups (geometry) to differential structures in some cases.

Related to this is the question that naturally arises concerning the given global topological coordinates, $\{p^\alpha\}$, which define the topological manifold $\mathbb{R}^4$, and their relationship to the local smooth coordinates given by the coordinate patch functions, ϕ_U^α. Both provide maps from an abstract $\mathbf{p} \in \mathbb{R}^4$, into $\mathbb{R}^4$ itself. Clearly the global topological coordinates cannot themselves be smooth everywhere since otherwise they would provide a diffeomorphism of $\mathbb{R}^4_\Theta$ onto standard $\mathbb{R}^4$. But can they be locally smooth? This is answered in the affirmative by

- **Fact** There exists a smooth copy of each $\mathbb{R}^4_\Theta$ for which the global C^0 coordinates are smooth in some neighborhood. That is, there exists a smooth copy, $\mathbb{R}^4_\Theta = \{(p^\alpha)\}$, for which $p^\alpha \in C^\infty$ for $|\mathbf{p}| < \epsilon$.

The implied obstruction to continuing the $\{p^\alpha\}$ as smooth beyond the ϵ limit presents a challenging source for further investigation. Related to this is a the defining feature of the early discovery work of $\mathbb{R}^4_\Theta$'s, namely the non-existence of arbitrarily large smoothly embedded three-spheres.

There are also certain natural "topological but not smooth" decompositions. For example,

- **Fact** $\mathbb{R}^4_\Theta$ is the topological, but not smooth, product, $\mathbb{R}^1 \times_\Theta \mathbb{R}^3$ is not smoothly equivalent (diffeomorphic) to $\mathbb{R} \times \mathbb{R}^3$.

Many interesting examples can be constructed using Gompf's "end-sum" techniques [69]. In this construction topological "ends" of non-compact smooth manifolds are glued together smoothly, $X \cup_{end} Y$. If one of the manifolds, say X, is also topological $\mathbb{R}^4$, the topology of the resultant space is unchanged, that is $\mathbb{R}^4 \cup_{end} Y$ is homeomorphic to Y. However, if X is an $\mathbb{R}^4_\Theta$ which cannot be smoothly embedded in standard $\mathbb{R}^4$, then neither can the the end sum. Thus,

- **Gompf's end sum result** If $X = \mathbb{R}^4_\Theta$ cannot be smoothly embedded in standard $\mathbb{R}^4$, but Y can be, then $\mathbb{R}^4 \times_\Theta \cup_{end} Y$ is homeomorphic, but not diffeomorphic to Y.

This technique will be used further below.

To do geometry we need a metric of the appropriate signature. It is a well known fact that any smooth manifold can be endowed with a smooth Riemannian metric, g_0. This follows from basic bundle theory [70]. Similarly, if the Euler number of X vanishes a globally non-zero smooth tangent vector, u exists. g_0 and u can be combined then to construct a global smooth metric of Lorentz signature, $(-, +, +, +)$, in dimension four. A generalization of this result follows also from standard bundle theory [70].

- **Theorem** If M is any smooth connected 4-manifold and A is a closed submanifold for which $H^4(M, A; \mathbf{Z}) = 0$, then any smooth time-orientable Lorentz signature metric defined over A can be smoothly continued to all of M.

One immediate conclusion about certain geometries on $\mathbb{R}^4_\Theta$ can be drawn from an investigation of the exponential map of the tangent space at some point, which is standard $\mathbb{R}^4$, onto the range of the resulting geodesics. The Hadamard-Cartan theorem guarantees that this map will be a diffeomorphism onto the full manifold if it is simply connected, the geometry has nonpositive curvature and is geodesically complete [71]. Thus,

- **Theorem** There can be no geodesically complete Riemannian metric with non-positive sectional curvature on $\mathbb{R}^4_{\Theta}$.[8]

The apparent lack of localization of the "exoticness" means that it must extend to infinity in some sense as illustrated by the lack of arbitrarily large smooth three-spheres. However, it turns out to be possible that the exoticness can be localized in a *spatial* sense as follows:

- **Theorem** There exists smooth manifolds which are homeomorphic but not diffeomorphic to $\mathbb{R}^4$ and for which the global topological coordinates (t, x, y, z) are smooth for $x^2 + y^2 + z^2 \geq \epsilon^2 > 0$, but not globally. Smooth metrics exists for which the boundary of this region is timelike, so that the exoticness is spatially confined.

The details of the construction of such manifolds are given in [73]. First, Gompf's end-sum technique is used to produce a $\mathbb{R}^4_{\Theta}$ for which the global topological coordinates are smooth outside of the cylinder, that is, in the closed set $c_0 = \{(t, x, y, z) | x^2 + y^2 + z^2 \geq \epsilon\}$ described in the first part of the theorem. Next, a Lorentz signature metric is constructed on c_0. This metric can even be a vacuum Einstein metric. For more details for the possibilities of smoothly exotic Einstein metrics on $\mathbb{R}^4$ see the paper of Kotschick [74]. The only condition is that the $\partial/\partial t$ be time like on c_0. The cross section continuation result with $A = c_0$ then guarantees the extension of the metric over the full space consistent with the conditions of the theorem. What makes the complement of c_0 exotic is the fact that the (x, y, z, t) cannot be continued as smooth functions over all of it. This result leads to

- **Conjecture** This localized exoticness can act as a source for some externally regular field, just as matter or a wormhole can.

Another set of interesting physical possibilities arise in a cosmological context inspired by the exotic product, $X = \mathbb{R} \times_{\Theta} S^3$, which arises from a puncturing of $\mathbb{R}^4_{\Theta}$. It is not hard to apply the same techniques used above to show that this product can be the standard smooth one for a finite, or semi-infinite range of the first variable, say t. The resulting manifold could then be endowed with a standard cosmological metric. This metric, and even the variable t itself, cannot be continued as globally smooth indefinitely, because of the exotic smoothness obstruction. Recall, however, that X is still a globally smooth manifold, with some globally smooth Lorentz-signature metric on it. Other interesting topological but not smooth products can be constructed by use of the end-sum construction. One interesting example is exotic

---

[8]However there are explicit metrics on Milnor's exotic S^7 [72], and it is known that a Riemannian metric exists on any smooth $\mathbb{R}^4$. The Lorentz signature case is different however, since the existence of a nowhere zero timelike vector would result in a smooth foliation of the manifold which would then reduce it to standard, so $\mathbb{R} \times \mathbb{R}^3 = \mathbb{R}^4$, and thus any such metric must have a singularity.

Kruskal, $X_K = \mathbb{R}^2 \times_\Theta S^2$. Using the cross section continuation theorem above, the standard vacuum Kruskal metric can be imposed on some closed set, $A \subset X_K$, and then continued to some smooth metric over the entire space. However, it cannot be continued as Kruskal, since otherwise X_K would then be standard $\mathbb{R}^2 \times S^2$. In sum,

- **Theorem** On some smooth manifolds which are topologically $\mathbb{R}^2 \times S^2$, the standard Kruskal metric cannot be smoothly continued over the full range, $u^2 - v^2 < 1$.

Finally, we close this section with a brief mention of the possible physical significance of Milnor's exotic seven-spheres, denoted here by Σ^7 (there are seven distinct ones). Recall that the standard S^7 as a Hopf bundle is the underlying bundle space for Yang-Mills ($SU(2)$) connections over S^4, which is compactified $\mathbb{R}^4$. On the other hand, Σ^7 as a bundle is no longer a principle $SU(2)$ bundle, but one associated to a principle $Spin(4)$ bundle. This could be regarded as some sort of generalized or exotic Yang-Mills bundle. It might prove interesting to investigate the possible physical ramifications of this.

Of course, there are many who are skeptical that this particular venture into differential topology will ever produce any result useful for physics. Typical reasons, and possible replies can be listed

- **No explicit coordinate patch** So there is not now, and may not be ever, any explicit coordinate expression for the metric

  - **Reply** But nevertheless certain general results can be proposed.

- **Necessity of singularities** Because an exotic $\mathbb{R}^4_\Theta$ cannot be globally smoothly foliated, there must necessarily be some type of singularity in any Lorentz signature metric.

  - **Reply** But the same is true with many if not most metrics investigated in manifolds with standard smoothness including those on manifolds with standard smoothness, such as collapse to a point, the origin of cosmology, etc.

1.10   Summary and Continuing Work

I had planned to end with some sort of a "conclusion," but the exotica work is ongoing, mostly by Torsten and others. So I will end with a slightly altered summary of significant work in the field, old and new. This has been provided to me by Torsten Asselmeyer-Maluga. Perhaps there is a motto along the lines: "If we cannot assume standard geometry (flat) for our entire universe, why should we assume standard smoothness?"

Perhaps the first paper trying to connect differential topology with physical models of spacetime was by Duane Randall and me [75]. Duane is a topolgist who patiently helped me understand the mathematics. This was followed by other papers [76–78]. In the meantime Sładkowski [79–81] started to investigate a possible relationship between exotic smoothness and particle physics and quantum gravity. About this time it occurred to me we are clearly making an assumption when we continue our spacetime observations, obviously limited to a single coordinate patch, to the entire manifold. Thus it may be that metric information cannot be continued in a model with exotic smoothness in the same way as for standard smoothness. For example, if we locally see the geometry of the Schwarzschild metric, we must make an assumption about the global smoothness structure $S^2 \times \mathbb{R}^2$ before describing the Kruskal diagram. If we extrapolate from our local coordinate patch and assume standard smoothness on $S^2 \times \mathbb{R}^2$, we find either a singularity or a material source. However our hidden assumption of standard smoothness need not be valid, since there are now known to be an infinity of others, none of which are physically equivalent to the standard one which is chosen without a priori justification. In other words it could be that the exotic structure might provide the source of what is seen in our local coordinate patch as a Schwarzschild metric. This speculation is not entirely well-formed and still needs further clarification, but became known as the *Brans conjecture*: exotic smoothness can serve as an additional source of gravity. This was further explored for compact manifolds by Asselmeyer [82] and for the exotic $\mathbb{R}^4$ by Sładkowski [68, 83]. Later this conjecture was extended in [84] to the possible generation for all forms of known energy, especially dark matter and dark energy. Exploring this in the context of dark energy resulted in a partial success [85] where the expectation value of an embedded surface was calculated. This value showed an inflationary behavior with a cosmological constant having a realistic value (in agreement with the Planck satellite results).

In 2004, Pfeiffer [86] discussed a strong relation between a smoothness structure and quantum gravity [87]. In the period 2004 to 2006, Król was able to find a connection to topos [88, 89] and model theory [90, 91] which uncovers some interesting relations to other areas like noncommutative geometry and quantum gravity. He continues to work on these ideas with Asselmeyer-Maluga. Their main focus is the relation of exotic smoothness to quantum gravity, using the theory of foliations [92]. Also, noncommutative geometry provides another tool to study these foliations and get relations to quantum field theory (QFT). For instance, the von Neumann algebra of a codimension-one foliation of an exotic $\mathbb{R}^4$ must contain a factor of type III_1 which is the local observable algebra in local algebraic QFT used to describe the vacuum [93–95]. But why is the foliation of an exotic 4-manifold so complicated? As an example consider the exotic $S^3 \times \mathbb{R}$. Clearly, there is always a topologically embedded 3-sphere but there is no smoothly embedded one. Let us assume the well-known hyperbolic metric of the spacetime $S^3 \times \mathbb{R}$ using the trivial foliation into leaves $S^3 \times \{t\}$ for all $t \in \mathbb{R}$. Suppose $S^3 \times \mathbb{R}$ carries an exotic smoothness structure. Then we will get only topologically, but not smoothly, embedded 3-spheres, within leaves $S^3 \times \{t\}$ (otherwise one obtains the standard smoothness structure, see [96] for instance). These topologically embedded 3-spheres are also known as wild 3-spheres. In [97], a relation to quantum D-branes was presented. Finally it was proved in [98] that the deformation quantization of a tame embedding (the usual embedding) is a wild embedding. Furthermore a geometric interpretation of quantum states naturally presents itself: wild embedded submanifolds are quantum states. This construction depends essentially on the continuum, wild embedded submanifolds always admit infinite triangulations. This approach opens a way to quantize a theory using geometric methods.

The inclusion of matter is also one of the main problems in this theory. For a special class of compact 4-manifolds it was shown in [99] that exotic smoothness can generate fermions and gauge fields using the so-called knot surgery of Fintushel and Stern [100]. Here, the knot is directly related to the appearance of an exotic smoothness structure. This provides a stable but fixed structure of fermions and gauge fields contradicting the results of QFT with a variable number of fermions and gauge fields[9] (where the (virtual) fermions and gauge fields will be generated or destroyed). The paper [101] presents an approach using the exotic $\mathbb{R}^4$ solving the difficulties of [99]. Cosmology is another application of exotic smoothness since the first work [84]. Here a smoothly exotic black hole [102] was constructed to show the absence of a singularity in the interior. The assumption of a smooth spacetime is a strong restriction as shown in [103]. The wild embedding has also a strong impact on the initial state of the universe: as shown in [104], a wild embedded 3-sphere (as quantum state) will pass to smooth 3-manifold by decoherence. This process leads to an exponential increase like inflation [85]. Notable are also the interesting relations to quantum gravity and string theory [105–108].

Appendix

Here I will summarize a selected list of topics from my thesis, finished in 1960, but formally presented and accepted by Princeton in May 1961.

- **Mach's Principle** I reviewed what I knew of it, and especially what I thought Dicke assumed. This of course required a careful look at the question of how a backgrund metric would affect the motion of particles, both point and extended. So this led to the next point.
- **Equations of Motion** During this time the questions associated with the equations of motion of both point and extended (fluid type) particles had been extensively studied by Einstein, Infeld, Papapetrou and others. So, in trying to confirm that I was using an operationally significant procedure for measuring and comparing inertial and gravitational mass for a given background metric. As I recall, Einstein and infeld were primarily concerned with point masses, while Papapetrou looked at fluid type stress-energy tensors.
- **How to measure "mass"** I tried to settle this for inertial mass by looking at motion in an external electric field.
- **Equivalence Principles** I explored both strong and weak as presented to me by Dicke.
- **Variational principles and field equations** I had already presented and explored the formalism Bob and I would use to arrive at equations satisfying the oxymoronic

---

[9]Here we do not distinguish between a fermionic quantum field and a fermion.

phrase "varying ...constant." Then, hearing of the work of Jordan and his group, I first reviewed their formalisms.

- **Static, spherically symmetric vacuum** I looked at the solution in Jordan's formulation known as the Heckmann solution, and then did the same for our formalism but in isotropic coordinates. This led to careful analysis of four qualitatively different metric forms.
- **Does G vary?** I then used various idealized operational procedure to look at what could be said about the dependence of $G \sim 1/\phi$ on the matter distribution in the universe.
- **Boundary Conditions** Of course these had to be carefully defined and were eventually defined in the usual way in terms of some "going to zero or other constant at infinity" procedure.
- **Cosmology** I could find no exact analog to the FRW metric of the Einstein equations, so I simple did some very extensive power series expansions for each of various ranges of arbitrary constants. A very few of you may recall that at that time we were amazed to have a human driven mechanical computer that could both multiply and divide long numbers within 10 s or so.

References

1. L. Steen, J. Seebach, *Counterexamples in Topology* (Holt, Rinehart and Winston, 1970)
2. C. Brans, On Unified Field Theories. Undergraduate thesis, Loyola University (1957) (Unpublished)
3. P. Dirac, Proc. Roy. Soc. (London) **A165**, 199 (1938)
4. J. Barbour, H. Pfister (ed.), *Mach's Principle* v. 6 of Einstein Studies (Birkauser, 1995)
5. D. Sciama, Month. Not. Royal Astron. Soc. **113**, 34 (1953)
6. C. Brans, Mach's principle and the locally measured gravitational constant in general relativity. Phys. Rev. **125**, 388 (1962)
7. P. Roll, R. Krotkov, R. Dicke, Ann. Phys. **26**, 442 (1964)
8. C. Brans, Mach's principle and a relativistic theory of gravitation, II. Phys. Rev. **125**, 2194 (1962)
9. P. Jordan, *Schwerkrft und Weltall* (Vieweg, 1955)
10. C. Brans, R.H. Dicke, Mach's principle and a relativistic theory of gravitation. Phys. Rev. **124**, 925 (1961)
11. E. Poisson, C. Will, *Gravity* (Cambridge, 2014)
12. C. Brans, Mach's Principle and a Varying Gravitational Constant. Ph.D. thesis, Princeton University (1961) (Unpublished)
13. R.H. Dicke, Phys. Rev. **125**, 2163 (1962)
14. R.H. Dicke, *The Theoretical Significance of Experimental Relativity* (Gordon and Breach, 1964)
15. C. Brans, C.H. Brans, Varying Newton's constant: a personal history of scalar-tensor theories, in Einstein Online, vol. 4, 1002 (2010)
16. C. Brans, Jordan-Brans-Dicke (2014), http://www.scholarpedia.org/article/Jordan-Brans-Dicke-Theory
17. E. Schucking, Phys. Today **52**, 26 (1999)
18. C. Brans, Invariant approach to the geometry of spaces in general relativity. J. Math. Phys. **6**, 94 (1965)
19. A.Z. Petrov, *Einstein Spaces* (Pergamon Press) English translation from Russian (1969)

20. A. Petrov, Sci. Not. Kazan **114**, 55 (1954)
21. C. Brans, A computer program for the non-numerical testing and reduction of sets of algebraic partial differential equations. J. Assoc. Comp. Mach. **14**, 45 (1967)
22. C. Brans, Invariant representation of all analytic Petrov type III solutions to the Einstein equations. J. Math. Phys. **11**, 1210 (1970)
23. H-J. Schmidt, Consequences of the non-compactness of the Lorentz group. Int. J. Theo. Phys. **37**, 691 (1998)
24. C. Brans, Complex two-form representation of the Einstein equations, the Petrov type III solutions. J. Math. Phys. **12**, 1616 (1971)
25. C. Brans, Complex bundle structure and the Einstein equations. Bull. A.P.S. **19**, 508 (1974)
26. C. Brans, Complex structures and representations of the Einstein equations. J. Math. Phys. **15**, 1559 (1974)
27. J. Thorpe, J. Math. Phys. **10**, 1 (1969)
28. C. Brans, Singularities in bootstrap gravitational geons. Phys. Rev. **140B**, 1174 (1965)
29. A. Komar, Phys. Rev. **137**, B462 (1965)
30. C. Brans, Some restrictions on algebraically general vacuum metrics. J. Math. Phys. **16**, 1008 (1975)
31. C. Brans, Complete integrability conditions of the Einstein-Petrov equations, type I. J. Math. Phys. **18**, 1378 (1977)
32. R.H. Dicke, P. Peebles, P.G. Roll, D.T. Wilkinson, Ap. J. **142**, 414 (1965)
33. A.A. Penzias, R.W. Wilson, Ap. J. **142**, 419 (1965)
34. C. Brans, Propagations of electromagnetic polarization effects in anisotropic cosmologies. Ap. J. **197**, 1 (1975)
35. A. Einstein, *The Meaning of Relativity* (Princeton, 1950)
36. C. Brans, Absence of inertial induction in general relativity. Phys. Rev. Lett. **39**, 856 (1977)
37. V. Varadarajan, *Geometry of Quantum Theory* (Van Nostrand, 1978)
38. A. Marlow (ed.), *Mathematical Foundations of Quantum Theory* (Academic Press, 1978)
39. C. Brans, D.R. Stewart, Unaccelerated-returning-twin paradox in flat space-time. Phys. Rev. D **8**, 1662 (1973)
40. J. Bell, Physics **1**, 195 (1964)
41. C. Brans, Bell's theorem does not eliminate fully causal hidden variables. Int. J. Theor. Phys. **27**, 219 (1988)
42. D. Mermin, Rev. Mod. Phys. **65**, 803 (1993)
43. K. deRaedt et al., Eur. Phys. J. B **53**, 139 (2006)
44. C. Brans, Consistency of field equations in 'self-creation' cosmologies. Gen. Relat. Grav. **19**, 949 (1987)
45. G. Barber, Gen. Rel. Grav. **14**, 117 (1982)
46. M. Ferraris et al., Class. Quant. Grav. **5**, L95 (1988)
47. G. Magnan, Gen. Rel. Grav. **19**, 465 (1987)
48. C. Brans, Non-linear Lagrangians and the significance of the metric. Class. Quantum Grav. **5**, L197 (1988)
49. C. Brans, Gravity and the tenacious scalar field, in *On Einstein's Path*, Essays in honor of Engelbert Schücking, ed. by A. Harvey. (Springer, Berlin, 1998), pp. 121–138, arXiv:gr-qc/9705069
50. C. Brans, The roots of scalar-tensor theory: an approximate history, in *Proceedings of International Workshop on Gravitation and Cosmology*, Santa Clara, Cuba (2004), arXiv:gr-qc/0506063
51. Y. Fujii, K-I. Maeda, *The Scalar-Tensor Theory of Gravitation* (Cambridge, 2003)
52. R. Dicke, Rev. Mod. Phys. **29**, 363 (1957)
53. P. Bergmann, Ann. Math. **49**, 255 (1948)
54. P. Bergmann, Int. J. Theo. Phys. **1**, 25 (1968)
55. T. Asselmeyer, C. Brans, **Book:***Exotic Structures and Physics: Differential Topology and Spacetime Models* (World Scientific Press, 2007)

56. R. Gompf, A. Stipsicz, *4-Manifolds and Kirby Calculus* (American Mathematical Society, 1999)
57. A. Scorpan, *The Wild World of 4-Manifolds* (American Mathematical Society, 2005)
58. M. Freedman, Not. Am. Mat. Soc. **31**, 3 (1984)
59. Daniel S. Freed, Karen K. Uhlenbeck, *Instantons and Four-Manifolds* (Springer, New York, 1984)
60. S. Donaldson, J. Diff. Geom. **18**, 269 (1983)
61. C. Brans, Gen. Rel. Grav. **34**, 1767 (2002)
62. C. Brans, Exotic Black Holes? arXiv:gr-qc/9303035
63. J. Milnor, Ann. Math. **64**, 399 (1956)
64. M. Jammer, *Concepts of Space: The History of Theories of Space in Physics*, 3rd edn. (Dover, 1993)
65. R.E. Gompf, J. Diff. Geom. **37**, 199 (1993)
66. M. Freedman, L. Taylor, J. Diff. Geom. **24**, 69 (1986)
67. N. Seiberg, E. Witten, Nucl. Phys. **B 426**, 19 (1994)
68. J. Sładkowski, Gravity on exotic $\mathbb{R}^4$ with few symmetries. Int. J. Mod. Phys. D **10**, 311–313 (2001)
69. R.E. Gompf, J. Diff. Geom. **21**, 283 (1985)
70. Norman Steenrod, *The Topology of Fiber Bundles* (Princeton University Press, Princeton, 1951)
71. S. Gallot, D. Hulin, J. Lafontaine, *Riemannian Geometry* (Springer, Berlin, 1990)
72. D. Gromoll, W. Mayer, Ann. Math. **100**, 401 (1974)
73. C. Brans, Localized exotic smoothness. Class. Quantum Grav. **11**, 1785 (1994)
74. D. Kotschick, Geom. Topol. **2**, 1 (1998)
75. C.H. Brans, D. Randall, Exotic differentiable structures and general relativity. Gen. Rel. Grav. **25**, 205 (1993)
76. C. Brans, Absolulte spacetime: the twentieth century ether. Gen. Rel. Grav. **31**, 597 (1999)
77. C.H. Brans, Exotic smoothness and physics. J. Math. Phys. **35**, 5494–5506 (1994)
78. C.H. Brans, Localized exotic smoothness. Class. Quant. Grav. **11**, 1785–1792 (1994)
79. J. Sładkowski, Exotic smoothness and particle physics. Acta Phys. Polon. **B 27**, 1649–1652 (1996)
80. J. Sładkowski. Exotic smoothness, fundamental interactions and noncommutative geometry (1996), arXiv:hep-th/9610093
81. J. Sładkowski, Exotic smoothness, noncommutative geometry and particle physics. Int. J. Theor. Phys. **35**, 2075–2083 (1996)
82. T. Asselmeyer, Generation of source terms in general relativity by differential structures. Class. Quant. Grav. **14**, 749–758 (1996)
83. J. Sładkowski, Strongly gravitating empty spaces (1999). Preprint arXiv:gr-qc/9906037
84. T. Asselmeyer-Maluga, C.H. Brans, Cosmological anomalies and exotic smoothness structures. Gen. Rel. Grav. **34**, 1767–1771 (2002)
85. T. Asselmeyer-Maluga, J. Król, Inflation and topological phase transition driven by exotic smoothness. Adv. HEP Volume 2014:867460 (article ID) (2014), http://dx.doi.org/10.1155/2014/867460, arXiv:1401.4815
86. H. Pfeiffer, Quantum general relativity and the classification of smooth manifolds. Report number: DAMTP **2004–32**, (2004)
87. T. Asselmeyer-Maluga, Exotic smoothness and quantum gravity. Class. Q. Grav. **27**, 165002 (2010), arXiv:1003.5506
88. J. Król, A model for spacetime II. The emergence of higher dimensions and field theory/strings dualities. Found. Phys. **36**, 1778 (2006)
89. J. Król, A model for spacetime: the role of interpretation in some Grothendieck topoi. Found. Phys. **36**, 1070 (2006)
90. J. Król, Background independence in quantum gravity and forcing constructions. Found. Phys. **34**, 361–403 (2004)

91. J. Król, Exotic smoothness and non-commutative spaces. The model-theoretic approach. Found. Phys. **34**, 843–869 (2004)

92. T. Asselmeyer-Maluga, J. Król, Abelian gerbes, generalized geometries and foliations of small exotic R^4 (2015), arXiv: 0904.1276v5 (subm. to Lett. Math. Phys.)

93. T. Asselmeyer-Maluga, J. Król, Exotic smooth $\mathbb{R}^4$ (2010), noncommutative algebras and quantization. arXiv:1001.0882

94. T. Asselmeyer-Maluga, J. Król. Constructing a quantum field theory from spacetime ()2011, arXiv:1107.3458

95. T. Asselmeyer-Maluga, R. Mader, Exotic R^4 and quantum field theory, in *7th International Conference on Quantum Theory and Symmetries (QTS7)*, ed. by C. Burdik et al. (IOP Publishing. Bristol, 2012), p. 012011, doi:10.1088/1742-6596/343/1/012011, arXiv:1112.4885

96. V. Chernov, S. Nemirovski, Cosmic censorship of smooth structures. Comm. Math. Phys. **320**, 469–473 (2013), arXiv:1201.6070

97. T. Asselmeyer-Maluga, J. Król, Topological quantum d-branes and wild embeddings from exotic smooth R^4. Int. J. Mod. Phys. A **26**, 3421–3437 (2011), arXiv:1105.1557

98. T. Asselmeyer-Maluga, J. Król, Quantum geometry and wild embeddings as quantum states. Int. J. Geom. Methods Mod. Phys. **10**(10), (2013) (will be published in November 2013), arXiv:1211.3012

99. T. Asselmeyer-Maluga, H. Rosé, On the geometrization of matter by exotic smoothness. Gen. Rel. Grav. **44**, 2825–2856 (2012), doi:10.1007/s10714-012-1419-3, arXiv:1006.2230

100. R. Fintushel, R. Stern, Knots, links, and 4-manifolds. Inv. Math. **134**, 363–400 (1998), arXiv:dg-ga/9612014

101. T. Asselmeyer-Maluga, C.H. Brans, How to include fermions into general relativity by exotic smoothness. Gen. Relat. Grav. **47**, 30 (2015), doi:10.1007/s10714-015-1872-x, arXiv:1502.02087

102. T. Asselmeyer-Maluga, C.H. Brans, *Smoothly Exotic Black Holes*. Space Science, Exploration and Policies (NOVA Publishers, 2012), pp. 139–156

103. T. Asselmeyer-Maluga, J. Król, On topological restrictions of the spacetime in cosmology. Mod. Phys. Lett. A **27**, 1250135 (2012), arXiv:1206.4796

104. T. Asselmeyer-Maluga, J. Król, Decoherence in quantum cosmology and the cosmological constant. Mod. Phys. Lett. A **28**, 1350158 (2013), doi:10.1142/S0217732313501587, arXiv:1309.7206

105. T. Asselmeyer-Maluga, J. Król, Small exotic smooth R^4 and string theory, in *International Congress of Mathematicians ICM 2010 Short Communications Abstracts Book*, ed. by R. Bathia (Hindustan Book Agency, 2010), p. 400

106. T. Asselmeyer-Maluga, J. Król, Exotic smooth R^4 and certain configurations of NS and D branes in string theory. Int. J. Mod. Phys. A **26**, 1375–1388 (2011), arXiv: 1101.3169

107. T. Asselmeyer-Maluga, J. Krol, Quantum D-branes and exotic smooth $\mathbb{R}^4$. Int. J. Geom. Methods Mod. Phys. **9**, 1250022 (2012), arXiv:1102.3274

108. T. Asselmeyer-Maluga, J. Król, Higgs potential and confinement in Yang-Mills theory on exotic $\mathbb{R}^4$ (2013), arXiv:1303.1632

Part I

Scalar-Tensor Theories (Brans–Dicke Theory)

"...Part II is mainly concerned with the introduction of a varying gravitational constant into the framework of general relativity, violating the strong while preserving the weak principle of equivalence (i.e. geodesics for uncharged test particles). To this end a scalar field, ϕ roughly corresponding to κ^{-1} (κ Gravitational constant) is added to the variational principle of general relativity. ..." (Ph.D. Thesis, Abstract).

"...The possibility of a varying gravitational constant has been discussed by Dirac, Jordan, and particularly with respect to Mach's principle by Dicke. The idea is to weaken the strong principle of equivalence through the effective gravitational constant. ... In choosing a variational principle violating the strong principle of equivalence by the introduction of a varying gravitational "constant," it seems desirable to satisfy at least two conditions. First, the variational principle must be similar to the standard Einstein principle, In other words, since the Einstein equations do agree with the observed data fairly well, any extension of the theory might be expected to be formally similar. Second, the variational principle must be consistent with the weak principle of equivalence which is just a generalization of the results of the Eötvös experiment. To satisfy this second condition it will be required that the operational definition of inertial mass be prescribed in a manner formally independent of the structure of the universe. The stress tensor of ponderable matter will be identified formally and interpretatively with that of general relativity. ...

...The variational principle will be thus taken to be

$$\delta_m \int d^4x \sqrt{-g} \left(\phi R + L_m + \omega \frac{\phi_{,\mu}\phi_{,\nu}g^{\mu\nu}}{\phi} \right)$$

Here ϕ has the dimensions of reciprocal gravitational constant, ω is a dimensionless constant number. The field equations associated with this principle become

$$\delta \int d^4x \sqrt{-g}\, L_m = 0$$

$$\phi S_{\alpha\beta} = (T_m)_{\alpha\beta} + \phi_{;\alpha;\beta} - g_{\alpha\beta}\Box\phi - \frac{\omega}{\phi}\left(\phi_{,\alpha}\phi_{,\beta} - \frac{1}{2}g_{\alpha\beta}\phi_{,\lambda}\phi^{,\lambda}\right)$$

$$\omega\left(\frac{2\Box\phi}{\phi} - \frac{\phi_{,\lambda}\phi^{,\lambda}}{\phi^2}\right) = R$$

in which δ_m signifies variation with respect to pertinent matter variable. ... ($S_{\alpha\beta}$ is the Einstein tensor)"

   Carl H. Brans: Mach's Principle and a varying Gravitational Constant, Ph.D. Thesis, Princeton University 1961

Chapter 2
Nonminimal Couplings in the Early Universe: Multifield Models of Inflation and the Latest Observations

David I. Kaiser

Abstract Models of cosmic inflation suggest that our universe underwent an early phase of accelerated expansion, driven by the dynamics of one or more scalar fields. Inflationary models make specific, quantitative predictions for several observable quantities, including particular patterns of temperature anistropies in the cosmic microwave background radiation. Realistic models of high-energy physics include many scalar fields at high energies. Moreover, we may expect these fields to have nonminimal couplings to the spacetime curvature. Such couplings are quite generic, arising as renormalization counterterms when quantizing scalar fields in curved spacetime. In this chapter I review recent research on a general class of multifield inflationary models with nonminimal couplings. Models in this class exhibit a strong attractor behavior: across a wide range of couplings and initial conditions, the fields evolve along a single-field trajectory for most of inflation. Across large regions of phase space and parameter space, therefore, models in this general class yield robust predictions for observable quantities that fall squarely within the "sweet spot" of recent observations.

2.1 Introduction

I first met Carl Brans about twenty years ago, in the mid-1990s, when I was a graduate student. Carl invited me to visit him at Loyola University in New Orleans, and he and his wife Anna kindly hosted me in their beautiful home. Our first meeting has always stood out in my mind: Carl picked me up at the airport, drove me straight to his office, and handed me a piece of chalk. I was to give him a lecture, right there at the blackboard, about cosmic inflation. I launched in, as best I could, and after a fun discussion Carl announced that it was time to pause and get some seafood gumbo; after all, we were in New Orleans. Ever since my first visit, I have found it

D.I. Kaiser (✉)
Department of Physics, Center for Theoretical Physics,
Massachusetts Institute of Technology, 77 Massachusetts Avenue,
Cambridge, MA 02139, USA
e-mail: dikaiser@mit.edu

© Springer International Publishing Switzerland 2016
T. Asselmeyer-Maluga (ed.), *At the Frontier of Spacetime*,
Fundamental Theories of Physics 183, DOI 10.1007/978-3-319-31299-6_2

terrifically inspiring to talk with Carl and to try to sharpen my own ideas in the face of his excellent questions, which he has always delivered in a gentle and encouraging way.[1]

Carl pursued what has become known as the "Brans-Dicke" theory of gravitation for his Ph.D. dissertation at Princeton, working closely with his advisor Robert Dicke [1–3]. Previous physicists had explored various ideas for scalar-tensor theories of gravity, including Pascual Jordan's well-known work, though none of the prior efforts had nearly the same galvanizing influence on the physics community as the Brans-Dicke work [4–8]. Brans and Dicke were motivated to try to incorporate Mach's principle in a relativistic theory of gravitation more consistently than Einstein had done in his general theory of relativity.[2]

The key insight in Brans and Dicke's work was to couple a scalar field directly to the Ricci spacetime curvature scalar in the action, thereby replacing Newton's constant, G, with an effective strength of gravity that could vary over space and time. Since Brans and Dicke introduced their formative work, several distinct theoretical motivations have emerged for such nonminimal couplings, beyond consideration of Mach's principle, including everything from dimensional compactification of higher-dimensional theories to effective couplings in supergravity and beyond. (For recent discussions, see [10–13].)

Perhaps the most mundane motivation for such nonminimal couplings today—but for me, the most compelling—is that nonminimal couplings arise as necessary counterterms when quantizing a self-interacting scalar field in curved spacetime. Even if the bare coupling is set to zero, quantum corrections will induce a nonzero coupling [14–20]. Moreover, the nonminimal coupling typically rises with energy scale under renormalization-group flow, with no ultraviolet fixed point [18]. It therefore makes sense to consider models with sizable nonminimal couplings at high energies, at or above the GUT scale—and hence to consider nonminimal couplings when thinking about the early universe.

2.2   Nonminimal Couplings and Inflation

Models of cosmic inflation suggest that our observable universe underwent an early phase of accelerated expansion, driven by the dynamics of one or more scalar fields [21]. (For reviews, see [22, 23].) There is by now a long history of building models of early-universe inflation incorporating nonmiminal couplings. Early models such as "induced-gravity inflation" [24], for example, built directly on work by Lee Smolin [25] and Anthony Zee [26], who had aimed to combine Brans-Dicke gravitation with a Higgs-like spontaneous symmetry breaking potential, in order to account for why the strength of gravity is so much weaker than the other fundamental forces. "Extended inflation" [27] likewise combined a Brans-Dicke field with

---

[1] Preprint MIT-CTP-4740.

[2] On Einstein's changing considerations of Mach's principle, see [9] and references therein.

a simple potential to drive accelerated expansion. Others considered more general nonmiminal couplings, in which the effective gravitational coupling G_{eff} arose as a combination of a bare coupling constant plus contributions from a scalar field coupled to the Ricci curvature scalar [28]. Among the most prominent recent examples is "Higgs inflation" [29]. In such models, the scalar field is expected to settle into a minimum of its potential near the end of inflation, leading to an effectively constant gravitational coupling for most of cosmic history. Hence such models present no tension with Solar System constraints on scalar-tensor gravity.

Realistic models of particle physics, relevant for inflationary energy scales, include many scalar fields [30]. The renormalization arguments alone suggest that each of these scalar fields should have a nonminimal coupling. So together with several students and collaborators, I have enjoyed exploring in recent years multifield models of inflation that incorporate nonminimal couplings [31–36].

The action for the original Brans-Dicke theory may be written

$$S_{BD} = \int d^4x\sqrt{-\tilde{g}}\left[\Phi\tilde{R} - \frac{\omega}{\Phi}\tilde{g}^{\mu\nu}\partial_\mu\Phi\partial_\nu\Phi\right], \tag{2.1}$$

where ω is a dimensionless constant and $\tilde{g}_{\mu\nu}(x)$ is the spacetime metric. (Greek letters label spacetime indices, $\mu, \nu = 0, 1, 2, 3$.) In $(3 + 1)$ spacetime dimensions, the Brans-Dicke field Φ has dimensions $(mass)^2$. Since high-energy theorists typically consider scalar fields that have dimension $mass$ in $(3 + 1)$ spacetime dimensions, we may rescale the Brans-Dicke field as $\Phi \to \phi^2/(8\omega)$. In terms of the rescaled field ϕ, the action of Eq. (2.1) may be written

$$S_{BD} = \int d^4x\sqrt{-\tilde{g}}\left[f_{BD}(\phi)\tilde{R} - \frac{1}{2}\tilde{g}^{\mu\nu}\partial_\mu\phi\partial_\nu\phi\right]. \tag{2.2}$$

The nonminimal coupling function takes the form

$$f_{BD}(\phi) = \frac{1}{2}\xi\phi^2, \tag{2.3}$$

where the dimensionless coupling constant ξ is related to the original Brans-Dicke parameter as $\xi = 1/(8\omega)$. Such a quadratic term is precisely the form in which quantum corrections arise for scalar fields in curved spacetime, and hence the form that appropriate counterterms must assume [14–20].

In Brans and Dicke's original formulation, the local strength of gravity, $G_{\text{eff}}(x)$, varies with the field $\phi(x)$: $G_{\text{eff}}(x) = 1/(8\pi\xi\phi^2)$. One may generalize such a coupling to include a bare (constant) mass, M_0, within the function $f(\phi)$:

$$f(\phi) = \frac{1}{2}\left[M_0^2 + \xi\phi^2\right], \tag{2.4}$$

with $(16\pi G_{\text{eff}})^{-1} = f(\phi)$. And this form, in turn, may be generalized to models
with N scalar fields:

$$f(\phi^I) = \frac{1}{2}\left[M_0^2 + \sum_{I=1}^{N} \xi_I\left(\phi^I\right)^2\right]. \tag{2.5}$$

We therefore consider models for which the action may be written

$$S = \int d^4x \sqrt{-\tilde{g}}\left[f(\phi^I)\tilde{R} - \frac{1}{2}\delta_{IJ}\tilde{g}^{\mu\nu}\partial_\mu\phi^I\partial_\nu\phi^J - \tilde{V}(\phi^I)\right]. \tag{2.6}$$

Here capital Latin letters label field-space indices, $I, J = 1, 2, ..., N$, and tildes
denote quantities in the so-called Jordan frame, in which the nonminimal couplings,
$f(\phi^I)\tilde{R}$, remain explicit in the action.

Because we are interested in comparing predictions from this family of models with recent astrophysical observations—especially high-precision measurements
of the cosmic microwave background radiation (CMB)—it is convenient to work
in the so-called Einstein frame, for which physicists have established a powerful
gauge-invariant formalism for treating gravitational perturbations.[3] (For reviews,
see [22, 38].)

In order to bring the gravitational portion of the action of Eq. (2.6) to the familiar Einstein-Hilbert form, we perform a conformal transformation, much as Dicke
described early in the study of Brans-Dicke gravitation [39]. We rescale the space-
time metric tensor, $\tilde{g}_{\mu\nu}(x) \to g_{\mu\nu}(x) = \Omega^2(x)\tilde{g}_{\mu\nu}(x)$. The conformal factor $\Omega^2(x)$
is positive definite and is related to the nonminimal coupling function that appears
in Eq. (2.6) as

$$\Omega^2(x) = \frac{2}{M_{\text{pl}}^2}f(\phi^I(x)), \tag{2.7}$$

where $M_{\text{pl}} \equiv 1/\sqrt{8\pi G} = 2.43 \times 10^{18}$ GeV is the reduced Planck mass, related to
Newton's gravitational constant, G. Upon performing this conformal transformation,
the action of Eq. (2.6) is transformed to [40]

$$S = \int d^4x \sqrt{-g}\left[\frac{M_{\text{pl}}^2}{2}R - \frac{1}{2}\mathcal{G}_{IJ}(\phi^K)g^{\mu\nu}\partial_\mu\phi^I\partial_\nu\phi^J - V(\phi^I)\right]. \tag{2.8}$$

The conformal transformation induces a field-space manifold whose metric, in the
Einstein frame, is given by

---

[3]We have bracketed, for now, the important and rather subtle question of whether there remains
any significant "frame dependence" for predictions from such multifield models. It seems clear
that one may map predictions for observables from one frame to another in the case of single-field
models [11, 12]. But making that mapping between frames in the presence of entropy (or isocurvature) perturbations—which can only arise in multifield models—seems to raise new subtleties [37].

$$\mathcal{G}_{IJ}(\phi^K) = \frac{M_{\text{pl}}^2}{2f(\phi^K)} \left[\delta_{IJ} + \frac{3}{f(\phi^K)} f_{,I} f_{,J} \right], \quad (2.9)$$

where $f_{,I} = \partial f/\partial \phi^I$.

We encounter an interesting feature when performing this conformal transformation for models with multiple scalar fields: unlike the well-studied case of a single-field model, in general there does not exist a rescaling of the scalar fields ϕ^I that can bring the gravitational portion of the action into Einstein-Hilbert form while also yielding canonical kinetic terms for the scalar fields. In particular, for $M_0 \neq 0$ and $N \geq 2$ scalar fields, the conformal transformation induces a field-space manifold whose metric, $\mathcal{G}_{IJ}(\phi^K)$, is not conformal to flat [40].[4] Instead, following the conformal transformation, models within this family assume the form of nonlinear sigma models [41].

The potential is also stretched by the conformal factor upon transformation to the Einstein frame. In particular, we find

$$V(\phi^I) = \frac{M_{\text{pl}}^4}{[2f(\phi^I)]^2} \tilde{V}(\phi^I). \quad (2.10)$$

This is the generalization of Dicke's original finding that masses of particles depend on the Brans-Dicke field following the conformal transformation [39]. In the context of simple inflationary models, this conformal stretching of the potential leads to important changes to the inflationary dynamics, compared to models with minimally coupled fields. The most important change is the emergence of strong single-field attractor behavior, which we discuss in Sect. 2.4.

Building on pioneering work on multifield inflation [44, 46], we developed in [32–36] a doubly covariant formalism with which to address dynamics in models that include multiple scalar fields with nonminimal couplings—that is, covariant with respect to both ordinary gauge transformations ($x^\mu \to x'^\mu$) as well as reparameterizations of the field-space coordinates ($\phi^I \to \phi'^I$). We consider perturbations around a Friedmann-Lemaître-Robertson-Walker spacetime metric, which we take to be spatially flat for convenience; the radius of curvature is stretched exponentially quickly during the first few efolds of inflation, so that a spatially flat background provides an excellent approximation for later dynamics. We then have

$$ds^2 = g_{\mu\nu}dx^\mu dx^\nu$$
$$= -(1 + 2A)dt^2 + 2a(\partial_i B)dx^i dt + a^2 \left[(1 - 2\psi)\delta_{ij} + 2\partial_i\partial_j E \right] dx^i dx^j, \quad (2.11)$$

where $a(t)$ is the scale factor, and $A(x^\mu)$, $B(x^\mu)$, $\psi(x^\mu)$, and $E(x^\mu)$ characterize the scalar degrees of freedom of the metric perturbations. Given the symmetries of the spacetime, to background order the fields can only depend on time:

---

[4]In the case of Brans-Dicke-like couplings, with $M_0 = 0$, one may rescale the fields ϕ^I to bring $\mathcal{G}_{IJ} \to \delta_{IJ}$, and hence restore canonical kinetic terms, only for $N \leq 2$. For $N > 2$, even with $M_0 = 0$, one again finds that $\mathcal{G}_{IJ}$ is not conformal to flat [40].

$$\phi^I(x^\mu) = \varphi^I(t) + \delta\phi^I(x^\mu). \tag{2.12}$$

The magnitude of the velocity vector for the background fields is given by

$$|\dot{\varphi}^I| \equiv \dot{\sigma} = \sqrt{\mathcal{G}_{IJ}\dot{\varphi}^I\dot{\varphi}^J}, \tag{2.13}$$

where overdots denote derivatives with respect to cosmic time, t. The background fields obey the equation of motion [32, 46]

$$\mathcal{D}_t\dot{\varphi}^I + 3H\dot{\varphi}^I + \mathcal{G}^{IJ}V_{,J} = 0, \tag{2.14}$$

where $H \equiv \dot{a}/a$ is the Hubble parameter, and we have introduced a (covariant) directional derivative for vectors A^I on the field-space manifold:

$$\mathcal{D}_t A^I \equiv \dot{\varphi}^J D_J A^I = \dot{A}^I + \Gamma^I_{JK}A^J\dot{\varphi}^K. \tag{2.15}$$

The Christoffel symbols Γ^I_{JK} are constructed from the field-space metric $\mathcal{G}_{IJ}$. The Friedmann equations (to background order) take the form [32, 46]

$$H^2 = \frac{1}{3M_{\text{pl}}^2}\left[\frac{1}{2}\dot{\sigma}^2 + V(\varphi^I)\right],$$
$$\dot{H} = -\frac{1}{2M_{\text{pl}}^2}\dot{\sigma}^2. \tag{2.16}$$

Equations (2.14) and (2.16) yield self-consistent inflationary solutions, with $|\dot{H}| \ll H^2$, across wide ranges of coupling constants and initial conditions [32–35].

The scale of H during inflation is constrained by recent observations. In particular, the present upper bound on the ratio of primordial tensor-to-scalar power spectra, r, requires $H_* \leq 3.4 \times 10^{-5} M_{\text{pl}}$ [45], where the asterisk indicates the value of H at the time when cosmologically relevant perturbations first crossed outside the Hubble radius during inflation. In simple, single-field models of chaotic inflation, one must fine-tune parameters, such as the quartic self-coupling $\lambda \sim \mathcal{O}(10^{-12})$, in order to accommodate this bound on H_*. In models with nonminimal couplings, however, the magnitude of H_* depends on both the Jordan-frame couplings (such as masses, m_I, and quartic self-couplings, λ_I, in $\tilde{V}(\phi^I)$), as well as the nonminimal coupling constants ξ_I, due to the conformal stretching of the potential in Eq. (2.10). Hence one may accommodate the observational constraint on H_* without exponentially fine-tuning the parameters [29, 32–35].

In order to study the behavior of the fluctuations, we may generalize the gauge-invariant Mukhanov-Sasaki variable to the multifield case, defining a vector of perturbations $Q^I(x^\mu)$ as a linear combination of the field fluctuations, $\delta\phi^I$, and the metric perturbation, ψ [32][5]:

$$Q^I \equiv \delta\phi^I + \frac{\dot{\varphi}^I}{H}\psi.$$ (2.17)

To linear order, the fluctuations Q^I satisfy the equation of motion [32, 46]

$$\mathcal{D}_t^2 Q^I + 3H\mathcal{D}_t Q^I + \left[\frac{k^2}{a^2}\delta^I{}_J + \mathcal{M}^I{}_J\right]Q^J = 0,$$ (2.18)

where we have performed a Fourier transform, $\nabla^2 Q^I = -k^2 Q^I$ with comoving wavenumber k, and the mass-squared matrix is given by

$$\mathcal{M}^I{}_J \equiv \mathcal{G}^{IK}\left(\mathcal{D}_J \mathcal{D}_K V\right) - \mathcal{R}^I{}_{LMJ}\dot{\varphi}^L\dot{\varphi}^M - \frac{1}{a^3 M_{\text{pl}}^2}\mathcal{D}_t\left(\frac{a^3}{H}\dot{\varphi}^I\dot{\varphi}_J\right).$$ (2.19)

Here $\mathcal{R}^I{}_{LMJ}$ is the Riemann tensor of the field-space manifold, constructed from $\mathcal{G}_{IJ}$ (and calculated to background order in the fields, φ^I); we raise and lower field-space indices with $\mathcal{G}_{IJ}$. The fluctuations thus acquire three distinct contributions to their effective mass: a term arising from the second derivative of the potential, akin to simple single-field models; a term (proportional to $\mathcal{R}^I{}_{LMJ}$) arising from the curvature of the field-space manifold; and a term (proportional to $1/M_{\text{pl}}^2$) arising from the coupled metric perturbations.

2.3 Predictions for Observables

Even to linear order, Eq. (2.18) couples fluctuations Q^I with Q^J and so on. The presence of several interacting degrees of freedom can lead to new observational features in multifield models, with no correlates in simple, single-field models. Two of the most important and best studied examples include the amplification of non-Gaussianities in the primordial power spectrum of curvature perturbations, and the amplification of isocurvature perturbations in addition to adiabatic modes. Non-Gaussianities are generically suppressed in single-field models [42, 43], and isocurvature modes do not arise at all in models with only a single scalar degree of freedom [22, 44]. Given tight constraints on primordial non-Gaussianities and isocurvature perturbations from the latest measurements of the CMB [45], many types of multifield models may therefore be in tension with the latest observations.

---

[5]Because the field-space manifold is curved, one must work with a representation of the field fluctuations that is covariant with respect to reparameterizations of the field-space coordinates, as discussed in [32] and references therein. That form reduces to Eq. (2.17) to linear order in the field fluctuations, which will suffice for our purposes here.

In order to quantify these multifield features, we build on techniques developed in [22, 44, 46] and introduce covariant measures with which to study the perturbation spectra [32–36]. We introduce a unit vector

$$\hat{\sigma}^I \equiv \frac{\dot{\varphi}^I}{\dot{\sigma}} \tag{2.20}$$

which points in the direction of the background fields' evolution. The directions in field space orthogonal to $\hat{\sigma}^I$ are spanned by

$$\hat{s}^{IJ} \equiv \mathcal{G}^{IJ} - \hat{\sigma}^I \hat{\sigma}^J. \tag{2.21}$$

We may then project the perturbations Q^I into components along the direction of the background fields' motion (the adiabatic direction) and orthogonal to that motion (the isocurvature directions):

$$Q_\sigma \equiv \hat{\sigma}_I Q^I, \quad \delta s^I \equiv \hat{s}^I{}_J Q^J. \tag{2.22}$$

The gauge-invariant curvature perturbation, $\mathcal{R}_c$, is defined as [22, 38],

$$\mathcal{R}_c \equiv \psi - \frac{H}{(\rho + p)} \delta q, \tag{2.23}$$

where ρ and p are the background-order energy density and pressure, respectively, and δq is the momentum flux of the perturbed fluid, $T^0_i = \partial_i \delta q$. Given the form of the action in Eq. (2.8), one may show that [32]

$$\mathcal{R}_c = \frac{H}{\dot{\sigma}} Q_\sigma. \tag{2.24}$$

Primordial curvature perturbations, $\mathcal{R}_c(x)$, lead to temperature anisotropies in the CMB. Photons that hail from regions of space that had a slightly greater-than-average gravitational potential will be slightly redshifted, upon expending a bit of extra energy to climb out of the potential well, compared to photons from regions of space that had a slightly less-than-average gravitational potential [22, 23, 38]. Hence the statistical properties of the tiny temperature anisotropies of the CMB provide a snapshot of primordial inhomogeneities, which in turn help to constrain models of early-universe inflation.

A critical insight [44, 46] is that Q_σ and δs^I are coupled only if the background fields *turn* in field space. Hence features like non-Gaussianities and isocurvature perturbations can be amplified in multifield models if the turn-rate, ω^I, is nonvanishing during the late stages of inflation (typically within the last 60 efolds of inflation). The covariant turn-rate may be defined as [32]

$$\omega^I \equiv \mathcal{D}_t \hat{\sigma}^I. \tag{2.25}$$

In multifield models, ω^I need not remain small during inflation, which can amplify features that are not observed in the CMB.

Consider the limit $\omega^I \rightarrow 0$ first, in which case the perturbations Q_σ and δs^I remain decoupled. The effective masses for the perturbations take the form [32]

$$\mathcal{M}_{\sigma\sigma} \equiv \hat{\sigma}_I \hat{\sigma}^J \mathcal{M}^I{}_J, \quad \mathcal{M}_{ss} \equiv \hat{s}_I{}^J \mathcal{M}^I{}_J. \tag{2.26}$$

In the limit $|\mathcal{M}_{\sigma\sigma}|, |\mathcal{M}_{ss}| \ll H^2$, each perturbation will evolve during inflation as a (nearly) massless scalar field in (quasi-) de Sitter space, and hence we may expect each perturbation to develop an amplitude of order [16, 19, 22]

$$\mathcal{P}_Q \simeq \left(\frac{H}{2\pi}\right)^2, \tag{2.27}$$

where the power spectrum is defined as $\mathcal{P}_Q \equiv (2\pi)^{-2} k^3 |Q_\sigma|^2$, which we have evaluated for modes of order the Hubble scale, $k \simeq aH$; likewise for $\mathcal{P}_S$, the power spectrum associated with the conventionally normalized isocurvature perturbations $S^I \equiv (H/\dot{\sigma})\delta s^I$. Upon using Eqs. (2.16), (2.24), and the usual definition of the slow-roll parameter,

$$\epsilon \equiv -\frac{\dot{H}}{H^2}, \tag{2.28}$$

we therefore expect an amplitude of curvature perturbations during inflation

$$\mathcal{P}_R \simeq \frac{1}{2M_{\text{pl}}^2 \epsilon} \left(\frac{H}{2\pi}\right)^2 \tag{2.29}$$

and similarly for $\mathcal{P}_S$.

The background fields $\varphi^I(t)$ evolve slowly during inflation, and hence neither $H(t)$ nor $\epsilon(t)$ will remain constant. That means that when modes of various comoving wavenumbers k cross outside the Hubble radius, with $k = aH$, they do so with slightly different amplitudes, $\mathcal{P}_R(k)$. Hence with a little more work, one may calculate the spectral tilt of the curvature perturbations [22, 32, 44, 46]:

$$n_s \equiv 1 + \frac{\partial \ln \mathcal{P}_R}{\partial \ln k} = 1 - 6\epsilon + 2\eta_{\sigma\sigma}, \tag{2.30}$$

where

$$\eta_{\sigma\sigma} \equiv M_{\text{pl}}^2 \frac{\mathcal{M}_{\sigma\sigma}}{V} \tag{2.31}$$

is the generalization of the second slow-roll parameter, for motion along the adiabatic direction. In the limit $\omega^I \rightarrow 0$, therefore, the amplitude and spectral tilt of primordial curvature perturbations in the multifield case look quite similar to the predictions from single-field models—with one important difference. If $|\mathcal{M}_{ss}| \ll H^2$, then such multifield models may amplify a sizable fraction of isocurvature modes, with $\beta_{\mathrm{iso}}(k) \equiv \mathcal{P}_S(k)/[\mathcal{P}_R(k) + \mathcal{P}_S(k)] \sim \mathcal{O}(1)$ at relevant wavenumbers k, which would be in significant tension with the latest observations [45].

Even greater deviations from the single-field case emerge if the fields turn in field space during inflation, $\omega^I \neq 0$. In that case, there can be a transfer of power from the isocurvature to the adiabatic modes, and the amplitude and tilt of $\mathcal{P}_R(k)$ will be affected. In particular, one may relate the power spectrum at time t_* (say, 50 or 60 efolds before the end of inflation) to its value at some later time, t, by means of a transfer function $T_{RS}(t_*, t)$ [32, 44, 46]:

$$
\begin{aligned}
\mathcal{P}_R(k) &= \mathcal{P}_R(k_*)\left[1 + T_{RS}^2(t_*, t)\right], \\
n_s &= n_s(t_*) + \frac{1}{H}\left(\frac{\partial T_{RS}}{\partial t_*}\right)\sin(2\Delta),
\end{aligned}
\tag{2.32}
$$

where $\Delta \equiv \arccos(T_{RS}/\sqrt{1 + T_{RS}^2})$. Even a modest transfer of power from the isocurvature to the adiabatic modes could push multifield models out of agreement with the latest high-precision measurements of quantities like n_s. Moreover, since T_{RS} is scale-dependent, such processes effectively couple modes of different wavenumber, k, and hence can amplify non-Gaussianities, pushing the coefficient of the bispectrum $f_{NL} \gg \mathcal{O}(1)$ [32, 46].[6]

In the Einstein frame, there is no anisotropic pressure to leading order in the perturbations ($\Pi^i_j \propto T^i_j \sim 0$ for $i \neq j$), and hence the tensor perturbations h_{ij} evolve just as in single-field models. Around the pivot scale k_*, the power spectrum thus obeys $\mathcal{P}_T \simeq 128(H^2/M_{\mathrm{pl}}^2)$ [34, 44, 46], which yields a prediction for the tensor-to-scalar ratio, r,

$$
r \equiv \frac{\mathcal{P}_T}{\mathcal{P}_R} = \frac{16\epsilon}{[1 + T_{RS}^2]}.
\tag{2.33}
$$

Just as the case for n_s and f_{NL}, predictions for r can deviate strongly from the usual single-field predictions in the case of significant transfer of power from isocurvature to adiabatic modes.

The exact form of T_{RS} for multifield models with nonminimal couplings may be found in [32]; the important point is that $T_{RS} \propto |\omega^I|$. In general, when significant turning occurs and $T_{RS} \geq \mathcal{O}(10^{-1})$, one finds both n_s and f_{NL} pulled significantly outside the 2σ bounds from the latest observations [32, 35].

---

[6]To calculate f_{NL} properly, one must go beyond linear order in the fluctuations and calculate the genuine bispectrum, $\langle \mathcal{R}_c(\mathbf{k}_1)\mathcal{R}_c(\mathbf{k}_2)\mathcal{R}_c(\mathbf{k}_3)\rangle$ [32, 42, 43]; upon performing the full calculation, we find a strong correlation between nonzero T_{RS} and sizable f_{NL} [32].

2.4 Single-Field Attractor

For multifield models with nonminimal couplings, the turn rate generically remains negligible. Therefore the types of observational consequences that may arise in multifield models, such as the overproduction of isocurvature modes or the amplification of significant non-Gaussianities, typically do not arise for this class of models. The reason comes from the conformal stretching of the potential, $V(\phi^I)$ of Eq. (2.10).

For simplicity, consider a two-field model, with $\phi^I = (\phi, \chi)$. Then for a generic, renormalizable potential in the Jordan frame,

$$\tilde{V}(\phi, \chi) = \frac{1}{2}m_\phi^2\phi^2 + \frac{1}{2}m_\chi^2\chi^2 + \frac{1}{2}g\phi^2\chi^2 + \frac{\lambda_\phi}{4}\phi^4 + \frac{\lambda_\chi}{4}\chi^4, \qquad (2.34)$$

and a nonminimal coupling function $f(\phi, \chi)$ as in Eq. (2.5), we find from Eq. (2.10) the potential in the Einstein frame[7]

$$V(\phi, \chi) = \frac{M_{\rm pl}^4}{4} \frac{(2m_\phi^2\phi^2 + 2m_\chi^2\chi^2 + 2g\phi^2\chi^2 + \lambda_\phi\phi^4 + \lambda_\chi\chi^4)}{[M_{\rm pl}^2 + \xi_\phi\phi^2 + \xi_\chi\chi^2]^2}. \qquad (2.35)$$

Whereas the potential in the Jordan frame, $\tilde{V}(\phi^I)$, grows as ϕ and/or χ becomes large, in the Einstein frame the potential $V(\phi^I)$ flattens out to long plateaus for large field values. (See Fig. 2.1.) That is, generically, the potential in the Einstein frame develops ridges (local maxima) and valleys (local minima), becoming flat along a given direction for asymptotically large field values. Both the ridges and valleys satisfy $V > 0$, and hence the system will inflate (albeit at different rates) whether the fields evolve along a ridge or a valley during inflation.

The ridge-valley structure of the potential leads to strong single-field attractor behavior during inflation, across a wide range of couplings and initial conditions [32–36]. If the fields happen to begin evolving along the top of a ridge, they will eventually fall into a neighboring valley at a rate that depends on the local curvature of the potential. Once the fields fall into a valley, Hubble drag quickly damps out any transverse motion in field space, after which the system evolves with virtually no turning for the remainder of inflation. (See Fig. 2.2.) In [36], we demonstrate that the strong attractor behavior persists in the limit $0 < \xi_I \leq 1$ as well as in the limit $\xi_I \gg 1$.

In the limit of strong nonminimal couplings, $\xi_I \gg 1$, the fields rapidly fall into a single-field attractor (within the first few efolds of inflation) unless one fine-tunes the ratio of couplings *and* the fields' initial conditions to exponential accuracy. Such attractor behavior is therefore a generic feature of multifield models with

---

[7]We have set $M_0 = M_{\rm pl}$ in $f(\phi^I)$, since for $\tilde{V}(\phi^I)$ in Eq. (2.34), the global minimum of the potential occurs at $\phi = \chi = 0$ rather than at any nonzero vacuum expectation value. Hence at the end of inflation, once ϕ and χ settle into the global minimum of the potential, $f(\phi^I) \to M_{\rm pl}^2/2$, recovering the usual gravitational coupling for general relativity.

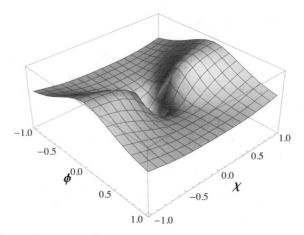

Fig. 2.1 The potential $V(\phi, \chi)$ in the Einstein frame, Eq. (2.35), for $\xi_\phi = 100, \xi_\chi = 80, \lambda_\phi = 10^{-2},$ $\lambda_\chi = 1.25 \times 10^{-2}, g = 0.8 \times 10^{-2}, m_\phi = 10^{-4} M_{\text{pl}}$, and $m_\chi = 1.5 \times 10^{-4} M_{\text{pl}}$. The field values are shown in units of M_{pl}

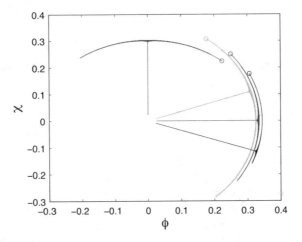

Fig. 2.2 Field trajectories for various couplings and initial conditions. *Open circles* indicate fields' initial values (in units of M_{pl}). We set $\dot{\phi}_0 = \dot{\chi}_0 = m_I = 0, \xi_\phi = 10^2, \lambda_\phi = 10^{-2}$, and vary the other parameters $\{\xi_\chi, \lambda_\chi, g, \theta_0\}$: $\{1.2\xi_\phi, 0.75\lambda_\phi, \lambda_\phi, \pi/4\}$ *(red)*; $\{0.8\xi_\phi, \lambda_\phi, \lambda_\phi, \pi/4\}$ *(blue)*; $\{0.8\xi_\phi, \lambda_\phi, 0.75\lambda_\phi, \pi/3\}$ *(green)*; $\{0.8\xi_\phi, 1.2\lambda_\phi, 0.75\lambda_\phi, \pi/6\}$ *(black)*. Here $\theta_0 \equiv \arctan(\chi_0/\phi_0)$. See also [34, 36]

nonminimal couplings, and subsumes the class of "α attractors" that has recently been identified [47].

The lack of turning in field space means that, generically, models in this family yield predictions very similar to those of simple single-field models of "plateau" inflation. With $\omega^I \simeq 0$, there is virtually no transfer of power from the isocurvature to the adiabatic modes, $T_{RS} \simeq 0$. Moreover, the effective mass of the isocurvature modes, $\mathcal{M}_{ss}$, remains *large* during inflation, while the fields evolve within a valley of the potential: $\mathcal{M}_{ss} \gg H^2$. Hence $\beta_{\mathrm{iso}}(k) \sim \mathcal{O}(10^{-5})$, well in keeping with recent observational constraints [35, 45].

Even better, within a single-field attractor and in the limit $\xi_I \gg 1$, one may integrate the equations of motion for the background fields within a slow-roll approximation, taking $|\ddot{\varphi}^I| \ll |H\dot{\varphi}^I|$; as demonstrated in [34], the resulting analytic expressions provide a remarkably close match for the exact numerical solutions within a given single-field attractor. In particular, we find [34]

$$\frac{\xi_\phi \phi_*^2}{M_{\mathrm{pl}}^2} \simeq \frac{4}{3} N_*, \tag{2.36}$$

where N_* is the number of efolds before the end of inflation, and we have considered (in this case) couplings such that the direction $\chi = 0$ is a local minimum of the potential. (We arrive at comparable expressions for other choices of couplings such that the local minimum lies along some angle $\theta = \arctan(\chi/\phi)$ in field space.) In that limit, we find expressions for the slow-roll parameters that are *independent* of the couplings:

$$\epsilon \simeq \frac{3}{4N_*^2}, \quad \eta_{\sigma\sigma} \simeq -\frac{1}{N_*}\left(1 - \frac{3}{4N_*}\right). \tag{2.37}$$

Returning to Eqs. (2.30), (2.32), and (2.33) with $\omega^I \simeq 0$ and hence $T_{RS} \simeq 0$, we then find [34]

$$n_s \simeq 1 - \frac{2}{N_*} - \frac{3}{N_*^2}, \quad r \simeq \frac{12}{N_*^2}, \tag{2.38}$$

independent of the values of the couplings and the fields' initial conditions. For typical reheating scenarios, one expects $50 \le N_* \le 60$ to correspond to the time during inflation when perturbations of a given comoving wavenumber first crossed outside the Hubble radius, which later re-entered the Hubble radius around the time the CMB was emitted [48]. Selecting $50 \le N_* \le 60$ in Eq. (2.38) yields

$$\begin{aligned} 0.959 \le n_s \le 0.966, \\ 0.003 \le r \le 0.005. \end{aligned} \tag{2.39}$$

This value of the spectral index, n_s, is in excellent agreement with the latest measurement by the Planck collaboration, $n_s = 0.968 \pm 0.006$ [45], while the predictions for r remain comfortably below the present upper bound of $r < 0.09$ [45]. Moreover, predictions for the running of the spectral index, $\alpha = dn_s/d\ln k$, satisfy $\alpha < 10^{-3}$ [34], likewise consistent with the latest observational estimates (which themselves are consistent with no observable running) [45]. And with $\omega^I \simeq 0$ and hence $T_{RS} \simeq 0$, these models predict $f_{NL} \sim \mathcal{O}(10^{-1})$ [32], again perfectly consistent with the latest observational constraints [45].

Lastly, one may study post-inflation reheating in this family of models [36]. The single-field attractor persists after the end of inflation, at least during times when the perturbations may be treated to linear order. The lack of turning in field space leads to efficient transfer of energy from the inflation condensate to coupled fluctuations, in contrast to multifield models with minimal couplings, in which "dephasing" of the background fields' oscillations typically suppresses resonances [48, 49]. Hence reheating in these models should be efficient, with an effective equation of state $w = p/\rho$ that interpolates between $w \simeq 0$ and $w \simeq 1/3$ within the first few efolds after the end of inflation [36].

2.5 Conclusions

More than half a century after Brans and Dicke introduced their scalar-tensor theory of gravitation, the study of scalar fields with nonminimal couplings continues to flourish. The number of compelling theoretical motivations for considering such nonminimal couplings has grown, and the relevance of such models for understanding the earliest moments of cosmic history is stronger than ever.

Brans and Dicke introduced their work at a time when Solar System tests of gravitation were still rare, and before the CMB had even been detected! It is an amazing testament to Carl's curiosity and physical insights that work stemming from his dissertation continues to inspire investigations of our cosmos to this day.[8] Congratulations to Carl on his 80th birthday, with admiration and gratitude.

Acknowledgments It is a great pleasure to thank Carl Brans for his kind encouragement over the years. I would also like to thank Torsten Asselmeyer-Maluga for editing this Festschrift in honor of Carl's 80th birthday. I am indebted to Evangelos Sfakianakis for pursuing the research reviewed here with me, along with our collaborators Matthew DeCross, Ross Greenwood, Edward Mazenc, Audrey (Todhunter) Mithani, Anirudh Prabhu, Chanda Prescod-Weinstein, and Katelin Schutz. This research has also benefited from discussions with Bruce Bassett and Alan Guth over the years. This work was conducted in MIT's Center for Theoretical Physics and supported in part by the U.S. Department of Energy under grant Contract Number DE-SC0012567.

References

1. C. H. Brans, Mach's principle and a varying gravitational constant. Ph.D. dissertation, Princeton University, 1961
2. C.H. Brans, R.H. Dicke, Mach's principle and a relativistic theory of gravitation. Phys. Rev. **124**, 925 (1961)
3. C.H. Brans, Mach's principle and a relativistic theory of gravitation, II. Phys. Rev. **125**, 2194 (1962); C.H. Brans, Mach's principle and the locally measured gravitational constant in general relativity. Phys. Rev. **125**, 388 (1962)

---

[8]Beyond his work on scalar-tensor gravitation, I also learned of Carl's interest in quantum entanglement and Bell's theorem [50] during my early visits with him—work that has also inspired some of my own recent research [51].

4. C. Will, *Was Einstein Right? Putting General Relativity to the Test*, 2nd ed. (Basic Books, New York, 1993 [1986])
5. J.D. Norton, Einstein, Nordström, and the early demise of Lorentz-covariant, scalar theories of gravitation. Arch. Hist. Exact Sci. **45**, 17 (1992)
6. D.I. Kaiser, When fields collide. Sci. Am. **296**, 62 (2007)
7. C.H. Brans, Varying Newton's constant: a personal history of scalar-tensor theories. Einstein Online **04**, 1002 (2010)
8. H. Goenner, Some remarks on the genesis of scalar-tensor theories. Gen. Rel. Grav. **44**, 2077 (2012). arXiv:1204.3455 [gr-qc]
9. M. Janssen, Of pots and holes: Einstein's bumpy road to general relativity. Ann. Phys. (Leipzig) **14**(Supplement), 58 (2005)
10. Y. Fujii, K.-I. Maeda, *The Scalar-Tensor Theory of Gravitation* (Cambridge University Press, New York, 2003)
11. V. Faraoni, *Cosmology in Scalar-Tensor Gravity* (Springer, New York, 2004)
12. S. Capozziello, V. Faraoni, *Beyond Einstein Gravity: A Survey of Gravitational Theories for Cosmology and Astrophysics* (Springer, New York, 2011)
13. S. Nojiri, S.D. Odintsov, Unified cosmic history in modified gravity: from $F(R)$ theory to Lorentz non-invariant models. Phys. Rep. **505**, 59 (2011). arXiv:1011.0544 [gr-qc]
14. C.G. Callan Jr., S.R. Coleman, R. Jackiw, A new improved energy-momentum tensor. Annals Phys. **59**, 42 (1970)
15. T.S. Bunch, P. Panangaden, L. Parker, On renormalization of $\lambda\phi^4$ field theory in curved spacetime, I. J. Phys. A **13**, 901 (1980); T.S. Bunch, P. Panangaden, On renormalization of $\lambda\phi^4$ field theory in curved spacetime, II. J. Phys. A **13**, 919 (1980)
16. N.D. Birrell, P.C.W. Davies, *Quantum Fields in Curved Space* (Cambridge University Press, New York, 1982)
17. S.D. Odintsov, Renormalization group, effective action and Grand Unification Theories in curved spacetime. Fortsh. Phys. **39**, 621 (1991)
18. I.L. Buchbinder, S.D. Odintsov, I.L. Shapiro, *Effective Action in Quantum Gravity* (Taylor and Francis, New York, 1992)
19. L.E. Parker, D.J. Toms, *Quantum Field Theory in Curved Spacetime* (Cambridge University Press, New York, 2009)
20. T. Markkanen, A. Tranberg, A simple method for one-loop renormalization in curved spacetime. J. Cosmol. Astropart. Phys. **08**, 045 (2013). arXiv:1303.0180 [hep-th]
21. A.H. Guth, The inflationary universe: A possible solution to the horizon and flatness problems. Phys. Rev. D **23**, 347 (1981); A.D. Linde, A new inflationary universe scenario: a possible solution of the horizon, flatness, homogeneity, isotropy, and primordial monopole problems. Phys. Lett. B **108**, 389 (1982); A. Albrecht, P.J. Steinhardt, Cosmology for Grand Unified Theories with radiatively induced symmetry breaking. Phys. Rev. Lett. **48**, 1220 (1982)
22. B.A. Bassett, S. Tsujikawa, D. Wands, Inflation dynamics and reheating. Rev. Mod. Phys. **78**, 537 (2006). arXiv:astro-ph/0507632
23. A.H. Guth, D.I. Kaiser, Inflationary cosmology: exploring the universe from the smallest to the largest scales. Science **307**, 884 (2005). arXiv:astro-ph/0502328; D.H. Lyth, A.R. Liddle, The Primordial Density Perturbation: Cosmology, Inflation, and the Origin of Structure (Cambridge University Press, New York, 2009); D. Baumann, TASI Lectures on Inflation. arXiv:0907.5424 [hep-th]; J. Martin, C. Ringeval, V. Vennin, Encyclopedia inflationaris. arXiv:1303.3787 [astro-ph.CO]; A.H. Guth, D.I. Kaiser, Y. Nomura, Inflationary paradigm after Planck 2013. Phys. Lett. B **733**, 112 (2014). arXiv:1312.7619 [astro-ph.CO]; A.D. Linde, Inflationary cosmology after Planck 2013. arXiv:1402.0526 [hep-th]
24. B.L. Spokoiny, Inflation and generation of perturbations in broken-symmetric theory of gravity. Phys. Lett. B **147**, 39 (1984); F.S. Accetta, D.J. Zoller, M.S. Turner, Induced-gravity inflation. Phys. Rev. D **31**, 3046 (1985); F. Lucchin, S. Matarrese, M.D. Pollock, Inflation with a nonminimally coupled scalar field. Phys. Lett. B **167**, 163 (1986); R. Fakir, W.G. Unruh, Induced-gravity inflation. Phys. Rev. D **41**, 1792 (1990); D.I. Kaiser, Constraints in the context of induced-gravity inflation. Phys. Rev. D **49**, 6347 (1994). arXiv:astro-ph/9308043; D.I.

Kaiser, Induced-gravity inflation and the density perturbation spectrum. Phys. Lett. B **340**, 23 (1994). arXiv:astro-ph/9405029; J.L. Cervantes-Code, H. Dehnen, Induced gravity inflation in the Standard Model of particle physics. Nucl. Phys. B **442**, 391 (1995). arXiv:astro-ph/9505069

25. L. Smolin, Towards a theory of spacetime structure at very short distances. Nucl. Phys. B **160**, 253 (1979)

26. A. Zee, Broken-symmetry theory of gravity. Phys. Rev. Lett. **42**, 417 (1979)

27. D. La, P.J. Steinhardt, Extended inflationary cosmology. Phys. Rev. Lett. **62**, 376 (1989); P.J. Steinhardt, F.S. Accetta, Hyperextended inflation. Phys. Rev. Lett. **64**, 2740 (1990); R. Holman, E. W. Kolb, Y. Wang, Gravitational couplings of the inflaton in extended inflation. Phys. Rev. Lett. **65**, 17 (1990); R. Holman, E.W. Kolb, S.L. Vadas, Y. Wang, Extended inflation from higher-dimensional theories. Phys. Rev. D **42**, 995 (1991)

28. T. Futamase, K. Maeda, Chaotic inflationary scenario of the universe with a nonminimally coupled 'inflaton' field. Phys. Rev. D **39**, 399 (1989); D.S. Salopek, J.R. Bond, J.M. Bardeen, Designing density fluctuation spectra in inflation. Phys. Rev. D **40**, 1753 (1989); R. Fakir, S. Habib, W.G. Unruh, Cosmological density perturbations with modified gravity. Astrophys. J. **394**, 396 (1992); R. Fakir, W.G. Unruh, Improvement on cosmological chaotic inflation through nonminimal coupling. Phys. Rev. D **41**, 1783 (1990); N. Makino, M. Sasaki, The density perturbation in the chaotic inflation with nonminimal coupling. Prog. Theor. Phys. **86**, 103 (1991); D.I. Kaiser, Primordial spectral indices from generalized Einstein theories. Phys. Rev. D **52**, 4295 (1995). arXiv:astro-ph/9408044; S. Mukaigawa, T. Muta, S.D. Odintsov, Finite Grand Unified Theories and inflation. Int. J. Mod. Phys. A **13**, 2839 (1998). arXiv:hep-ph/9709299; E. Komatsu, T. Futamase, Complete constraints on a nonminimally coupled chaotic inflationary scenario from the cosmic microwave background. Phys. Rev. D **59**, 064029 (1999). arXiv:astro-ph/9901127; A. Linde, M. Noorbala, A. Westphal, Observational consequences of chaotic inflation with nonminimal coupling to gravity. J. Cosmol. Astropart. Phys. **1103**, 013 (2011). arXiv:1101.2652 [hep-th]

29. F.L. Bezrukov, M.E. Shaposhnikov, The Standard Model Higgs boson as the inflaton. Phys. Lett. B **659**, 703 (2008). arXiv:0710.3755 [hep-th]

30. D.H. Lyth, A. Riotto, Particle physics models of inflation and the cosmological density perturbation. Phys. Rept. **314**, 1 (1999). arXiv:hep-ph/9807278; A. Mazumdar, J. Rocher, Particle physics models of inflation and the curvaton scenarios. Phys. Rep. **497**, 85 (2011). arXiv:1001.0993 [hep-ph]; V. Vennin, K. Koyama, D. Wands, Encyclopedia curvatonis. arXiv:1507.07575 [astro-ph.CO]

31. D.I. Kaiser, A.T. Todhunter, Primordial perturbations from multifield inflation with nonminimal couplings. Phys. Rev. D **81**, 124037 (2010). arXiv:1004.3805 [astro-ph.CO]

32. D.I. Kaiser, E.A. Mazenc, E.I. Sfakianakis, Primordial bispectrum from multifield inflation with nonminimal couplings. Phys. Rev. D **87**, 064004 (2013). arXiv:1210.7487 [astro-ph.CO]

33. R.N. Greenwood, D.I. Kaiser, E.I. Sfakianakis, Multifield dynamics of Higgs inflation. Phys. Rev. D **87**, 044038 (2013). arXiv:1210.8190 [hep-ph]

34. D.I. Kaiser, E.I. Sfakianakis, Multifield inflation after Planck: the case for nonminimal couplings. Phys. Rev. Lett. **112**, 011302 (2014). arXiv:1304.0363 [astro-ph.CO]

35. K. Schutz, E.I. Sfakianakis, D.I. Kaiser, Multifield inflation after Planck: Isocurvature modes from nonminimal couplings. Phys. Rev. D **89**, 064044 (2014). arXiv:1310.8285 [astro-ph.CO]

36. M.P. DeCross, D.I. Kaiser, A. Prabhu, C. Prescod-Weinstein, E.I. Sfakianakis, Preheating after multifield inflation with nonminimal couplings, I: covariant formalism and attractor behavior. arXiv:1510.08553 [hep-ph]

37. J. White, M. Minamitsuji, M. Sasaki, Curvature perturbation in multifield inflation with nonminimal coupling. J. Cosmol. Astropart. Phys. **07**, 039 (2012). arXiv:1205.0656 [astro-ph.CO]; J. White, M. Minamitsuji, M. Sasaki, Nonlinear curvature perturbation in multifield inflation models with nonminimal coupling. J. Cosmol. Astropart. Phys. **09**, 015 (2013). arXiv:1406.6186 [astro-ph.CO]; A. Yu. Kamenshchik, C.F. Steinwachs, Question of quantum equivalence between Jordan frame and Einstein frame. Phys. Rev. D **91**, 084033 (2015). arXiv:1408.5769 [gr-qc]

38. H. Kodama, M. Sasaki, Cosmological perturbation theory. Prog. Theor. Phys. Suppl. **78**, 1 (1984); V.F. Mukhanov, H.A. Feldman, R.H. Brandenberger, Theory of cosmological perturbations. Phys. Rep. **215**, 203 (1992); K.A. Malik, D. Wands, Cosmological perturbations. Phys. Rep. **475**, 1 (2009). arXiv:0809.4944 [astro-ph]
39. R.H. Dicke, Mach's principle and invariance under transformation of units. Phys. Rev. **125**, 2163 (1962)
40. D.I. Kaiser, Conformal transformations with multiple scalar fields. Phys. Rev. D **81**, 084044 (2010). arXiv:1003.1159 [gr-qc]
41. S.V. Ketov, *Quantum Nonlinear Sigma Models* (Springer, New York, 2000)
42. J.M. Maldacena, Non-Gaussian features of primordial fluctuations in single field inflationary models. J. High Energy Phys. **0305**, 013 (2003). arXiv:astro-ph/0210603
43. N. Bartolo, E. Komatsu, S. Matarrese, A. Riotto, Non-Gaussianity from inflation: theory and observations. Phys. Rep. **402**, 103 (2004). arXiv:astro-ph/0406398; X. Chen, Primordial non-Gaussianities from inflation models. Adv. Astron., 638979 (2010). arXiv:1002.1416 [astro-ph]
44. C. Gordon, D. Wands, B.A. Bassett, R. Maartens, Adiabatic and entropy perturbations from inflation. Phys. Rev. D **63**, 023506 (2001). arXiv:astro-ph/0009131; D. Wands, N. Bartolo, S. Matarrese, A. Riotto, An observational test of two-field inflation. Class. Quant. Grav. **19**, 613 (2002). arXiv:hep-ph/0205253
45. P.A.R. Ade et al. (Planck collaboration), Planck 2015 results, XIII: cosmological parameters. arXiv:1502.01589 [astro-ph.CO]
46. M. Sasaki, E.D. Stewart, A general analytic formula for the spectral index of the density perturbations produced during inflation. Prog. Theor. Phys. **95**, 71 (1996). arXiv:astro-ph/9507001; D. Wands, Multiple field inflation. Lect. Notes Phys. **738**, 275 (2008). arXiv:astro-ph/0702187; D. Langlois, S. Renaux-Petel, Perturbations in generalized multifield inflation. J. Cosmol. Astropart. Phys. 0804 (2008), 017. arXiv:0801.1085 [hep-th]; C.M. Peterson, M. Tegmark, Testing multifield inflation: a geometric approach. arXiv:1111.0927 [astro-ph.CO]; J.-O. Gong, T. Tanaka, A covariant approach to general field space metric in multifield inflation. J. Cosmol. Astropart. Phys. **1103**, 015 (2011). arXiv:1101.4809 [astrod-ph.CO]
47. R. Kallosh, A. Linde, Nonminimal inflationary attractors. J. Cosmol. Astropart. Phys. **1310**, 033 (2013). arXiv:1307.7938 [hep-th]; J.J.M. Carrasco, R. Kallosh, A. Linde, Cosmological attractors and initial conditions for inflation. arXiv:1506.00936 [hep-th], and references therein
48. M.A. Amin, M.P. Hertzberg, D.I. Kaiser, J. Karouby, Nonperturbative dynamics of reheating after inflation: a review. Int. J. Mod. Phys. D **24**, 1530003 (2015). arXiv:1410.3808 [hep-ph]
49. N. Barnaby, J. Braden, L. Kofman, Reheating the universe after multifield inflation. J. Cosmol. Astropart. Phys. **1007**, 016 (2010). arXiv:1005.2196 [hep-th]; T. Battefeld, A. Eggemeier, J.T. Giblin, Jr., Enhanced preheating after multifield inflation: on the importance of being special. J. Cosmol. Astropart. Phys. **11**, 062 (2012). arXiv:1209.3301 [astro-ph.CO], and references therein
50. C.H. Brans, Bell's theorem does not eliminate fully causal hidden variables. Int. J. Theor. Phys. **27**, 219 (1988)
51. J. Gallicchio, A.S. Friedman, D.I. Kaiser, Testing Bell's inequality with cosmic photons: closing the setting-independence loophole. Phys. Rev. Lett. **112**, 110405 (2014). arXiv:1310.3288 [quant-ph]

Chapter 3
A New Estimate of the Mass
of the Gravitational Scalar Field
for Dark Energy

Yasunori Fujii

Abstract A new estimate of the mass of the pseudo dilaton is offered by following the fundamental nature that a massless Nambu-Goldstone boson, called a dilaton, in the Einstein frame acquires a nonzero mass through the loop effects which occur with the Higgs field in the relativistic quantum field theory as described by poles of D, spacetime dimensionality off the physical value $D = 4$. Naturally the technique of dimensional regularization is fully used to show this pole structure to be suppressed to be finite by what is called a Classical-Quantum-Interplay, to improve our previous attempt. Basically the same analysis is extended to derive also the coupling of a pseudo dilaton to two photons.

3.1 Introduction

We have developed our own version of the Scalar-Tensor theory (STT) [1, 2] due originally to Jordan [3], also to Brans and Dicke [4], but now with the unique feature that we are then allowed to be free from the possible fine-tuning problem in understanding the small size of a cosmological constant (CC), or the dark energy (DE), fitted to the observed accelerating universe [5]. Today's value of CC $\sim t_0^{-2}$ with t_0 the present age of the universe, is this small simply because we are this old cosmologically.[1]

---

[1]We use the reduced Planckian unit system defined by $c = \hbar = M_P(= (8\pi G)^{-1/2}) = 1$. The units of length, time and energy in conventional units are given by 8.10×10^{-33}cm, 2.70×10^{-43}s, 2.44×10^{18} GeV, respectively. As an example of the converse of the last entry, we find $1\,\text{GeV} = 2.44^{-1} \times 10^{-18}$ in units of the Planck energy. In the same way, the present age of the universe $t_0 \approx 1.37 \times 10^{10}$y is $10^{60.2}$ in units of the Planck time.

Y. Fujii (✉)
Advanced Research Institute for Science and Engineering,
Waseda University, Okubo, Shinjuku-ku, Tokyo 169-8555, Japan
e-mail: fujiitana@gmail.com

© Springer International Publishing Switzerland 2016
T. Asselmeyer-Maluga (ed.), *At the Frontier of Spacetime*,
Fundamental Theories of Physics 183, DOI 10.1007/978-3-319-31299-6_3

59

    This approach is an outgrowth, in retrospect, of a conceptual attempt based on
a simple Λ cosmology for the radiation-dominated universe [6]. According to an
attractor and asymptotic solution, the Jordan frame (JF), with a variable G, describes
unrealistically a static universe, while the Einstein frame (EF, with subscript $*$), with
G_* kept constant, provides fortunately with an expanding universe, $a(t_*) \sim t_*^{1/2}$,
hence accepted as the physical frame. Also the scalar field density interpreted as
dark energy density falls off like t_*^{-2}, from which follows the scenario of a decaying
cosmological constant, as emphasized above. On the other hand, however, starting
off by assuming a conventional mass term in JF, we come to finding that the microscopic fundamental particles, including an electron, with their masses which *fall
off* like $\sim a_*(t_*)^{-1} \sim t_*^{-1/2}$ in EF. This is totally in conflict with today's view on
the astronomical measurements; cosmological size is measured in reference to the
microscopic units of length provided by the inverse mass of the microscopic particles. Also we have no way of detecting any variation of units themselves, implying
their constancy in the physical frame, as we once called the *Own-unit-insensitivity
principle* (OUIP) [7], which ultimately derives the receding speeds of distant objects
in terms of the red-shifts of the observed atomic spectra.

    From these arguments we view the theory of STT to face a serious flaw when
it meets the microscopic physics. Sometime ago we came across [1, 2] that this
flaw can be avoided miraculously in terms of global scale invariance broken spontaneously with the field ϕ playing the role of a massless Nambu-Goldstone (NG) boson
[8–10], called *dilaton*, which, like many other examples of NG bosons, would acquire
a nonzero mass hence a *pseudo dilaton*. We further suggested a tentative estimate of
its mass-squared; $\mu^2 \sim m_q^2 M_{ss}^2 / M_P^2 \sim (10^{-9} \text{ eV})^2$, with m_q for the averaged quark
mass, while $M_{ss} \sim 10^3$ GeV for the supersymmetry mass scale to be prepared for
the quadratic cutoff of the self-energy of a scalar field.[2]

    Now in the current article, we are going to replace quarks by the Higgs field as an
origin for the masses of fundamental particles, also with an improved technique for
the mass acquisition mechanism applied uniquely to the dilaton. The new numerical
estimate based on the Standard Model (SM) turns out to result in μ somewhat heavier
than our previous estimate, still basically more or less in the same range of the order
of magnitude, remaining responsible for the experimental searches for the DE [12]
of the accelerating universe through $\gamma\gamma$ scattering.

    On the theoretical side, as we also point out, the two different concepts, scaling
behavior and infinities in the qantized field theory, are described by a single common
variable, the spacetime dimensionality assumed to be continuous off the physical
value 4. On the basis of dimensional regularization (DR) technique [13], the formulation is then not only simple and straightforward but also continued smoothly from
the *first half* of the spontaneously broken scale invariance.

---

[2]This mass corresponds approximately to the force-range ~ 100 m, related to the suggested non-
Newtonian force, as was discussed in [11], for example.

The same mechanism applies also to the coupling of a pseudo dilaton to two photons. These experiences prompt us to present our phenomenological results even, for the moment, only with a preliminary study of the effects of the conventional renormalization procedure.

3.2 Basic Equations

We start with the basic Lagrangian

$$\mathcal{L} = \begin{cases} \sqrt{-g}\left(\frac{1}{2}\xi\phi^2 R - \frac{1}{2}\epsilon g^{\mu\nu}\partial_\mu\phi\partial_\mu\phi + L_{\text{matter}} - \Lambda\right), \\ \sqrt{-g_*}\left(\frac{1}{2}R_* - \frac{1}{2}g_*^{\mu\nu}\partial_\mu\sigma\partial_\mu\sigma + L_{*\text{matter}} - \Lambda\exp(-4\hat{\zeta}\sigma)\right), \end{cases} \tag{3.1}$$

for $\Lambda > 0$ and $\xi > 0$ by avoiding antigravity, expressed both in JF and EF, respectively, where the factor Ω for the conformal transformation satisfies[3]

$$g_{\mu\nu} = \Omega(x)^{-2}g_{*\mu\nu}, \quad \phi = \hat{\xi}^{-1/2}\Omega, \quad \text{with} \quad \Omega = \exp(\hat{\zeta}\sigma), \tag{3.2}$$

where

$$\hat{\xi} = \xi M_{\text{P}}^{-2}, \quad \hat{\zeta} = \zeta M_{\text{P}}^{-1}, \quad \text{with} \quad \zeta^{-2} \equiv 6 + \epsilon\xi^{-1} > 0. \tag{3.3}$$

The symbol ϕ implies the gravitational scalar field in JF with $\epsilon = \pm 1$, or 0.[4] Notice that ϕ, even with an apparently ghost nature with $\epsilon = -1$, has been brought to be mixed to the spinless portion of $g_{\mu\nu}$ through the nonminimal coupling term, the first term on RHS of the upper line of (3.1), to emerge as a canonical nonghost scalar field σ in EF under the condition specified at the last of (3.3).

From (3.1) we derive the cosmological equations in each of radiation-dominated JF and EF, also in the standard spatially flat Robertson-Walker metric with the matter density approximated by a uniform distribution, finding the attractor and asymptotic solutions.[5] This is the way we reach the static universe in JF and the expanding universe in EF, as stated in Sect. 3.1.

During the calculation, we derived the asymptotic relation for the matter density ρ in JF;

$$\rho = -3\Lambda\frac{2\xi + \epsilon}{6\xi + \epsilon}. \tag{3.4}$$

Using $\Lambda > 0$ and $\xi > 0$ as stated immediately following (3.1) before also the last condition of (3.3), we find that the obvious condition $\rho > 0$ results only if

---

[3] Symbols φ, ω used in the original Refs. [3, 4] are now re-expressed by our more convenient ones; $\varphi = (1/2)\xi\phi^2$, $4\omega = \epsilon\xi^{-1}$.

[4] $\epsilon = 0$ implies $\zeta^2 = 1/6$ corresponding to the coefficient $a = 1/3$ of the scalar component of the combined potential [11, 14]. We choose $\eta_{00} = \eta^{00} = -1$.

[5] For details see [1, 2, 15].

$$\epsilon = -1, \quad \text{hence} \quad \frac{1}{6} < \xi < \frac{1}{2}, \quad \text{and} \quad \frac{1}{4} < \zeta^2 < \infty, \tag{3.5}$$

as will be used later. We also come to find

$$\phi \to t, \quad \text{or} \quad \Omega \to t, \quad \text{as} \quad t \to \infty. \tag{3.6}$$

Many of the important features in these approximate solutions are taken over to the more exact numerical solutions, which include such a unique complication like the occasional step-like falling-off behavior of the scalar-field density, as was discussed with the help of a supporting assumption, in Sects. 5.4.1–5.4.2 of [1] and Sects. 6.1–6.2 of [2], thus allowing the wording, a *decaying* cosmological *constant*, acceptable semantically. But more urgently, we re-emphasize briefly how OUIP is observed by the spontaneously broken scale invariance.

The crucial point is that we are supposed to start with the *interaction* term in JF;

$$- \mathcal{L}_\mathrm{I} = \frac{1}{2}\sqrt{-g}\, h\phi^2 \Phi^2, \tag{3.7}$$

instead of the conventional mass term $-\mathcal{L}_m = \sqrt{-g}(1/2)m^2\Phi^2$, applied to an example of the real scalar-field matter Φ, where h is a dimension*less* coupling constant, hence indicating *scale invariance* in JF. The conformal transformation to EF yields

$$- \mathcal{L}_\mathrm{I} = \frac{1}{2}\sqrt{-g_*}\,\Omega^{-4}h\hat{\xi}^{-1}\Omega^2\Omega^2\Phi_*^2 = \frac{1}{2}\sqrt{-g_*}\,m_*^2\Phi_*^2, \tag{3.8}$$

$$\text{with} \quad \Phi = \Omega\Phi_*, \quad m_*^2 = h\hat{\xi}^{-1}.$$

Notice that all of the Ω's cancel each other, hence leaving a truly *constant* m_*. The last result might be combined with $\xi \sim \mathcal{O}(1)$ as obtained from the second of (3.5) to find $h \sim \mathcal{O}(m_*^2/M_\mathrm{P}^2)$, which will be inherited basically to the more realistic but more complicated model for the Higgs field as will be developed soon later.

In this connection, we also notice that (3.7) fails to observe a premise that ϕ be decoupled from the matter Lagrangian, as emphasized by Brans and Dicke [4] who realized this to be the simplest way to implement Weak Equivalence Principle (WEP) as one of the most important features in the macroscopic and classical gravity.[6]

Now facing the microscopic and cosmological gravity, we might try an attempt beyond their premise. This is the reason why we come to (3.7), showing remarkably the global scale invariance *taking ϕ into account*. In a sense, we exploit the scale invariance of the whole STT terms in JF Lagrangian except for Λ.

From none of the mass in (3.7) we have created the mass *spontaneously*. In fact the mass dimension has been smuggled through $\hat{\xi}^{-1/2}$ as a VEV of ϕ, following the second of (3.2). The spontaneous nature might also be better interpreted by computing

---

[6]The amplitude through the nonminimal coupling term observes the same tensor coupling as in General Relativity.

the Noether current of the scale transformation[7];

$$\delta g_{\mu\nu} = 2\ell g_{\mu\nu}, \quad \delta\phi = -\ell\phi, \quad \delta\Phi = -\ell\Phi, \tag{3.9}$$

which derives,

$$J^\mu = \frac{1}{2}\sqrt{-g}\,g^{\mu\nu}\partial_\nu\left(\xi\zeta^{-2}\phi^2 + \Phi^2\right), \tag{3.10}$$

an exact result, as detailed in Appendix M of [1]. We then re-express RHS now in EF. In this process, we obtain $\Box_*\Phi_*^2$ on RHS, then by using the field equation of Φ_* coming to find the contribution from $\Lambda_* = \exp(-4\hat\zeta\sigma)\Lambda$. In the rest of the present article, however, we focus upon the epochs in which Λ_* remains rather smaller than the ordinary matter as was demonstrated by our realistic solutions in [1, 2], in the reasonable past from today. Then we may ignore Λ_* approximately, also impose the conservation law $\partial_\mu J_*^\mu = 0$, thus obtaining

$$\zeta^{-1}\Box_*\sigma = -\left(m_*^2\Phi_*^2 + g_*^{\mu\nu}\partial_\mu\Phi_*\partial_\nu\Phi_*\right), \tag{3.11}$$

where LHS comes directly from the first term on RHS of (3.10), hence reaching the massless nature of σ, to be called a *dilaton*, precisely as had been shown by Nambu [8]. The occurrence of the massless dilaton is expected to survive the approximations mentioned above.

We now leave the *first half* of the scenario of spontaneously broken scale invari-nace, entering its *second half* in which the massless dilaton grows into a *massive pseudo dilaton*. For this purpose, we start with the spacetime dimensionality D off, but close to, the physical value 4. We also re-interpret the above Φ now as the Higgs field supposed to provide with the origin of the masses of all the microscopic fields. In fact the familiar *Mexican-Hat* potential is shown to inherit the core of the 4-dimensional scale-invariance within the realm of STT.

We then extend (3.7) to

$$-\mathcal{L}_H = \sqrt{-g}\left(\frac{1}{2}h\phi^2\Phi^2 + \frac{\lambda}{4!}\Phi^4\right) = \sqrt{-g_*}\Omega^{D-4}\left(\frac{1}{2}\tilde{m}^2\Phi_*^2 + \frac{\lambda}{4!}\Phi_*^4\right), \tag{3.12}$$

$$\text{with} \quad \tilde{m}^2 = h\hat\xi^{-1},$$

where Ω in the second of (3.2) and (3.8) are both replaced by Ω^{d-1} with $d = D/2$. Then together with $\sqrt{-g} = \Omega^{-D}\sqrt{-g_*}$, we come to find an overall multiplier on RHS of (3.12);

$$\Omega^{-D}\left(\Omega^{d-1}\right)^4 = \Omega^{-D+2D-4} = \Omega^{D-4}. \tag{3.13}$$

---

[7]The special form of the first in the following, corresponding to $g_{\mu\nu} \to \Omega^2 g_{\mu\nu}$, has been selected, because, unlike another type of the coordinate transformation $x^\mu \to \Omega x^\mu$ or $\delta x^\mu = \ell x^\mu$, making it straightforward to be applied to the expanding universe with the 3-space simply uniform without any particular origin. See [16], for example, on introducing $\delta x^\mu/x^2 = \beta^\mu$.

Obviously, adding the quartic self-coupling of Φ^4 to (3.7) leaves our experiences of the scaling behaviors and spontaneous symmetry breaking nearly unchanged.

Following the well-known procedure, we shift the origin by the VEV v;

$$\Phi_* = v + \tilde{\Phi}. \tag{3.14}$$

By requiring the absence of the term linear in $\tilde{\Phi}$, we arrive finally at[8]

$$-\mathcal{L}_\mathrm{H} = \exp\left(2\hat{\zeta}(d-2)\sigma\right)\sqrt{-g_*}\mathcal{V}, \quad \text{with} \quad \mathcal{V} = \frac{1}{2}m^2\tilde{\Phi}^2 + \frac{1}{2}m\sqrt{\frac{\lambda}{3}}\tilde{\Phi}^3 + \frac{\lambda}{4!}\tilde{\Phi}^4, \tag{3.15}$$

where m defined by

$$m^2 = -2\tilde{m}^2 = \frac{\lambda}{3}v^2, \tag{3.16}$$

is the observed mass of the Higgs field, 1.26×10^2 GeV [17].

In this way we define the dynamics of $\tilde{\Phi}$ and σ, where the exponential factor comes from (3.13), substituted from the third of (3.2), while $\mathcal{V}$ is the Higgs potential.[9] We have ignored the possible vacuum terms as well as part of CC, in accordance with what we stated before between (3.10) and (3.11). As we consider, the current process of mass creation and the pseudo dilaton belongs to a *local physics* expected not to be affected seriously by CC, or DE.

To be noticed more explicitly, the field σ occurs *only in association* with $d - 2 = (D - 4)/2$ separated from what is called the Higgs potential $\mathcal{V}$. In other words, the dilaton σ might appear to be present only for $D \neq 4$. This by no means implies that σ is entirely outside our realistic concern at $D = 4$, because another part $\mathcal{V}$ contains infinities at $D = 4$, as will be shown shortly. We should take the same attitude as in DR that we keep $d \neq 2$ during the computation until we come back to the physical value $d = 2$ only at the very end of the calculation.

For more details of the calculation, we apply the expansion

$$\exp(2\hat{\zeta}(d-2)\sigma) \approx 1 - 2\hat{\zeta}(2-d)\sigma. \tag{3.17}$$

On the other hand, the potential $\mathcal{V}$, representing a collection of the field $\tilde{\Phi}$, might develop certain Feynman diagrams which may happen to induce closed loops of $\tilde{\Phi}$ then exhibiting divergences. The simplest 1-loop divergence is described by

---

[8]The symbol m in the second equation is the mass in EF, to be better denoted by m_*. To avoid too much notational complications in the following equations, however, we continue to use m without the subscript $*$ for the observed mass for the Higgs mass in the whole subsequent part of the article.

[9]This potential $\mathcal{V}$ is shown to agree with the relevant part of the SM. See (87.3) of [18], for example. His $V(\varphi) = (\lambda_r/4)(\varphi^\dagger\varphi - (v^2/2))^2$ is reproduced precisely by our (3.15) by choosing $\sigma = 0$, $\lambda_r/4 = \lambda/4!$, and $\varphi = (1/\sqrt{2})(v + \tilde{\Phi})$ only for the single component, with another component vanishing, corresponding to his (87.13). Notice also that a special relation chosen between the two terms in the parenthesis on RHS of $V(\varphi)$ above has the same effect of the term linear in $\tilde{\Phi}$ removed, which we required in the sentence just prior to the foregoing footnote 8.

$$\Gamma(2-d) \sim (2-d)^{-1}, \quad \text{as} \quad d \to 2, \tag{3.18}$$

according to the technique of DR.

By substituting (3.17) and (3.18) into the first of (3.15), we find a product like $\zeta(d-2)\sigma\Gamma(2-d)$, for which we apply the relation

$$(2-d)\Gamma(2-d) \to 1, \quad \text{as} \quad d \to 2, \tag{3.19}$$

to be called a *Classical-Quantum Interplay* (CQI), which connects a classical factor $2-d$ to $\Gamma(2-d)$ obviously representing a quantum nature. The above result implies that σ happens to pick up a $\tilde{\Phi}$-loop, which thus plays a major role in the same way as the cutoff conjectured by Nambu and Jona-Lasinio [10] for the study of the pion-nuleon system, thus re-discovered in a somewhat different but closely related context.

Notice also that we no longer suffer from infinities, as far as the relevant processes are dominated by CQI. This might be, however, related to the Brans-Dicke premise as was discussed following (3.8). The pole structure (3.18) should apply only to such fundamental fields like Φ, or quarks and leptons, but not to composite particles like hadrons. In this sense, the WEP violation effect due to the occurrence of ϕ in the matter Lagrangian tends to be smaller, but might need more detailed analysis before reaching the final comparison with the observations.

In the next section we are going to discuss how this crucial relation, CQI, can be used naturally in calculating the mass of the pseudo dilaton.

3.3 Computing the Mass of the Pseudo Dilaton

In order to derive the mass μ of σ, we first consider the simplest form of the σ self-energy (SE) part, as shown in the upper line of Fig. 3.1, where the dotted and solid lines are for σ and $\tilde{\Phi}$, respectively. Each vertex is read out from the second term on RHS of (3.17) times the first term in $\mathcal{V}$ of (3.15), deriving the effective vertex part;

$$g_0 = -2\hat{\zeta}(2-d)m^2. \tag{3.20}$$

The loop integral of $\tilde{\Phi}$ gives[10,11]

$$\int d^D k \frac{1}{(k^2+m^2)^2} = i\pi^2 \left(m^2\right)^{d-2} \frac{\Gamma(d)\Gamma(2-d)}{\Gamma(2)} \approx i\pi^2 \Gamma(2-d). \tag{3.21}$$

---

[10]Some of the details of the required integrals will be found in Appendix N of [1].

[11] Strictly speaking, the denominators should be $((k+q/2)^2+m^2)((k-q/2)^2+m^2)$, where q is for the momentum of the size of $\sim\mu$. Since we finally find μ negligibly smaller than m as in (3.49), we might justify the approximate computation as in (3.21). The same kind of approximation applies to almost any of the loop integrals to be encountered in the following of the present article.

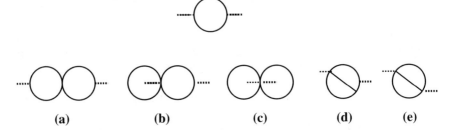

Fig. 3.1 In the *upper line* we show the simplified 1-loop SE part of σ, to start with. *Dotted* and *solid lines* are for the σ and the shifted Higgs field $\tilde{\Phi}$, respectively. In the *lower line*, we illustrate the 2-loop amplitudes to which CQI can be applied. They are different from each other in the ways the *dotted lines* couple to different parts of the three terms in the potential $\mathcal{V}$ on RHS of (3.15)

The whole contribution is

$$\sim g_0^2 i \pi^2 \Gamma(2-d) \sim 4m^4 \zeta^2 (2-d)^2 \Gamma(2-d) \sim 2-d \to 0, \quad \text{as} \quad d \to 2,$$
(3.22)

since the pole $(2-d)^{-1}$ has been over-cancelled by $(d-2)^2$ in accordance with CQI. As a lesson, a finite nonzero result will occur only if the $\tilde{\Phi}$ in $\mathcal{V}$ must include 2-loop divergences.

We now try to extend the argument to construct 2-loop amplitudes for the σ mass term, as illustrated by (a)–(e) in the lower line of Fig. 3.1. First in the diagram (a), we show a simply doubled 1-loop diagrams, in fact by connecting the simplest one to another diagram formed by the first term and the third term of $\mathcal{V}$ in (3.15). By computing explicitly, we do reach two poles resulting in the nonzero result;

$$\mathcal{J}_a = \frac{[-i(2\pi)^4]^3}{[i(2\pi)^4]^4} \lambda m^4 \left[2\hat{\zeta}(2-d)\sigma \right]^2 \int d^D k \frac{1}{(k^2+m^2)^2} \int d^D k' \frac{1}{(k'^2+m^2)^2}$$

$$= i(2\pi)^{-4} (i\pi^2)^2 \lambda \hat{\zeta}^2 4m^4 \sigma^2 = -i\frac{1}{4}\lambda \hat{\zeta}^2 m^4 \sigma^2, \quad \text{as} \quad d \to 2.$$
(3.23)

The combination in (b) is basically the same as in (a), different only where one of the dilaton lines reaches $\mathcal{V}$, thus causing a *linear* denominator. Still due to

$$\int d^D k' \frac{1}{k'^2+m^2} = i\pi^2 (m^2)^{d-1} \frac{\Gamma(d)\Gamma(1-d)}{\Gamma(3-d)} = -i\pi^2 m^2 \Gamma(2-d), \quad (3.24)$$

somewhat different from (3.21), though, we find the result, which happens to be the same as (3.23);

$$\mathcal{J}_b = \frac{[-i(2\pi)^4]^2}{[i(2\pi)^4]^3}\lambda m^2 \left[2\hat{\zeta}^2(2-d)\sigma\right]^2 \int d^D k' \frac{1}{k'^2+m^2} \int d^D k \frac{1}{(k^2+m^2)^2}$$

$$= i\frac{1}{4}\pi^{-4}\lambda m^4 (i\pi^2)^2\hat{\zeta}^2\sigma^2 = -i\frac{1}{4}\lambda\hat{\zeta}^2 m^4\sigma^2 = \mathcal{J}_a. \tag{3.25}$$

The same result will follow for the diagram in (c);

$$\mathcal{J}_c = -i(2\pi)^4\left[i(2\pi)^4\right]^{-2}\lambda\left[2\hat{\zeta}(2-d)\sigma\right]^2 \int d^D k'\frac{1}{k'^2+m^2}\int d^D k\frac{1}{k^2+m^2}$$

$$= i(2\pi)^{-4}\lambda 4\hat{\zeta}^2\sigma^2\left[i\pi^2 m^2\frac{2-d}{\Gamma(2-d)}\right]^2 = \mathcal{J}_a. \tag{3.26}$$

We then face another diagram in (d) due to the trilinear term on RHS of (3.15);

$$\mathcal{J}_d = \frac{[-i(2\pi)^4]^3}{[i(2\pi)^4]^4}\lambda m^2\left[2\hat{\zeta}(2-d)\sigma\right]^2\left(\frac{3!m}{2\sqrt{3}}\right)^2$$

$$\times \int d^D k \int d^D k'\frac{1}{k^2+m^2}\frac{1}{(k'^2+m^2)^2}\frac{1}{(k-k')^2+m^2}, \tag{3.27}$$

where the last term under the double integral is ovelapping not separable into the functions of k alone, and k' alone, respectively, corresponding to the inclined line in (d) in Fig. 3.1.

We have two alternative ways, carrying out (i) the k' integral first, or (ii) the k integral first. This also corresponds to including the overlapping integral as part of k' integral, or k integral, or expressed more explicitly as

$$\mathcal{J}_d^{(i)} = K_d(2-d)^2 \int d^D k\frac{1}{k^2+m^2}\left(\int d^D k'\frac{1}{(k'^2+m^2)^2}\frac{1}{(k-k')^2+m^2}\right), \tag{3.28}$$

$$\mathcal{J}_d^{(ii)} = K_d(2-d)^2 \int d^D k'\frac{1}{(k'^2+m^2)^2}\left(\int d^D k\frac{1}{k^2+m^2}\frac{1}{(k-k')^2+m^2}\right), \tag{3.29}$$

where

$$K_d = \frac{[-i(2\pi)^4]^3}{[i(2\pi)^4]^4}4\hat{\zeta}^2\sigma^2\left(\frac{m}{2}\sqrt{\frac{\lambda}{3}}3!\right)^2 = 3i\frac{\lambda m^2}{4\pi^4}\hat{\zeta}^2\sigma^2. \tag{3.30}$$

We are going to show first that $\mathcal{J}_d^{(i)}$ vanishes due to the k' integration which does not behave divergently, by analyzing only the relevant portion of (3.28);

$$\hat{\mathcal{J}}_d^{(i)} = (2-d)\int d^D k'\frac{1}{(k'^2+m^2)^2}\frac{1}{(k-k')^2+m^2}$$

$$= -2(2-d) \int_0^1 x dx \int d^D \tilde{k}' \frac{1}{(\tilde{k}'^2 + \mathcal{M}_{(i)}^2)^3}$$

$$= -i\pi^2(2-d) \int x dx (\mathcal{M}_{(i)}^2)^{d-3} \Gamma(3-d)$$

$$\sim -i\pi^2 \frac{2-d}{\mathcal{M}_{(i)}^2} \to 0, \quad \text{as} \quad d \to 2, \tag{3.31}$$

in which the gamma function $\Gamma(3-d)$ is left convergent at $d = 2$, hence (3.31), with the factor $2-d$, vanishes in the same way as in (3.22), despite that $\mathcal{M}_{(i)}^2$ defined by

$$\mathcal{M}_{(i)}^2 = m^2 + x(1-x)k^2, \quad \text{with} \quad \tilde{k}' = k' - k. \tag{3.32}$$

is not a pure constant.

We then move on to $\mathcal{J}_d^{(ii)}$. Also using (3.21), we find

$$\mathcal{J}_d^{(ii)} = K_d (2-d)^2 \int_0^1 dx \int d^D k' \frac{1}{(k'^2 + m^2)^2} \int d^D \tilde{k} \frac{1}{(\tilde{k}^2 + \mathcal{M}_{(ii)}^2)^2}$$

$$= K_d (2-d) i \pi^2 \int d^D k' \frac{1}{(k'^2 + m^2)^2} (\mathcal{M})_{(ii)}^{d-2} (2-d) \Gamma(2-d)$$

$$= K_d (2-d) i \pi^2 \int d^D k' \frac{1}{(k'^2 + m^2)^2}$$

$$= K_d (i\pi^2)^2 (2-d) \Gamma(2-d) = K_d (i\pi^2)^2$$

$$= -3i \frac{1}{4\pi^4} \lambda \hat{\zeta}^2 m^4 \sigma^2 = 3\mathcal{I}_a, \tag{3.33}$$

3 times as large as the last term of (3.23) for $\mathcal{J}_a$, where

$$\mathcal{M}_{(ii)}^2 = m^2 + x(1-x)k'^2, \quad \text{with} \quad \tilde{k} = k - k', \tag{3.34}$$

by the repeated use of CQI. Notice that the fact that $\mathcal{M}_{(ii)}^2$ is not purely constant does not affect the conclusion in the limit $d \to 2$.

Basically the same analysis can be applied to the diagram (e). From a visual inspection, we readily identify the difference in the equations; we only replace the double-pole terms of the type $(k^2 + m^2)^{-2}$ by the single-pole terms $(k^2 + m^2)^{-1}$, where k might be k'. This removes the difference between (3.28) and (3.29), leaving us to consider only (3.33), with the double-pole term at the very end of the second line replaced by the single-pole term, implying a sign change as we notice between (3.21) and (3.24). This is offset by the difference between $[-i(2\pi)^4]^3/[i(2\pi)^4]^4$ at the top of (3.27) and the corresponding factor $[-i(2\pi)^4]^2/[i(2\pi)^4]^3$ supposed to occur in $\mathcal{J}_e$. In this way we reach the simple result

$$\mathcal{J}_e = \mathcal{J}_d. \tag{3.35}$$

We then take the sum

$$\mathcal{J} = \mathcal{J}_a + 2\mathcal{J}_b + \mathcal{J}_c + 2\mathcal{J}_d + 2\mathcal{J}_e = -i\frac{16}{4}\lambda m^4 \hat{\zeta}^2 \sigma^2 = -4i\lambda m^4 \hat{\zeta}^2 \sigma^2, \quad (3.36)$$

where we have doubled the contribution from (b) and (d), to recover the right-left symmetry, while another doubling in (e) has been applied because the diagrams displayed in Fig. 3.1 are only half of the two possible choices of the dotted lines.

We are then allowed to compare our result with the effective mass term of σ;

$$\mathcal{J} = -L_\mu = \frac{1}{2}\mu^2 \sigma^2, \quad (3.37)$$

thus allowing us to identify (3.36) with $-i(2\pi)^4 \mu^2 \sigma^2$, hence

$$\mu^2 = \frac{4}{16\pi^4}\hat{\zeta}^2 \lambda m^4, \quad \text{or} \quad \mu = \frac{1}{2\pi^2}\sqrt{\lambda}\hat{\zeta} m^2. \quad (3.38)$$

In this way we have come to conclude that a massless dilaton in the presence of quantum loops does acquire a nonzero mass, thus becoming a pseudo dilaton, almost automatically in accordance with a simple interpretation of the exponential factor in (3.15). It seems even amusing to find that the same Higgs field plays indispensable roles in creating masses both of the pseudo dilaton, part of gravity, and the rest of the fundamental particles.

In spite of this desired result in deriving the mass μ, we still admit some uncertainty which might arise from the contribution without being derived from the CQI process. Consider a typical example of an additional $\tilde{\Phi}$ loop inserted between the two loops in the diagram (a) of Fig. 3.1, for example, also connected to the neighboring loops through the λ term in $\mathcal{V}$ of (3.15).[12] In the absence of $(d-2)\sigma$ we must appeal to the conventional procedure in DR;

$$\mathcal{R} \equiv -i(2\pi)^4 \lambda^2 [i(2\pi)^4]^{-2} \int d^D k \frac{1}{(k^2 + m^2)^2} \approx -\frac{\lambda^2}{16\pi^2}\Gamma(-\delta)\exp\left(-\delta/\delta_m\right), \quad (3.39)$$

where $\delta = d - 2$, $(m^2)^\delta = \exp(\delta \ln m^2)$, and $\lambda \sim \mathcal{O}(1)$, also

$$\delta_m \equiv \frac{1}{-\ln m^2} \approx \frac{1}{75} \approx 0.013. \quad (3.40)$$

The foregoing CQI calculations would be left undisturbed if $|\mathcal{R}|$ is kept well below the order unity in the reduced Planckian units.

The relation (3.39) is evaluated at $\delta \to 0+$. The divergence from $\Gamma(-\delta) \sim -\delta^{-1}$ requires a re-regularization, which is expected to convert a pole to a cutoff $\delta_c > 0$, for example. In the absence of any general way of fixing δ_c, also without related

---

[12]Basically the same type of analysis can be applied to another example of a loop attached to the side of the left loop in Fig. 3.1b, for example.

physical observables,[13] we might choose it to be δ_m defined by (3.40), which is not only unique to the current model with the occurrence of m^2, but also happens to be reasonably small. In this way we have now reached;

$$|\mathcal{R}(\delta_c)| = |\mathcal{R}(\delta_m)| \approx \frac{-\ln m^2}{16\pi^2} e^{-1} \approx \frac{0.48}{2.72} \approx 0.18 \lesssim \mathcal{O}(1), \qquad (3.41)$$

which turns out to be barely a lower end of $\mathcal{O}(1)$.

This result achieved by a simple but natural approach appears to be an encouraging sign for a theoretically favored idea of the CQI-dominance, at least in certain physical situations, though more details are yet to be scrutinized, probably in the future. In the next section, we then enter the final step of our numerical analysis on the basis of the expected CQI computation.

3.4 Estimating μ

In the second equation of (3.38), we first re-express $m^2 = (1.26 \times 10^2)^2 \, (\text{GeV})^2$ into the Planckian unit system;

$$m^2 = (1.26 \times 10^2)^2 \times (2.44^{-1} \times 10^{-18})^2. \qquad (3.42)$$

However, $\hat{\zeta}$ has the mass dimension -1, so that the product $\hat{\zeta} m^2$ has the mass dimension $+1$, in agreement with μ on LHS of the second of (3.38), and is going to be re-expressed in units of GeV instead of $(\text{GeV})^2$;

$$\hat{\zeta} m^2 = \hat{\zeta}(1.26 \times 10^2)^2 \times (2.44^{-1} \times 10^{-18}) = \zeta \frac{1.26^2 \times 10^4}{2.44} \times 10^{-18} \, \text{GeV}$$

$$= 0.651\zeta \times 10^{-14} \, \text{GeV} = 6.51\zeta \, \mu\text{eV}. \qquad (3.43)$$

In order to know λ, we first use (3.16). We then go through the two well-known steps in SM;

$$m_W = -\frac{gv}{2}, \quad \text{and} \quad \frac{g^2}{m_W^2} = \frac{8}{\sqrt{2}} G_F, \qquad (3.44)$$

where $m_W = 80.2 \, \text{GeV}$ is the mass of the W meson interpreted as a Higgs mechanism for the mass of a gauge field, while $G_F = 1.17 \times 10^{-5} \, \text{GeV}^{-2}$ is the Fermi constant for the weak interaction. We then use the second equation of (3.44) to derive[14]

---

[13] For the self-masses in QED or QCD, divergences are removed simply to fit the observed values.
[14] The effect of renormalized fields might have been ignored at this moment, given the crude approximation to be allowed in (3.49). For details see [18], for example.

$$g^2 = \frac{8}{\sqrt{2}} m_W^2 \, G_F = 0.426, \quad \text{or} \quad g = 0.653, \tag{3.45}$$

which is then substituted into the first of (3.44), finding

$$v = -\frac{2}{g} m_W = 246 \, \text{GeV}, \tag{3.46}$$

finally substituted into (3.16) hence arriving at

$$\lambda = 3 \left(\frac{1.26}{2.46} \right)^2 = 0.787, \quad \text{or} \quad \sqrt{\lambda} = 0.887. \tag{3.47}$$

We also need an estimate of ζ. According to (3.5), we have $\epsilon = -1$ and $1/6 < \xi < 1/2$, hence $1/4 < \zeta^2 < \infty$ for the radiation-dominated universe, while $1/2 < \xi < 3/2$ and $3/16 < \zeta^2 < 1/4$ for the dust-dominated universe, as shown in Fig. 1 of [2, 7]. Our own fit to the accelerating universe yields $\zeta \sim 1.58$. as read in Fig. 5.8 of [1] and Fig. 9 of [2]. Also noted is $\xi = 1/4$ and $\zeta^2 = 1/2$, or $\zeta = 0.714$ from Super String Theory as indicated in (3.4.58) of [19]. In view of these findings, we suggest a tentative but still convenient bound of ζ somewhere between 0.5 and 2.0;

$$\zeta \approx (0.5 - 2.0). \tag{3.48}$$

Summarizing them all finally in the second of (3.38), we obtain

$$\mu \approx (0.15 - 0.59) \, \mu\text{eV} \sim (150 - 590) \, \text{neV}. \tag{3.49}$$

The far RHS suggests nearly two orders of magnitude over the previous estimate $\sim 10^{-9} \, \text{eV}$.[15]

3.5 Coupling of a Pseudo Dilaton to Two Photons

As an application of the current approach, we now try to re-derive the coupling term

$$-L_3 = A\sigma \frac{1}{4} F_{\mu\nu} F^{\mu\nu}, \tag{3.50}$$

which plays a pivotal role in the experimental searches for DE [12], hopefully on a wider perspective in exploiting the scale invariance than in the past attempts [1, 2].

---

[15]It still appears that the two estimates above are more or less close to each other from a wider point of view, probably because the two approaches share the same concept on the pseudo dilaton in some way or the other.

To emphasize the unique nature of the pseudo dilaton, we are going to study the photon SE part, or the vacuum polarization, represented by the diagram, the same type as shown on the upper line of Fig. 3.1, though the solid line, used to be for the neutral Higgs field, is now re-interpreted as a charged matter field, also with the dotted line used for the pseudo dilaton, replaced by the photon line. For simplicity for the moment, we assume a singly charged and massive Dirac field ψ, a representative of quarks and leptons.

We start off with the simple electromagnetic interaction of ψ first in JF, followed by moving to EF;

$$- \mathcal{L}_{\text{em}} = bie \left(\bar{\psi} b^{i\mu} \gamma_i \psi \right) A_\mu \rightarrow b_* \Omega^{-D} ie \Omega^{D-1} \left(\bar{\psi}_* b_*^{i\mu} \Omega \gamma_i \psi_* \right) \Omega^{d-2} A_{*\mu}$$
$$\equiv -b_* \Omega^{d-2} L_{*\text{em}}, \tag{3.51}$$

where e is the elementary charge chosen to be a pure constant, with $\alpha = e^2/(4\pi) \approx 1/137$;

$$- L_{*\text{em}} = ie \bar{\psi}_* b_*^{i\mu} \gamma_i \psi_* A_{*\mu}, \tag{3.52}$$

with b_μ^i the dimensionally extended tetrad with $b = \sqrt{-g}$, also the electromagnetic field transforming as $A_\mu = \Omega^{d-2} A_{*\mu}$. The occurrence of Ω^{d-2} on the far RHS of (3.51) indicates scale invariance in 4 dimensions, hence the same nature as generating σ as in the previous sections.

In EF, we approximate spacetime by locally Minkowskian to apply the ordinary Feynman rules based on (3.52) to the same type of the diagram as on the upper line of Fig. 3.1 in terms of DR, obtaining the gauge-invariant form[16]

$$\Pi_{\mu\nu}(k) = \frac{\alpha}{3\pi} \left(k_\mu k_\nu - k^2 \eta_{\mu\nu} \right) \Gamma(2 - d), \tag{3.53}$$

where we have reversed the overall sign due to the antisymmetric nature of ψ and $\bar{\psi}$.

Further re-installing Ω^{d-2} on the far RHS of (3.51), substituted from the third of (3.2), we reach the whole result

$$\sqrt{-g_*} \exp \left(2\hat{\zeta}(d - 2)\sigma \right) \frac{-\alpha}{3\pi} \left(k_\mu k_\nu - k^2 \eta_{\mu\nu} \right) \Gamma(2 - d). \tag{3.54}$$

We then pick up the term liner in σ, also using the CQI in the form of (3.19), comparing the result with (3.50) by using the relation

$$\frac{1}{4} F_{\mu\nu} F^{\mu\nu} = -\epsilon_f^\mu \left(k_\mu k_\nu - k^2 \eta_{\mu\nu} \right) \epsilon_i^\nu, \tag{3.55}$$

---

[16] See (6.174)–(6.181) of [1] for details simply for the scalar loop field. Extending to the Dirac field is tedious but straightforward.

with $\epsilon_f'^\mu$ and ϵ_i^ν for the polarization vectors of the final and initial photons, respectively, then identifying the constant A as

$$A = -\frac{\alpha}{3\pi}\hat{\zeta}. \tag{3.56}$$

In this way we come to determine A basically of the size of the inverse of the Planck mass, as expected to be.

For each of the quarks and the leptons, the result (3.56) adds up with the corresponding multiplicative factors for the electric charges squared. Notice also that a cancellation always takes place between the terms of the same m, the mass of the loop fields, to meet the gauge-invariance of the form (3.53), thus yielding A independnent of m. This has an advantage that we derive A basically $\sim \alpha\hat{\zeta}$, but, on the other hand, deprives us of a familiar procedure to suppress the contribution from much heavier and uncertain loop fields.

From this point of view, we may also consider the charged scalar fields, sharing the same charge-structure as indicated by supersymmetry, for which we develop obviously the pararell computations with ψ in (3.52) replaced by Φ together with the additional 4-point term $\sim e^2 \bar{\Phi}\Phi g^{\mu\nu} A_\mu A_\nu$,[17] without the sign change due to the antisymmetry of the fermionic field, ending up with a multiplicative factor $-1/4$ to (3.53),

$$A_{sc} = \frac{\alpha}{12\pi}\hat{\zeta}. \tag{3.57}$$

in place of (3.56).

The reduction factor $1/4$ can be interpreted by $(1/2)^2$ with 1 and 2 for the spin degrees of freedom for the scalar and the Dirac fields, respectively, while the squaring takes care of the occurrence of two lines in each of the main loop diagrams. In this sense, every 4 scalar fields offset the effect of 1 Dirac field, no matter how heavy they might be. Obviously, it appears too early to make an unambiguous prediction before we develop a more general survey on what the fundamental particles are to be included in the loop, also considering wider class of spin-statistics combinations, again left to future studies, at this time. An uncertainty of this kind still unavoidable at present might result in an adjustable parameter multiplied to $\Gamma^{1/2}$ with Γ for the decay width of σ into two photons in the formulation in [12].[18]

---

[17]This term contributes a term proportional to m^2 due to a simple 1-loop of $\langle 0|\Phi\bar{\Phi}|0\rangle \propto m^2$ in DR, thus cancelling the term of $\sim m^2$ in the main loop term as in the upper line of Fig. 3.1.

[18]The previous result Eq. (3.91) in [2], for example, can even be re-interpreted as our (3.56) multiplied by an "adjustable parameter" $B/A = \zeta^{-1}(2/3)\mathcal{Z}$.

3.6 Summary

We started out by assuming the scale-invariance of the Higgs potential in JF in STT in 4 dimensions, reaching, somewhat unexpectedly, a time-independent particle mass in EF in conformity with today's view on measuring cosmological size in units of the inverse of the mass of the microscopic particles. We then extended the spacetime dimensionality D off the physical value 4, allowing us to analyze the behavior of the pseudo NG boson, pseudo dilaton. We derived its mass by studying the σ SE part. By considering the 2-loop amplitudes, we obtained the nonzero and finite mass μ of the pseudo dilaton, by maximally exploiting the CQI relation, also utilizing SM, somewhere around μeV, which turns out approximately 2 orders of magnitude heavier than our previous tentative estimate. The difference is understood naturally because we now deal with a theoretical model quite different from our previous simple-minded one. It still seems helpful if we find any aspect more tractable on the non-CQI terms possibly in the future. As an extended idea, we tried also to re-derive the coupling strength of σ into 2γ, still short of the fully unique determination of the multiplier at present.

Acknowledgments The author expresses his sincere thanks to K. Homma, H. Itoyama, C.S. Lim, K. Maeda, T. Tada and T. Yoneya for many useful discussions.

References

1. Y. Fujii, K. Maeda, *The Scalar-Tensor Theory* (Cambridge University Press, Cambridge, 2003)
2. Y. Fujii, Entropy. **14**, 1997 (2012). www.mdpi.com/journal/entropy
3. P. Jordan, *Schwerkraft und Weltall* (Friedrich Vieweg und Sohn, Braunschweig, 1955)
4. C. Brans, R.H. Dicke, Phys. Rev. **124**, 925 (1961)
5. A.G. Rises et al., Astron. J. **116**, 1009 (1998); S. Perlmutter et al., Astrophys. J. **517**, 565 (1999); D.N. Spergel et al., Astrophys. J. **517**, 565 (1999); D.N. Spergel et al., Astrophys. J. Suppl. **148**, 175 (2003)
6. D. Dolgov, An attempt to get rid of the cosmological constant, in *The Very Early Universe, Proceedings of Nuffield Workshop*, ed. by G.W. Gibbons, S.T. Siklos (Cambridge Univeristy Press. Cambridge, 1982)
7. Y. Fujii, Prog. Theor. Phys. **118**, 983 (2007)
8. Y. Nambu, Phys. Rev. Lett. **4**, 380 (1960)
9. J. Goldstone, Nuovo Cim. **19**, 154 (1961)
10. Y. Nambu, G. Jona-Lasinio, Phys. Rev **122**, 345 (1961)
11. Y. Fujii, Nat. Phys. Sci. **234**, 5 (1971)
12. Y. Fujii, K. Homma, Prog. Theor. Phys. **126**, 531 (2011); Prog. Theor. Exp. Phys. **2014**, 089203, K. Homma *et al.* Prog. Theor. Exp. Phys. **2014**, 083C01
13. G. 't Hooft, M. Veltman, Nucl. Phys. **44**, 189 (1972); K. Chikashige, Y. Fujii, Prog. Theor. Phys. **57**, 623; 1038 (1977); Y. Fujii, Dimensional regularization and hyperfunctions, in *Particles and Fields*, ed. by D. Bohl, A. Kamal (Plenum Press, New York, 1977)
14. J. O'Hanlon, Phys. Rev. Lett. **29**, 137 (1972)
15. K. Maeda, Y. Fujii, Phys. Rev. D **79**, 084026 (2009)
16. A. Salam, J. Strathdee, Phys. Rev. **184**, 1760 (1969)

17. The ATLAS Collaboration, Phys. Lett. B **716**, 1 (2012) (The CMS Collaboration. Phys. Lett. B **716**, 30 (2012))
18. M. Srednicki, *Quantum Field Theory* (Cambridge University Press, Cambridge, 2007)
19. M.B. Green, J.H. Schwarz, E. Witten, *Superstring Theory* (Cambridge University Press, Cambridge, 1987)

Chapter 4
Axion and Dilaton + Metric Emerge Jointly from an Electromagnetic Model Universe with Local and Linear Response Behavior

Friedrich W. Hehl

Universally coupled, thus gravitational, scalar fields are still active players in contemporary theoretical physics. So, what is the relationship between the scalar of scalar-tensor theories, the dilaton and the inflaton? Clearly this is an unanswered and important question. The scalar field is still alive and active, if not always well, in current gravity research.

Carl H. Brans (1997)

Abstract We take a quick look at the different possible universally coupled scalar fields in nature. Then, we discuss how the gauging of the group of scale transformations (dilations), together with the Poincaré group, leads to a Weyl-Cartan spacetime structure. There the *dilaton* field finds a natural surrounding. Moreover, we describe shortly the phenomenology of the hypothetical *axion* field. In the second part of our essay, we consider a spacetime, the structure of which is exclusively specified by the premetric Maxwell equations and a fourth rank electromagnetic response tensor density $\chi^{ijkl} = -\chi^{jikl} = -\chi^{ijlk}$ with 36 independent components. This tensor density incorporates the permittivities, permeabilities, and the magneto-electric moduli of spacetime. No metric, no connection, no further property is prescribed. If we forbid birefringence (double-refraction) in this model of spacetime, we eventually end up with the fields of an axion, a dilaton, and the 10 components of a metric tensor with Lorentz signature. If the dilaton becomes a constant (the vacuum admittance) and the axion field vanishes, we recover the Riemannian spacetime of general relativity theory. Thus, the metric is encapsulated in χ^{ijkl}, it can be *derived* from it.

F.W. Hehl (✉)
Institute for Theoretical Physics, University of Cologne,
50923 Cologne, Germany
e-mail: hehl@thp.uni-koeln.de

F.W. Hehl
Department of Physics and Astronomy, University of Missouri,
Columbia, MO 65211, USA

© Springer International Publishing Switzerland 2016
T. Asselmeyer-Maluga (ed.), *At the Frontier of Spacetime*,
Fundamental Theories of Physics 183, DOI 10.1007/978-3-319-31299-6_4

4.1 Dilaton and Axion Fields

4.1.1 Scalar Fields

The Jordan-Brans[1]-Dicke *scalar*, the *dilaton*, the *axion*, the *inflaton*—scalar fields everywhere—and eventually even one, the scalar, that is, the spinless *Higgs* boson H^0, which has been found experimentally as heavy as some 134 protons. These different scalar fields[2] are not necessarily independent from each other, it could be, for example, that the JBD-scalar can be identified with the dilaton (see [30]) or the Higgs boson with the inflaton (see [6, 7]). Thus, the list of potentially existing universally coupled scalar fields could be somewhat smaller. For the history of the JBD-scalar, one can compare Brans [11] and Goenner [32] and, for the role of the inflaton in different models, Vennin et al. [97].

4.1.2 Einstein Gravity and the Energy-Momentum Current

As remarked by Brans [10] in the quotation above, if universally coupled, the scalar fields are intrinsically related to the gravitational field. In Einstein's theory of gravity, general relativity (GR), the gravitational potential is the metric g_{ij}, with $i, j = 0, 1, 2, 3$ as (holonomic) coordinate indices. As its source acts the symmetric energy-momentum tensor T_{ij} of matter. This is a second rank tensor, which is generated already in special relativity (SR) with the help of the group $T(4)$ of *translations* in space and in time. Together with the Lorentz transformations $SO(1, 3)$, the translations $T(4)$ build up the Poincaré group $P(1, 3)$ as a semi-direct product: $P(1, 3) = T(4) \rtimes SO(1, 3)$. This is the group of motion in the Minkowski space of special relativity, see [31].

Accordingly, if one desires to understand gravity from the point of view of the gauge principle, the $T(4)$ is an indispensable part of these considerations. However, being only *one* piece of the $P(1, 3)$, it is suggestive to gauge the complete $P(1, 3)$. This is exactly what Sciama and Kibble did during the beginning 1960s, see [73], [9, Chap. 4], and [16].

---

[1] Carl Brans is one of the pioneers of the scalar-tensor theory of gravitation. This essay is dedicated to Carl on the occasion of his 80th birthday with all best wishes to him and his family. During the year of 1998, we had the privilege to host Carl, as an Alexander von Humboldt awardee, for several months at the University of Cologne. I remember with pleasure the many lively discussions we had on scalars, on structures of spacetime, on physics in general, and on various other topics.

[2] We skip here the plethora of *scalar* mesons,

$$\pi^\pm, \pi^0, \eta, f_0(500), \eta'(958), f_0(980), a_0(980), ...,$$
$$K^\pm, K^0, K^0_S, K^0_L, K^*_0(1430), D^\pm, D^0, D^*_0(2400)^0, D^\pm_s, ...;$$

they are all composed of two quarks. Thus, the scalar mesons do not belong to the fundamental particles.

4.1.3 Einstein-Cartan Gravity: The Additional Spin Current

This gauging of the $P(1, 3)$ extends the geometrical framework of gravity. The 4 translational potentials $e_i{}^\alpha$ and the 6 Lorentz potentials $\Gamma_i{}^{\alpha\beta} = -\Gamma_i{}^{\beta\alpha}$ span a Riemann-Cartan spacetime, enriching the Riemannian spacetime of GR by the presence of Cartan's torsion; here $\alpha, \beta = 0, 1, 2, 3$ are (anholonomic) frame indices. Whereas the translational potentials couple to the canonical energy-momentum tensor of matter $\Sigma_\alpha{}^i$, the Lorentz potentials couple to the canonical spin current of matter $\tau_{\alpha\beta}{}^i = -\tau_{\beta\alpha}{}^i$.

The simplest version of the emerging Poincaré gauge models is the Einstein-Cartan theory (EC), a viable gravitational theory competing with GR if highest matter densities are involved. If l_{Planck} denotes the Planck length and λ_{Compton} the Compton wave length of a particle, then deviations of the Einstein-Cartan from Einstein's theory are expected at length scales of below $\sim(l_{\text{Planck}}^2 \lambda_{\text{Compton}})^{1/3}$; for protons, prevalent in the early cosmos, it is about 10^{-29} m. According to Mukhanov [65], it is exactly this order of magnitude down to which, according to recent cosmological data, GR is known to be valid.

From a gauge theoretical point of view, the EC-theory looks more reasonable than GR since the Einsteinian principles of how to heuristically derive a gravitational theory were followed closely: they were just applied to fermionic matter instead of to macroscopic point particles or Euler fluids or to classical electromagnetism, as Einstein did.

Incidentally, in the EC-theory and, more generally, in the Poincaré gauge theory, the Poincaré and, in particular, the Lorentz covariance are valid locally by construction, similar as in a $SU(2)$ Yang-Mills theory, we have local $SU(2)$ covariant. Kostelecký (priv. comm., Jan. 2016) agrees that the "Einstein-Cartan theory maintains local Lorentz invariance." Then the same is true for a Poincaré gauge theory, which acts likewise in a Riemann-Cartan spacetime with torsion and curvature. However, in an experimental set-up, according to Kostelecký [48], torsion must be considered as an external field and, according to the "standard lore for backgrounds," local Lorentz invariance is broken. By the same token, an external magnetic field in electrodynamics breaks local Lorentz invariance. This is, in my opinion, an abuse of language, which conveys the wrong message that the existence of a torsion field violates local Lorentz invariance.

If, for theoretical reasons, one wants to evade the emergence of the Riemann-Cartan spacetime, then one can manipulate, in the underlying Minkowski space, the intrinsic or spin part of the total angular momentum of matter in such a way that it vanishes on the cost of increasing the orbital part of it by the corresponding amount, see [63]. This procedure is called Belinfante-Rosenfeld symmetrization of the canonical energy-momentum current, which, in general, is defined as an asymmetric tensor by the Noether procedure. Accordingly, by symmetrization the energy-momentum current is made fit to act as a source of the Einstein field equation. In this way, one can effectively sweep the spin and the torsion under the rug and can live happily forever in the paradise of the Riemannian spacetime of GR.

Of course, in the end observations and/or experiments will decide which of the two theories, GR or EC, will survive. We opt for the latter.

4.1.4 Dilaton Field and Dilation Current

The dilaton field ϕ entered life as a Nambu-Goldstone boson of broken scale invariance, see Fujii in [30]. Thus, ϕ is related to dilat[at]ions or scale transformations in space and time. But the dilaton also occurs in theories of gravity (JBD) and in string theory, see Di Vecchia et al. [21]. The $P(1, 3)$, if multiplied (semi-directly) with the scale group, becomes the Weyl group $W(1, 3)$. This 11-parameter group is an invariance group for massless particles in special relativity. The translations, via Noether's theorem, generate the conserved energy-momentum tensor Σ_{ij}, the Lorentz transformations the conserved total angular momentum tensor $J_{ij}{}^k := \tau_{ijk} + x_{[i}\Sigma_{j]}{}^k = -J_{ji}{}^k$, and the dilation the conserved total dilation current $\Upsilon^k := \Delta^k + x^l\Sigma_l{}^k$,

$$\partial_k \Sigma_i{}^k = 0, \tag{4.1}$$

$$\partial_k J_{ij}{}^k = \partial_k \tau_{ij}{}^k - \Sigma_{[ij]} = 0, \tag{4.2}$$

$$\partial_k \Upsilon^k = \partial_k \Delta^k + \Sigma_k{}^k = 0, \tag{4.3}$$

see [47, 59] particularly for Υ^k. Thus, if a universal coupling is assumed, then ϕ should have the intrinsic dilation current Δ^k as its source; for theories in Weyl spaces in which Δ^k does *not* play a role, see Scholz [86, 87].

There are numerous field theoretical models under way which, if scale or dilation invariance is implemented, have conformal invariance as a consequence; for a more recent review see Nakayama [66]. Hence, jumping to conformal invariance, before one understood scale invariance, is probably not a very good strategy. For this reason we confine ourselves here to scale invariance, to the dilaton, and to the 11-parametric Weyl group. But it should be understood that the light cone is also invariant under the 15-parametric conformal group, see Barut and Rączka [5] and Blagojević [8] and, for a historical account, Kastrup [45].

Both currents, the intrinsic dilation current Δ^k and the energy-momentum current $\Sigma_i{}^k$ are related to *external* groups, to the dilation (scale) and to the translation groups, respectively. This is the reason for their universality.

4.1.5 The Weyl-Cartan Spacetime as a Natural Habitat of the Dilaton Field

We only tried to make a strong case in favor of the EC-theory in order to repeat the corresponding arguments for the dilation group. Gauging the Weyl group yields a Weyl-Cartan spacetime. The classical paper in that respect is the one of Charap and Tait, see [9, Chap. 8]. A universally coupled massless scalar field induces a *Weyl covector* Q_i as the corresponding dilation potential willy nilly. This is the type of spacetime Weyl used (with vanishing torsion) for his failed unified theory of 1918. Here the Weyl space with the connection $^W\Gamma$ is resurrected for the dilation current, instead of for the electric current, see [8]:

$$\overset{W}{\nabla}_i g_{jk} = -Q_i g_{jk}, \quad {}^W\Gamma_{ijk} = {}^{RC}\Gamma_{ijk} + \frac{1}{2}(Q_i g_{jk} + Q_j g_{ki} - Q_k g_{ij}); \quad (4.4)$$

$^{RC}\Gamma$ is the connection of the Riemann-Cartan space. Again, as in the case of the Lorentz group, one can manipulate the total dilation current Υ^k and can transform its intrinsic part into an orbital part by modifying in this case the *trace* $\Sigma_k{}^k$ of the energy-momentum current. Then, again, one can stay within the realm of the Riemannian space of GR, see Callan et al. [13].

As we mentioned already, the gauge theoretical answer was given by Charap and Tait [15]. Again, which approach will succeed is eventually a question to experimental verification.

We see, if the JBD-scalar is interpreted as a dilaton, then we would expect that the Weyl-Cartan spacetime is its arena. Clearly this does only provide the kinematics of the theory. The dynamics would depend on the exact choice of the dilaton Lagrangian.

Recently, Lasenby and Hobson [53] wrote an in-depth review of gauging the Weyl group and, moreover, formulated an "extended Weyl gauge theory." Also within their framework, the Weyl-Cartan space, and a straightforward extension of it, play an important role, see also Haghani et al. [33]. Definite progress has also been achieved in the study of equations of motion within the scalar tensor theories of gravity, see Obukhov and Puetzfeld [76, 81, 82]. The breaking of scale invariance in the more general approach of metric-affine gravity was studied in [34], for example; for somewhat analogous breaking mechanisms, see [60–62].

4.1.6 Axion Field

Dicke did not only introduce in 1961, together with Brans [12], a scalar field into gravity, but he also discussed, in 1964, and pseudoscalar or axial scalar field φ^2 in the context of gravitational theory, see [20, Appendix 4, p. 51, Eq. (7)].

Subsequently, in the early 1970s, Ni [67] investigated matter coupled to the gravitational field and to electromagnetism and looked for consistency with the equivalence principle. He found it possible to introduce in this context a new neutral *pseudo*scalar field accompanying the metric field, see also [4, 68–70]. Later, in the context of the vacuum structure of quantum chromodynamics, a light neutral pseudoscalar, subsequently dubbed "axion" was hypothesized, see Weinberg [98, pp. 458–461]. Similar as Ni's field, the axion couples also to the electromagnetic field, see Wilczek's paper [99] on "axion electrodynamics".

The axion field is of a similar universality as the gravitational field. In other words, the axion belongs to the universally coupled scalar fields. Let in electrodynamics, $\mathcal{H}^{ij} = (\mathbf{D}, \mathbf{H}) = -\mathcal{H}^{ji}$ and $F_{ij} = (\mathbf{E}, \mathbf{B}) = -F_{ji}$ denote the excitation and the field strength, respectively. The constitutive relation characterizing the axion field $\alpha(x)$

(in elementary particle terminology it is called A^0) reads [35],

$$\mathcal{H}^{ij} = \frac{1}{2}\alpha\epsilon^{ijkl}F_{kl} \quad \text{or} \quad \begin{cases} D^a = \alpha B^a, \\ H_a = -\alpha E_a, \end{cases} \tag{4.5}$$

see also [37] for the corresponding formalism; here ϵ is the totally antisymmetric Levi-Civita symbol with $\epsilon^{ijkl} = \pm 1$, moreover, $a = 1, 2, 3$. Clearly, the axion embodies the magnetoelectric effect par excellence. It is a pseudoscalar under 4-dimensional diffeomorphisms.

In electrotechnical terms, the axion behaves like the (nonreciprocal Tellegen) gyrator of network analysis, see [49, 100]; also the perfect electromagnetic conductor (PEMC) of Lindell and Sihvola [58, 93] represents an analogous structure. Metaphorically speaking, as we see from (4.5), the axion "rotates" the voltages $(\mathbf{B}, \mathbf{E})$ into the currents $(\mathbf{D}, \mathbf{H})$. In SI, we have the units $[B] = \text{Vs/m}^2$, $[E] = \text{V/m}$; $[D] = \text{As/m}^2$, $[H] = \text{A/m}$. Thus, $[\alpha] = 1/\text{ohm} = 1/\Omega$ carries the physical dimension of an admittance. Now, in the Maxwell Lagrangian, we find an additional piece $\sim \alpha(x)\epsilon^{ijkl}F_{ij}F_{kl} \sim \alpha(x)\mathbf{E}\cdot\mathbf{B}$, a term, which was perhaps first discussed by Schrödinger [89, pp. 25–26]. If α were a constant, the field equations would not change.

As we already remarked, α is a 4-dimensional pseudoscalar. The same is true for the von Klitzing constant $R_{\mathrm{K}} \approx 25\,813\,\Omega$. And this covariance is a prerequisite for its universal meaning. Phenomenologically, the quantum Hall effect (QHE) can also be described by a constitutive law of the type (4.5), see [35, Eq. (B.4.60)].

It is possible to apply the constitutive relation (4.5) directly to a solid, too. By the evaluation of experiments we have shown [37] that in the multiferroic $\mathrm{Cr_2O_3}$ (chromium sesquioxide) we have a nonvanishing axion piece of up to $\sim 10^{-3}\lambda_0$, where λ_0 is the vacuum admittance of about $1/377\,\Omega$. This fact demonstrates that there exist materials with a nonvanishing, if small, (pseudoscalar) axion piece. This may be considered as a plausibility argument in favor of a similar structure emerging in fundamental physics. If the A^0 were found, it would *not* be an unprecedented structure, see in this context also Ni et al. [72].

In matter-coupled $\mathcal{N} = 2$ supergravity models, there are examples in which a dilaton and an axion are contained simultaneously in the allowed particle spectrum, see Freeman and Van Proeyen [29, p. 451]. However, in the next section we will demonstrate that in a fairly simple classical model of an electromagnetic universe, the axion can emerge jointly with a dilaton and the metric.

More recently, there have been attempts to relate the axion field to the torsion of spacetime, see, for example Mielke et al. [64] and Castillo-Felisola et al. [14]. To us, this assumed link between the internal symmetry $U(1)$ of the axion with the external translation symmetry $T(4)$ related to the torsion appears to be artificial and not supported by physical arguments.

4.2 An Electromagnetic Model Universe

4.2.1 The Premetric Maxwell Equations

We consider a 4-dimensional differentiable manifold. The electromagnetic field is specified by its excitation $\mathcal{H}^{ij}$, a 2nd rank antisymmetric contravariant tensor density, and by its field strength F_{ij}, a 2nd rank antisymmetric covariant tensor; the electric current $\mathcal{J}^k$ is a contravariant vector density, see Post [80]. On this manifold, the Maxwell equations read

$$\partial_k \mathcal{H}^{ik} = \mathcal{J}^k, \qquad \partial_{[i} F_{jk]} = 0; \qquad (4.6)$$

the brackets $[\ ]$ denote antisymmetrization of the corresponding indices with 1/3! as a factor, see [88]; for the Tonti-diagram of (4.6), compare [95, p. 315].

In none of these equations the metric tensor g_{ij} nor the connection $\Gamma_{ij}{}^k$ are involved. Still, these equations are valid and are generally covariant in the Minkowski space of special relativity, in the Riemann space of general relativity, and in the Riemann-Cartan or Weyl-Cartan space of gravitational gauge theories. The Maxwell equations (4.6) as such, apart from a historical episode up to 1916, see [22, 23], have no specific relation to the Poincaré or the Lorentz group.

Perlick [78] has shown that the initial value problem in electrodynamics can be particularly conveniently implemented by means of the premetric form of the Maxwell equations.

In contrast to most textbook representations, no "comma goes to semicolon rule" is required. The Maxwell equations (4.6) are just universally valid for all forms of electrically charged matter. Incidentally, this represents also a simplifying feature for numerical implementations. The price one has to pay is to introduce, as Maxwell did, the excitation $\mathcal{H}^{ij}$, besides F_{ij}, as an independent field quantity and to note that it is a tensor density. From a phenomenological point of view, this is desirable anyway, since the excitation has an operational definition of its own, namely as charge/length^2 (**D**) and current/length (**H**), respectively, which is independent from the definition of the field strength as force/charge (**E**) and force/current (**B**). For a rendition in the calculus of exterior differential forms, one can compare with the axiomatic scheme in [18, 35], see also [19].

Let us stress additionally that $\mathcal{H}^{ij}$, F_{ij}, and $\mathcal{J}^k$ can be defined in a background independent way.

The Maxwell equations (4.6) are based on the conservation laws of electric charge $Q := \int d\sigma^{ijk} \varepsilon_{ijkl} \mathcal{J}^l$ (unit in SI "coulomb") and magnetic flux $\Phi := \int d\sigma^{ij} F_{ij}$ (unit in SI "weber"). Charge Q and flux Φ are 4-dimensional scalars. They induce the structure of the excitation $\mathcal{H}_{ij}$ and the field strength F_{ij}. In this context, the field strength is operationally defined via the Lorentz force density $\mathfrak{f}_i = F_{ij} \mathcal{J}^j$, the current being directly observable and the force and its measurement known from mechanics.

The charge and its conservation is the anchor of electrodynamics. Its current $\mathcal{J}^k$ defines, by means of the Lorentz force density $\mathfrak{f}_i$, the field strength F_{ij}, which allows to define the magnetic flux Φ. Faraday's induction law is an incarnation of magnetic flux conservation.

Some people have no intuition about the conservation of a quantity that is defined in 3 dimensions by integration over a 2-dimensional area $\sim \int d\sigma^a B_a$, since we usually associate conservation with a quantity won by 3-dimensional volume integration, namely $\sim \int dV \rho$. Some mathematics education about dimensions will enable us to understand the induction law as a "continuity equation."

Summing up: the premetric Maxwell equations are a close-knit structure, the 4-dimensional diffeomorphisms covariance holds it all together. Clearly, a metric as well as a connection are alien to the Maxwell equations.

4.2.2 A Local and Linear Electromagnetic Response

In order to fill the Maxwell equations with life, one has to relate F_{ij} to $\mathcal{H}^{ij}$:

$$\mathcal{H}^{ij} = \mathcal{H}^{ij}(F_{kl}) \tag{4.7}$$

If we assume this functional to be local, that is, $\mathcal{H}^{ij}(x)$ depends only on $F_{kl}(x)$, and linear homogeneously, then we find

$$\mathcal{H}^{ij} = \frac{1}{2} \chi^{ijkl} F_{kl} \quad \text{with} \quad \chi^{ijkl} = -\chi^{ijlk} = -\chi^{jikl}; \tag{4.8}$$

here the field $\chi^{ijkl}(x)$ represents the electromagnetic response tensor density of rank 4 and weight $+1$, with the physical dimension $[\chi] = 1/\text{resistance}$. An antisymmetric pair of indices corresponds, in 4 dimensions, to 6 independent components. Thus, χ^{ijkl} can be understood as a 6×6 matrix with 36 independent components.

We want to characterize the electromagnetic model spacetime by this response tensor field $\chi^{ijkl}(x)$ with 36 independent components.[3] This is the tensor density defining the *structure* of spacetime. It transcends the metric and/or the connection.

We decompose the 6×6 matrix into its 3 irreducible pieces. On the level of χ^{ijkl}, this induces [17, 35]

$$\chi^{ijkl} = {}^{(1)}\chi^{ijkl} + {}^{(2)}\chi^{ijkl} + {}^{(3)}\chi^{ijkl}. \tag{4.9}$$
$$36 = \quad 20 \quad \oplus \quad 15 \quad \oplus \quad 1.$$

---

[3] Schuller et al. [90] took the χ^{ijkl}-tensor density, which arises so naturally in electrodynamics, called the tensor proportional to it "area metric", and generalized it to n dimensions and to string theory. For reconstructing a volume element, they have, depending on the circumstances, two different recipes, like, for example, taking the *sixth* root of a determinant. From the point of view of 4-dimensional electrodynamics, the procedure of Schuller et al. looks contrived to us.

The third part, the *axion* part, is totally antisymmetric $^{(3)}\chi^{ijkl} := \chi^{[ijkl]} = \alpha\,\epsilon^{ijkl}$, with the pseudoscalar α, see also [83]. The *skewon* part is defined according to $^{(2)}\chi^{ijkl} := \frac{1}{2}(\chi^{ijkl} - \chi^{klij})$. Under reversible conditions, (4.8) can be derived from a Lagrangian, then $^{(2)}\chi^{ijkl} = 0$. The *principal* part $^{(1)}\chi^{ijkl}$ fulfills the symmetries $^{(1)}\chi^{ijkl} = {}^{(1)}\chi^{klij}$ and $^{(1)}\chi^{[ijkl]} = 0$.

The local and linear response relation now reads

$$\mathcal{H}^{ij} = \frac{1}{2}\left({}^{(1)}\chi^{ijkl} + {}^{(2)}\chi^{ijkl} + \alpha\,\epsilon^{ijkl}\right)F_{kl}, \tag{4.10}$$

and, split in space and time [35, 37],

$$D^a = (\varepsilon^{ab} - \epsilon^{abc}n_c)E_b + (\gamma^a{}_b + s_b{}^a - \delta^a_b s_c{}^c)B^b + \alpha\,B^a, \tag{4.11}$$

$$H_a = (\mu^{-1}_{ab} - \epsilon_{abc}m^c)B^b + (-\gamma^b{}_a + s_a{}^b - \delta^b_a s_c{}^c)E_b - \alpha\,E_a; \tag{4.12}$$

here $\epsilon^{abc} = \epsilon_{abc} = \pm1, 0$ are the 3-dimensional Levi-Civita symbols. The 6 permittivities $\varepsilon^{ab} = \varepsilon^{ba}$, the 6 permeabilities $\mu_{ab} = \mu_{ba}$ were already known to Maxwell. The 8 magnetoelectric pieces $\gamma^a{}_b$ (its trace vanishes, $\gamma^c{}_c = 0$) were found since 1961, see Astrov [2]. Eventually, the hypothetical skewon piece [35] carries 3 permittivities n_a, the 3 permeabilities m^a, and the 9 magnetoelectric pieces $s_a{}^b$. Equivalent response relations were formulated by Serdyukov et al. [91, p. 86] and studied in quite some detail, see also de Lange and Raab [52].

Suppose we have as *special case* a vacuum spacetime described by a Riemannian metric g_{ij}. Then the response tensor turns out to be

$$\chi^{ijkl} = {}^{(1)}\chi^{ijkl} = 2\lambda_0\sqrt{-g}\,g^{i[k}g^{l]j} \quad \text{and} \quad \mathcal{H}^{ij} = \lambda_0\sqrt{-g}\,F^{ij}, \tag{4.13}$$

with the vacuum admittance $\lambda_0 \approx 1/377\,\Omega$. Thus, we recover known structures, and we recognize that the relation (4.8) represents a natural generalization of the vacuum case. The metric g^{ij} can be considered as some kind of a square root of the electromagnetic response tensor χ^{ijkl}.

We should keep in mind that a local and homogeneous electromagnetic response like (4.8) can be, if the circumstances require it, generalized to nonlocal and/or to nonlinear laws. Examples of *nonlocal* laws have been proposed by Bopp and Podolsky[4] and by Mashhoon.[5] *Nonlinear* laws are due to Heisenberg and Euler,[6] Born and Infeld,[7] and Plebański.[8] Fresnel surfaces for the nonlinear case were found by Obukhov and Rubilar [77], for example. More recently, Lämmerzahl et al. [51] and Itin et al. [44] investigated electrodynamics in Finsler spacetimes. In the premetric

---

[4] See [28, Sect. 28-8].
[5] See [35, Sect. E.2.2].
[6] See [35, Sect. E.2.3].
[7] See [35, Sect. E.2.4].
[8] See [35, Sect. E.2.5].

framework, this corresponds to a nonlocal constitutive law, see [44, Eq. (3.29)], somewhat reminiscent of the Bopp-Podolsky scheme.

4.2.3 Propagation of Electromagnetic Disturbances

The obvious next step in evaluating the physics of our model of spacetime is to look how electromagnetic disturbances propagate in this spacetime. One can either consider the short wave-length limit of the electromagnetic theory, the WKB-approximation, or one can study, as we will do here, the propagation of electromagnetic disturbances with a technique developed by Hadamard; for a general outline, see [96, Chap. C].

Hadamard describes an elementary wave as a process that forms a wave surface. Across this surface, the electromagnetic field is continuous, but the derivative of the field has a jump. The direction of a jump is given by the wave covector. The subsequent integration produces the rays, with the wave vectors as tangents to rays, see for our case [35, 40, 46, 74]. In the meantime, our methods have been improved, see [3, 24, 27].

Out of the electromagnetic response tensor density we can define, with the help of the covariant Levi-Civita symbol $\epsilon_{ijkl} = \pm 1, 0$, the premetric "diamond" (single) dual and the diamond double dual, respectively:

$$\chi^{\diamond\,ij}{}_{kl} := \frac{1}{2}\chi^{ijcd}\epsilon_{cdkl}\,, \qquad {}^{\diamond}\chi^{\diamond}_{ijkl} := \frac{1}{2}\epsilon_{ijab}\chi^{\diamond\,ab}{}_{kl} = \frac{1}{4}\epsilon_{ijab}\chi^{abcd}\epsilon_{cdkl}\,. \qquad (4.14)$$

The covariant Levi-Civita symbol carries weight -1 and χ^{abcd} weight $+1$. Thus, the double dual has weight $+1$, too. Performing the double dual apparently corresponds to a lowering of all four indices of χ^{abcd}—and this is achieved without having access to a metric of spacetime.

After this preparation, it is straightforward to define the (premetric) 4th rank *Kummer tensor density*, which is cubic in χ, as [3]

$$\mathcal{K}^{ijkl}[\chi] := \chi^{aibj}\,{}^{\diamond}\chi^{\diamond}_{acbd}\chi^{ckdl}\,. \qquad (4.15)$$

It has weight $+1$ and obeys the symmetry $\mathcal{K}^{ijkl} = \mathcal{K}^{klij}$.

At each point in spacetime, the wave covectors $q_i = (\omega, \mathbf{k})$ of the electromagnetic waves span the Fresnel wave surfaces, which are quartic in the wave covectors according to

$$\mathcal{K}^{ijkl}[\chi]\,q_i q_j q_k q_l = \mathcal{K}^{(ijkl)}[\chi]\,q_i q_j q_k q_l = 0\,. \qquad (4.16)$$

The Tamm-Rubilar (TR) tensor density [35, 84], with the conventional factor 1/6, is defined by

$$\mathcal{G}^{ijkl}[\chi] := \frac{1}{6}\mathcal{K}^{(ijkl)}[\chi] = \frac{1}{6}\chi^{a(ij|b} \diamond \chi^{\diamond}_{acbd}\chi^{c|kl)d} .\qquad (4.17)$$

It is totally symmetric and carries 35 independent components. By straightforward algebra it can be shown that the axion field drops out from the TR-tensor:

$$\mathcal{G}^{ijkl}[\chi] = \mathcal{G}^{ijkl}[\,^{(1)}\chi + \,^{(2)}\chi];\qquad (4.18)$$

see in this connection also [39] and the references given there. The effect of the skewon piece on light propagation has been studied in [75]. Ni [68] was the first to understand that the axion field doesn't influence the light propagation in the geometrical optics limit. Note that $^{(1)}\chi + {}^{(2)}\chi$ has $20 + 15$ independent components, exactly as $\mathcal{G}$—probably not by chance.

Accordingly, the totally symmetric TR-tensor $\mathcal{G}^{ijkl}[\chi]$, with its 35 independent components, can, up to a factor, be observed by optical means, that is, the TR-tensor—in contrast to the Kummer tensor, as far as we know—has a direct operational interpretation.

4.2.4 Fresnel Wave Surface

The (generalized) Fresnel equation

$$\mathcal{G}^{ijkl}[\chi]\,q_iq_jq_kq_l = 0,\qquad (4.19)$$

determines a Fresnel wave surface. A trivial test for checking the correctness of (4.19) is to substitute the response tensor for the Maxwell-Lorentz vacuum electrodynamics $(4.13)_1$ into the TR-tensor of (4.19). One finds straightforwardly $(g^{ij}q_iq_j)^2 = 0$, that is, two light cones that collapse onto each other. The decomposition of (4.19) into space and time can be found in [35, (D.2.44)].

For illustration, following [3, 85], see also [41], we will display a classical example of such a surface. In Eqs. (4.11) and (4.12), we choose an anisotropic permittivity tensor with three different principal values and assume trivial vacuum permeability, whereas all magnetoelectric moduli—with the possible exception of the axion α—vanish,

$$(\varepsilon^{ab}) = \begin{pmatrix} \varepsilon^1 & 0 & 0 \\ 0 & \varepsilon^2 & 0 \\ 0 & 0 & \varepsilon^3 \end{pmatrix} \quad \text{and} \quad (\mu_{ab}^{-1}) = \mu_0^{-1}\begin{pmatrix} 1 & 0 & 0 \\ 0 & 1 & 0 \\ 0 & 0 & 1 \end{pmatrix}.\qquad (4.20)$$

Substitution into the Fresnel equation yield the quartic polynomial

$$\begin{aligned}&(\alpha^2x^2 + \beta^2y^2 + \gamma^2z^2)(x^2 + y^2 + z^2)\\ &- [\alpha^2(\beta^2 + \gamma^2)x^2 + \beta^2(\gamma^2 + \alpha^2)y^2 + \gamma^2(\alpha^2 + \beta^2)z^2] + \alpha^2\beta^2\gamma^2 = 0,\quad (4.21)\end{aligned}$$

with the 3 parameters[9] $\alpha := c/\sqrt{\varepsilon_1}$, $\beta := c/\sqrt{\varepsilon_2}$, $\gamma := c/\sqrt{\varepsilon_3}$, and with $c = 1/\sqrt{\varepsilon_0\mu_0}$ as the vacuum speed of light.

The corresponding surface is drawn in Fig. 4.1. As an example of a Fresnel surface for a more exotic material, we provide one for the so-called PQ-medium of Lindell [56]. It may turn out that this response tensor can only be realized with the help of a suitable metamaterial, see [92]. Corresponding investigations are underway by Favaro [25].

Let us shortly look back on what we have achieved so far: We have formulated the Maxwell equation in a premetric way. For the response tensor only local and linear notions are used, no distances or angles were mentioned nor implemented. Under such circumstances, electromagnetic disturbances propagate in a birefringent way in accordance with the Fresnel wave surfaces, such as presented in Figs. 4.1 and 4.2.

How can we now bring in distances and angles, which are concepts omnipresent in everyday life? The answer is obvious, we have to suppress birefringence.

4.2.5 Suppression of Birefringence: The Light Cone

Looking at the figures, it is clear that we have to take care that both shells in each Fresnel wave surface become identical spheres. Then light propagates like in vacuum. For this purpose, we can solve the quartic Fresnel equation (4.19) with respect to the frequency q_0, keeping the 3-covector q_a fixed. One finds four solutions, for the details please compare [38, 50]. To suppress birefringence, one has to demand *two conditions*. In turn, the quartic equation splits into a product of two quadratic equations proportional to each other. Thus, we find a light cone $g^{ij}(x)\,q_i q_j = 0$ at each point of spacetime.

Perhaps surprisingly, we derived also the *Lorentz signature,* see [35, 42, 43]. This can be traced back to the Lenz rule, which determines the relative sign of the two terms in the induction law, as compared to the relative sign in the Ampère-Maxwell law. The Lorentz signature can be understood on the level of classical electrodynamics, no appeal to quantum field theory, which is widespread in the literature, is necessary.

Globally in the cosmos, birefringence is excluded with high accuracy, see the observations of Polarbear [1] and the discussion of Ni [71].

4.2.6 Axion, Dilaton, Metric

At the premetric level of our framework, besides the principal piece, first the skewon and the axion fields emerged. Only subsequently the light cone was brought up. The skewon field was phased out by our insistence of the vanishing birefringence in the vacuum. Accordingly, the axion field and the light cone survived the suppression of the birefringence.

---

[9]Here, in this context, α is *not* the axion field!.

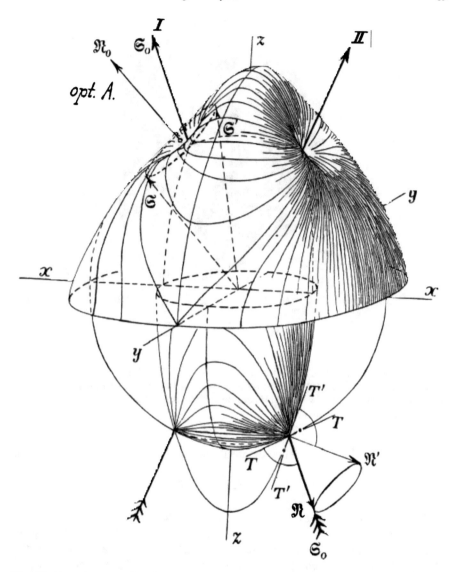

Fig. 4.1 Fresnel wave surface for the permittivities and the permeabilities of Eq. (4.20). It had been drawn by Jaumann for an optically biaxial crystal, see Schaefer [85, p. 485]. This crystal has the property of birefringence (or double refraction). The origin at $x = y = z = 0$ is the point in 3-dimensional space from where the wave covectors **k** originate. They end on the Fresnel wave surface. Their modulus is proportional to the reciprocal of the phase velocity ω/k. In other words, up to a sign, we have usually in one direction two different phase velocities. This is an expression of the birefringence. Only along the optical axes **I** and **II**, we have only one wave covector. The upper half depicts the exterior shell with the funnel shaped singularities, the lower half the inner shell. The two shells cross each other at four points forming cusps

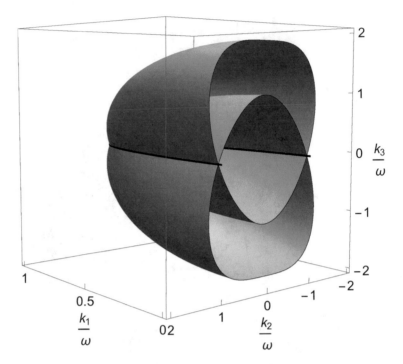

Fig. 4.2 Fresnel wave surface for a PQ-medium of Lindell [56, 57]. Using Lindell's dyadic version of the Fresnel equation [54, 55], Sihvola [94] drew the Fresnel wave surface by using Mathematica. Our image was later created by Favaro [25] in a similar way, again with Mathematica. For the wave covector, we have $q_i = (\omega, k_1, k_2, k_3)$

The light cone does not define the metric uniquely. Rather an arbitrary function $\lambda(x)$ is left over:

$$\lambda(x)\, g^{ij}(x)\, q_i q_j = 0. \tag{4.22}$$

The light cone is invariant under the 15-parametric conformal group. The 4 proper conformal transformation correspond to a reflection at the unit circle and, as such, are of a nonlocal nature. As a consequence, if two frames are related to each other by a proper conformal transformation and one frame is inertial, the other one is accelerated with respect to the former one. Accordingly, there is an operational distinction possible between a proper conformal and a dilation or scale transformation. Thus, only the 11 parameter Weyl subgroup of the 15 parameter conformal group is based on *local* transformations.

If we compare our result in (4.22) with vacuum response in (4.13), we recognize, not forgetting the axion field, that we find the following response equation for vanishing birefringence:

$$\mathcal{H}^{ij} = [\underbrace{\lambda(x)}_{\text{dilaton}} \sqrt{-g}\, g^{ik}(x)\, g^{jl}(x) + \underbrace{\alpha(x)}_{\text{axion}}\, \epsilon^{ijkl}\,]\, F_{kl}. \tag{4.23}$$

Because of the presence of the dilation within the Weyl group, it is natural to identify the function $\lambda(x)$ with the dilaton field.[10]

In the calculus of exterior differential forms, see [35], the twisted excitation 2-form $H = \frac{1}{2}\epsilon_{ijkl}\mathcal{H}^{kl}dx^i \wedge dx^j$ and the untwisted field strength 2-form $F = \frac{1}{2}F_{ij}dx^i \wedge dx^j$, together with the twisted current 3-form $J = \frac{1}{3!}\epsilon_{ijkl}\mathcal{J}^l dx^i \wedge dx^j \wedge dx^k$, obey the Maxwell equations $dH = J$ and $dF = 0$. By means of the metric, we can introduce the Hodge star * operator. Then the response relation (4.23) becomes even more compact [26, 36]:

$$H = [\lambda(x)^\star + \alpha(x)]F. \tag{4.24}$$

Equations (4.23) and (4.24) represent the end result of investigating an electromagnetic spacetime model with local and linear response and without birefringence. The three fields $\lambda(x)$, $g^{ij}(x)$, and $\alpha(x)$ come up together with a reasonable interpretation. At least in the way we defined them here, $\lambda(x)$, $g^{ij}(x)$, and $\alpha(x)$ are all three *descendants of electromagnetism.*

As we have argued in Sect. 1.5, the dilaton seems to be at home in the Weyl-Cartan spacetime. Our results (4.23) or (4.24) are consistent with this expectation, that is, we believe that these equations are valid in a Weyl-Cartan spacetime.

What are we told by experiments and observations? The axion A^0 has not been found so far, so we can provisionally put $\alpha = 0$. Moreover, under normal circumstances, the dilaton seems to be a constant field and thereby sets a certain scale, that is, $\lambda(x) = \lambda_0 = $ const, where λ_0 is the admittance of free space, the value of which is, in SI-units, $\approx 1/(377\ \Omega)$. Under these conditions, we are left with the response relation of conventional Maxwell-Lorentz electrodynamics,

$$\mathcal{H}^{ij} = \lambda_0\sqrt{-g}\,F^{ij} \quad \text{or} \quad H = \lambda_0{}^\star F. \tag{4.25}$$

The possible generalizations are apparent.

4.3 Discussion

Gravity, coupling to all objects carrying energy-momentum, is a truly universal interaction. Electromagnetism is only involved in electrically charged matter. What is curious and what we still do not understand is that the gravitational potential g^{ij} emerges in an electromagnetic context, that is, in studying electromagnetic disturbances, we can suppress birefringence, and then the light cone emerges. And the light cone is essentially involved in general relativity. In other words, we

---

[10]In the early 1980s, Ni [69] has shown the following: Suppressing the birefringence is a necessary and sufficient condition for a Lagrangian based constitutive tensor to be decomposable into metric+dilaton+axion in a weak gravitational field (weak violation of the Einstein equivalence principle), a remarkable result. Note that Ni assumed the existence of a metric. We, in (4.23), derived the metric from the electromagnetic response tensor density χ^{ijkl}.

cannot formulate a general-relativistic theory of gravity unless some electric charge is around: electromagnetic waves are a necessary tool for constructing general relativity.

Perlick is not concerned about it. He observes that [79] "...the vacuum Maxwell equations are but one example that have the light cones of the spacetime metric for their characteristics. The same is true of the Dirac equation, the Klein-Gordon equation and others...." Yes, this is true. However, if a metric is *not* prescribed, we cannot even formulate Dirac's theory. In contrast, in premetric electrodynamics, if a local and linear response tensor density is assumed, we can derive the metric, as we discussed above. In this sense, electrodynamics is distinguished from Dirac's theory—and in this, and only in this sense, the premetric Maxwell equations are more fundamental than the Dirac equation.

Accordingly, there seems to be a deep connection between electromagnetism and gravity, even though gravity is truly universal, in contrast to electrodynamics.

Acknowledgments This project was partly supported by the German-Israeli Research Foundation (GIF) by the grant GIF/No. 1078-107.14/2009. I am grateful to Claus Kiefer (Cologne) for helpful remarks re Higgs cosmology, to Yuri Obukhov (Moscow) re questions of torsion and of the Hadamard method, to Helmut Rumpf (Vienna) re the Mukhanov bound of 10^{-29} m, and to Volker Perlick (Bremen) re gravity versus electromagnetism and what is more fundamental. I would like to thank Ari Sihvola and Ismo Lindell (both Espoo/Helsinki) and Alberto Favaro (London) for the permission to use the image in Fig.2. Many useful comments on a draft of this paper were supplied by Hubert Goenner (Göttingen), Yakov Itin (Jerusalem), Claus Lämmerzahl (Bremen), Ecardo Mielke (Mexico City), Wei-Tou Ni (Hsinchu), Erhard Scholz (Wuppertal), Frederic Schuller (Erlangen), and Dirk Puetzfeld (Bremen). I am grateful to Alan Kostelecký (Bloomington) for a last-minute exchange of emails.

References

1. P.A.R. Ade et al. [Polarbear Collaboration], Polarbear constraints on cosmic birefringence and primordial magnetic fields. Phys. Rev. D **92**, 123509 (2015). arXiv.org:1509.02461
2. D.N. Astrov, Magnetoelectric effect in chromium oxide. Sov. Phys. JETP **13**, 729–733 (1961) (Zh. Eksp. Teor. Fiz. **40**, 1035–1041 (1961))
3. P. Baekler, A. Favaro, Y. Itin, F.W. Hehl, The Kummer tensor density in electrodynamics and in gravity. Annals Phys. (NY) **349**, 297–324 (2014). arXiv.org:1403.3467
4. A.B. Balakin, W.T. Ni, Non-minimal coupling of photons and axions. Class. Quant. Grav. **27**, 055003 (2010). arXiv.org:0911.2946
5. A.O. Barut, R. Rączka, *Theory of Group Representations and Applications* (PWN-Polish Scientific, Warsaw, 1977)
6. A.O. Barvinsky, A.Y. Kamenshchik, C. Kiefer, A.A. Starobinsky, C. Steinwachs, Asymptotic freedom in inflationary cosmology with a non-minimally coupled Higgs field. JCAP **0912**, 003 (2009). arXiv.org:0904.1698
7. F. Bezrukov, M. Shaposhnikov, Inflation, LHC and the Higgs boson. Comptes Rendus Physique **16**, 994–1002 (2015). http://dx.doi.org/10.1016/j.crhy.2015.08.005
8. M. Blagojević, *Gravitation and Gauge Symmetries* (IoP, Bristol, 2002)
9. M. Blagojević, F.W. Hehl (eds.), *Gauge Theories of Gravitation, a Reader with Commentaries* (Imperial College Press, London, 2013)

10. C.H. Brans, Gravity and the tenacious scalar field, in *On Einstein's Path, Essays in Honor of E. Schucking*, ed. by A. Harvey (Springer, New York, 1999), Chap. 9, pp. 121–138. arXiv.org:gr-qc/9705069

11. C.H. Brans, The roots of scalar-tensor theory: an approximate history, presented during the *1st Internat. Workshop on Gravitation and Cosmology, 31 May–4 Jun 2004, Santa Clara, Cuba 2004*. arXiv.org:gr-qc/0506063

12. C. Brans, R.H. Dicke, Mach's principle and a relativistic theory of gravitation. Phys. Rev. **124**, 925–935 (1961)

13. C.G. Callan Jr., S.R. Coleman, R. Jackiw, A new improved energy-momentum tensor. Annals Phys. (NY) **59**, 42–73 (1970)

14. O. Castillo-Felisola, C. Corral, S. Kovalenko, I. Schmidt, V.E. Lyubovitskij, Axions in gravity with torsion. Phys. Rev. D **91**, 085017 (2015). arXiv.org:1502.03694

15. J.M. Charap, W. Tait, A gauge theory of the Weyl group. Proc. R. Soc. Lond. A **340**, 249–262 (1974)

16. C.M. Chen, J.M. Nester, R.S. Tung, Gravitational energy for GR and Poincaré gauge theories: a covariant Hamiltonian approach. Int. J. Mod. Phys. D **24**, 1530026 (2015). arXiv.org:1507.07300

17. D.H. Delphenich, On linear electromagnetic constitutive laws that define almost-complex structures. Annalen Phys. (Berlin) **16**, 207–217 (2007). arXiv.org:gr-qc/0610031

18. D.H. Delphenich, *Pre-metric Electromagnetism*, electronic book, 410 pp. http://www.neo-classical-physics.info/ (2009)

19. D. Delphenich, Pre-metric electromagnetism as a path to unification. arXiv.org:1512.05183

20. R.H. Dicke, *The Theoretical Significance of Experimental Relativity* (Blackie & Son, London, 1964)

21. P. Di Vecchia, R. Marotta, M. Mojaza, J. Nohle, The story of the two dilatons (2015). arXiv.org:1512.03316

22. A. Einstein, *Eine neue formale Deutung der Maxwellschen Feldgleichungen der Elektrodynamik* (A New Formal Interpretation of Maxwell's Field Equations of Electrodynamics) (Sitzungsber. Königl. Preuss. Akad. Wiss., Berlin, 1916), pp. 184–188; see also A.J. Kox et al. (eds.), *The Collected Papers of Albert Einstein*, vol. 6 (Princeton University Press, Princeton, 1996), pp. 263–269

23. A. Einstein, *The Meaning of Relativity*, 5th edn. (Princeton University Press, Princeton, 1955) (first published in 1922)

24. A. Favaro, Recent advances in classical electromagnetic theory. Ph.D. thesis, Imperial College London (2012)

25. A. Favaro, Private communication (2015)

26. A. Favaro, L. Bergamin, The non-birefringent limit of all linear, skewonless media and its unique light-cone structure. Annalen Phys. (Berlin) **523**, 383–401 (2011). arXiv.org:1008.2343

27. A. Favaro, F.W. Hehl, Light propagation in local and linear media: Fresnel-Kummer wave surfaces with 16 singular points. Phys. Rev. A **93**, 013844 (2016). arXiv.org:1510.05566

28. R.P. Feynman, R.B. Leighton, M. Sands, *The Feynman Lectures on Physics, Vol.II: Mainly Electromagnetism and Matter* (Addison-Wesley, Reading, 4th Printing 1969)

29. D.Z. Freedman, A. Van Proeyen, *Supergravity* (Cambridge University Press, Cambridge, 2012)

30. Y. Fujii, K. Maeda, *The Scalar-Tensor Theory of Gravitation* (Cambridge University Press, Cambridge, 2003)

31. D. Giulini, The rich structure of Minkowski space, in *Minkowski Spacetime: A Hundred Years Later*, ed. by V. Petkov. Fundamental Theories of Physics (Springer), vol. 165, pp. 83–132 (2010). arXiv.org:0802.4345

32. H. Goenner, Some remarks on the genesis of scalar-tensor theories. Gen. Rel. Grav. **44**, 2077–2097 (2012). arXiv.org:1204.3455

33. Z. Haghani, N. Khosravi, S. Shahidi, *The Weyl-Cartan Gauss-Bonnet gravity*, Class. Quant. Grav. **32**, 215016 (2015). arXiv.org:1410.2412

34. F.W. Hehl, J.D. McCrea, E.W. Mielke, Y. Ne'eman, Progress in metric-affine gauge theories of gravity with local scale invariance. Found. Phys. **19**, 1075–1100 (1989)

35. F.W. Hehl, Yu.N Obukhov, *Foundations of Classical Electrodynamics: Charge, Flux, and Metric* (Birkhäuser, Boston, 2003)
36. F.W. Hehl, Yu.N. Obukhov, *Spacetime Metric from Local and Linear Electrodynamics: A New Axiomatic Scheme.* Lecture Notes in Physics (Springer), vol. 702, pp. 163–187 (2006). arXiv.org:gr-qc/0508024
37. F.W. Hehl, Yu.N. Obukhov, J.P. Rivera, H. Schmid, Relativistic nature of a magnetoelectric modulus of $Cr_2 O_3$ crystals: a four-dimensional pseudoscalar and its measurement. Phys. Rev. A **77**, 022106 (2008). arXiv.org:0707.4407
38. Y. Itin, No-birefringence conditions for spacetime. Phys. Rev. D **72**, 087502 (2005). arXiv.org:hep-th/0508144
39. Y. Itin, Photon propagator for axion electrodynamics. Phys. Rev. D **76**, 087505 (2007). arXiv.org:0709.1637
40. Y. Itin, On light propagation in premetric electrodynamics. Covariant dispersion relation. J. Phys. A **42**, 475402 (2009). arXiv.org:0903.5520
41. Y. Itin, Dispersion relation for anisotropic media. Phys. Lett. A **374**, 1113–1116 (2010). arXiv.org:0908.0922
42. Y. Itin, Y. Friedman, Backwards on Minkowski's road. From 4D to 3D Maxwellian electromagnetism. Annalen Phys. (Berlin) **17**, 769–786 (2008). arXiv.org:0807.2625
43. Y. Itin, F.W. Hehl, Is the Lorentz signature of the metric of space-time electromagnetic in origin? Annals Phys. (NY) **312**, 60–83 (2004). arXiv.org:gr-qc/0401016
44. Y. Itin, C. Lämmerzahl, V. Perlick, Finsler-type modification of the Coulomb law. Phys. Rev. D **90**, 124057 (2014). arXiv.org:1411.2670
45. H.A. Kastrup, On the advancements of conformal transformations and their associated symmetries in geometry and theoretical physics. Annalen Phys. (Berlin) **17**, 631–690 (2008). arXiv.org:0808.2730
46. R.M. Kiehn, G.P. Kiehn, J.B. Roberds, Parity and time-reversal symmetry breaking, singular solutions, and Fresnel surfaces. Phys. Rev. A **43**, 5665–5671 (1991)
47. W. Kopczyński, J.D. McCrea, F.W. Hehl, The Weyl group and its currents. Phys. Lett. A **128**, 313–317 (1988)
48. V.A. Kostelecký, N. Russell, J. Tasson, Constraints on torsion from bounds on Lorentz violation. Phys. Rev. Lett. **100**, 111102 (2008). arXiv.org:0712.4393
49. K. Küpfmüller, W. Mathis, A. Reibiger, *Theoretische Elektrotechnik*, 19th edn. (Springer Vieweg, Berlin, 2013)
50. C. Lämmerzahl, F.W. Hehl, Riemannian light cone from vanishing birefringence in premetric vacuum electrodynamics. Phys. Rev. D **70**, 105022 (2004). arXiv.org:gr-qc/0409072
51. C. Lämmerzahl, V. Perlick, W. Hasse, Observable effects in a class of spherically symmetric static Finsler spacetimes. Phys. Rev. D **86**, 104042 (2012). arXiv.org:1208.0619
52. O.L. de Lange, R.E. Raab, Multipole theory and the Hehl-Obukhov decomposition of the electromagnetic constitutive tensor. J. Math. Phys. **56**, 053502 (2015)
53. A. Lasenby, M. Hobson, Scale-invariant gauge theories of gravity: theoretical foundations (2015). arXiv.org:1510.06699
54. I.V. Lindell, Electromagnetic wave equation in differential-form representation. Progr. Electromagn. Res. (PIER) **54**, 321–333 (2005)
55. I.V. Lindell, *Multiforms, Dyadics, and Electromagnetic Media*, IEEE Press, Wiley, Hoboken, 2015)
56. I.V. Lindell, Plane-wave propagation in electromagnetic PQ medium. Progr. Electromagn. Res. (PIER) **154**, 23–33 (2015)
57. I.V. Lindell, A. Favaro, Decomposition of electromagnetic Q and P media. Progr. Electromagn. Res. B (PIER B) **63**, 79–93 (2015)
58. I.V. Lindell, A. Sihvola, Perfect electromagnetic conductor. J. Electromagn. Waves Appl. **19**, 861–869 (2005). arXiv.org:physics/0503232
59. J. Łopuszánski, *An Introduction to Symmetry and Supersymmetry in Quantum Field Theory* (World Scientific, Singapore, 1991)

60. E.W. Mielke, Spontaneously broken topological $SL(5, R)$ gauge theory with standard gravity emerging. Phys. Rev. D **83**, 044004 (2011)
61. E.W. Mielke, Weak equivalence principle from a spontaneously broken gauge theory of gravity. Phys. Lett. B **702**, 187–190 (2011)
62. E.W. Mielke, Symmetry breaking in topological quantum gravity. Int. J. Mod. Phys. D **22**, 1330009 (2013)
63. E.W. Mielke, F.W. Hehl, J.D. McCrea, Belinfante type invariance of the Noether identities in a Riemannian and a Weitzenböck spacetime. Phys. Lett. A **140**, 368–372 (1989)
64. E.W. Mielke, E.S. Romero, Cosmological evolution of a torsion-induced quintaxion. Phys. Rev. D **73**, 043521 (2006)
65. V. Mukhanov, *Physics Colloquium*, University of Bonn on 2013 May 17
66. Y. Nakayama, Scale invariance vs conformal invariance. Phys. Rep. **569**, 1–93 (2015). arXiv.org:1302.0884
67. W.-T. Ni, *A Non-metric Theory of Gravity*, (Montana State University, Bozeman, 1973). http://astrod.wikispaces.com/
68. W.-T. Ni, Equivalence principles and electromagnetism. Phys. Rev. Lett. **38**, 301–304 (1977)
69. W.-T. Ni, Equivalence principles and precision experiments, in *Precision Measurement and Fundamental Constants II*, ed. by B.N. Taylor, W.D. Phillips. Natl. Bur. Stand. (US) Spec. Publ. **617**, 647–651 (1984). http://astrod.wikispaces.com/
70. W.-T. Ni, Dilaton field and cosmic wave propagation. Phys. Lett. A **378**, 3413–3418 (2014). arXiv.org:1410.0126
71. W.T. Ni, Spacetime structure and asymmetric metric from the premetric formulation of electromagnetism. Phys. Lett. A **379**, 1297–1303 (2015). arXiv.org:1411.0460
72. W.-T. Ni, S.-S. Pan, H.-C. Yeh, L.-S. Hou, J. Wan, Search for an axionlike spin coupling using a paramagnetic salt with a dc SQUID. Phys. Rev. Lett. **82**, 2439 (1999)
73. Yu.N. Obukhov, Poincaré gauge gravity: Selected topics. Int. J. Geom. Meth. Mod. Phys. **3**, 95–138 (2006). arXiv.org:gr-qc/0601090
74. Y.N. Obukhov, T. Fukui, G.F. Rubilar, Wave propagation in linear electrodynamics. Phys. Rev. D **62**, 044050 (2000). arXiv.org:gr-qc/0005018
75. Yu.N. Obukhov, F.W. Hehl, On possible skewon effects on light propagation. Phys. Rev. D **70**, 125015 (2004). arXiv.org:physics/0409155
76. Yu.N. Obukhov, D. Puetzfeld, Equations of motion in scalar-tensor theories of gravity: a covariant multipolar approach. Phys. Rev. D **90**, 104041 (2014). arXiv.org:1404.6977
77. Yu.N. Obukhov, G.F. Rubilar, Fresnel analysis of the wave propagation in nonlinear electrodynamics. Phys. Rev. D **66**, 024042 (2002). arXiv.org:gr-qc/0204028
78. V. Perlick, On the hyperbolicity of Maxwell's equations with a local constitutive law. J. Math. Phys. **52**, 042903 (2011). arXiv:1011.2536
79. V. Perlick, Private communication (2015)
80. E.J. Post, *Formal Structure of Electromagnetics—General Covariance and Electromagnetics* (North Holland, Amsterdam, 1962) (Dover, Mineola, 1997)
81. D. Puetzfeld, Y.N. Obukhov, Equations of motion in metric-affine gravity: a covariant unified framework. Phys. Rev. D **90**, 084034 (2014). arXiv:1408.5669
82. D. Puetzfeld, Y.N. Obukhov, Equivalence principle in scalar-tensor gravity. Phys. Rev. D **92**, 081502(R) (2015). arXiv.org:1505.01285
83. R.E. Raab, A.H. Sihvola, On the existence of linear non-reciprocal bi-isotropic (NRBI) media. J. Phys. A **30**, 1335–1344 (1997)
84. G.F. Rubilar, Linear pre-metric electrodynamics and deduction of the light cone. Annalen Phys. (Berlin) **11**, 717–782 (2002). arXiv.org:0706.2193
85. C. Schaefer, *Einführung in Die Theoretische Physik*, vol.3, Part 1 (de Gruyter, Berlin, 1932)
86. E. Scholz, Higgs and gravitational scalar fields together induce Weyl gauge. Gen. Rel. Grav. **47**(7) (2015). arXiv.org:1407.6811
87. E. Scholz, MOND-like acceleration in integrable Weyl geometric gravity. Found. Phys. **46**, 176–208 (2016). arXiv.org:1412.0430

88. J.A. Schouten, *Tensor Analysis for Physicists,* corr. 2nd edn. (Clarendon Press, Oxford, 1959) (Dover, New York, 1989)
89. E. Schrödinger, *Space-Time Structure* (Cambridge University Press, Cambridge, 1954)
90. F.P. Schuller, C. Witte, M.N.R. Wohlfarth, Causal structure and algebraic classification of area metric spacetimes in four dimensions. Ann. Phys. (NY) **325**, 1853–1883 (2010). arXiv.org:0908.1016
91. A. Serdyukov, I. Semchenko, S. Tretyakov, A. Sihvola, *Electromagnetics of Bi-anisotropic Materials. Theory and Applications* (Gordon and Breach, Amsterdam, 2001)
92. E. Shamonina, L. Solymar, Metamaterials: How the subject started. Metamaterials **1**, 12–18 (2007)
93. A. Sihvola, I.V. Lindell, Perfect electromagnetic conductor medium. Ann. Phys. (Berlin) **17**, 787–802 (2008)
94. A. Sihvola, Private communication (2015)
95. E. Tonti, *The Mathematical Structure of Classical and Relativistic Physics: a general classification diagram* (Birkhäuser/Springer, New York, 2013)
96. C. Truesdell, R.A. Toupin, The classical field theories, in *Encyclopedia of Physics*, vol. III/1, ed. by S. Flügge (Springer, Berlin, 1960), pp. 226–793
97. V. Vennin, J. Martin, C. Ringeval, Cosmic inflation and model comparison. Comptes Rendus Physique **16**, 960–968 (2015)
98. S. Weinberg, *The Quantum Theory of Fields. Volume II: Modern Applications* (Cambridge University Press, Cambridge, 1996)
99. F. Wilczek, Two applications of axion electrodynamics. Phys. Rev. Lett. **58**, 1799–1802 (1987)
100. P. Russer, *Electromagnetics, Microware Circuit and Antenna Design for Communications Engineering*, 2nd edn. (Artech House, Norwood, MA, 2006)

Chapter 5
Gravitational Theories with Stable (anti-)de Sitter Backgrounds

Tirthabir Biswas, Alexey S. Koshelev and Anupam Mazumdar

Abstract In this article we will construct the most general torsion-free parity-invariant covariant theory of gravity that is free from ghost-like and tachyonic instabilities around constant curvature space-times in four dimensions. Specifically, this includes the Minkowski, de Sitter and anti-de Sitter backgrounds. We will first argue in details how starting from a general covariant action for the metric one arrives at an "equivalent" action that at most contains terms that are quadratic in curvatures but nevertheless is sufficient for the purpose of studying stability of the original action. We will then briefly discuss how such a "quadratic curvature action" can be decomposed in a covariant formalism into separate sectors involving the tensor, vector and scalar modes of the metric tensor; most of the details of the analysis however, will be presented in an accompanying paper. We will find that only the transverse and trace-less spin-2 graviton with its two helicity states and possibly a spin-0 Brans-Dicke type scalar degree of freedom are left to propagate in 4 dimensions. This will also enable us to arrive at the consistency conditions required to make the theory perturbatively stable around constant curvature backgrounds.

5.1 Introduction

5.1.1 A Personal Note from Tirtho

Initially, it felt a bit strange to me to write about attempts to modify General Theory of Relativity (GR) when we are celebrating one hundred very successful years of

T. Biswas (✉)
Loyola University, 6363 St. Charles Avenue, Box 92, New Orleans 70118, USA
e-mail: tbiswas@loyno.edu

A.S. Koshelev
Departamento de Física and Centro de Matemática e Aplicações, Universidade
da Beira Interior, 6200 Covilhã, Portugal
e-mail: alexey@ubi.pt

A. Mazumdar
Consortium for Fundamental Physics, Lancaster University, Lancaster LA1 4YB, UK
e-mail: a.mazumdar@lancaster.ac.uk

© Springer International Publishing Switzerland 2016
T. Asselmeyer-Maluga (ed.), *At the Frontier of Spacetime*,
Fundamental Theories of Physics 183, DOI 10.1007/978-3-319-31299-6_5

Einstein's master piece, but then I remembered one of the fundamental tenants of science, that we can never know whether a theory is correct, only that it is not yet wrong! So, it is not so surprising after all that in spite of its success in these hundred years, literally hundreds of attempts have been made to modify Einstein's theory of gravity. Having said that, GR has proved to be impossibly difficult to dislodge. Perhaps, there is an emotional component to it, after all we all fell in love with GR when we saw for the first time how a theory could replace the abstract notions of force and "action at a distance" with a physically intuitive and beautiful geometric picture that could explain gravity. And, it is always hard to extricate ourselves from something that we love. While we now know how to construct countless theories of gravity which preserves the same basic geometric structure and symmetries, for instance, that the force of gravity is encoded in curvatures of space-time that can be built from a metric, Einstein had also based his theory on deep philosophical ideas, such as the Equivalence principle, that are harder to preserve, not that we are obliged to. Theoretically speaking, if one at least wants to preserve general covariance without introducing any new fields beyond the metric, the modifications that one can consider must involve higher derivative terms which tend to be plagued by ghost-like or tachyonic instabilities. As if these were not enough impediments, experimentally, GR has been tested to unprecedented levels of accuracy, and it passes with flying colors. Indeed, GR has experienced success in explaining a multitude of different phenomena starting from purely astronomical observations, such as the bending of light near a massive object, to the cosmic expansion of our universe.

So, why do we keep searching for this elusive "better" theory so vigorously, why can't we just leave GR alone for a while? The answer is obvious to most theoretical physicists, notwithstanding its amazing success, GR is profoundly incomplete. It is plagued by classical singularities in the ultraviolet (UV), as seen inside the blackholes or at the Big Bang. GR suffers from quantum divergences that cannot be renormalized and constructing a consistent quantum theory of gravity remains one of the outstanding challenges of 21st century physics. To draw a contrast, while we may not be completely happy with the Standard Model of particle physics that describes the three other fundamental forces of nature, for instance, it suffers from the hierarchy problem, doesn't explain the origin of its twenty odd parameters, the fact of the matter is that it is a perfectly consistent theory that has till date explained/predicted experimental observations quite brilliantly.

On the infrared (IR) front we have also "recently" been greeted with a surprise, we have found out that our universe at the largest scales is apparently able to defy gravity and speed up its cosmic expansion. While this major inconvenience can be explained away without having to tamper with GR just by invoking a cosmological constant, albeit a disconcertingly small one, there is a school of thought that it is gravity that is perhaps becoming weak at cosmic scales thereby allowing our universe to accelerate.

Thus, today it has become especially fashionable to try to modify GR in ways that could address either the UV or IR problems/puzzles, but the reason, in my opinion, why Carl's work with Dicke is phenomenal is not just because they realized the importance of going beyond GR and constructed in comprehensive detail their scalar tensor model of gravity, but even more because they did so around a half a century ago! What is also remarkable is that this first attempt to modify GR is arguably still the most fruitful of the modifications that has been considered in the literature. Indeed, Brans-Dicke theories or generalizations thereof are what emerge from fundamental theories such as Kaluza Klein theories, supergravity models and string theory after compactifications of extra dimensions. It has been phenomenologically the most successful, finding applications in inflation theory and various dark energy models such as quintessence.

Unsurprisingly, I had worked on several different versions of Brans-Dicke theory before I actually met Carl during my job interview at Loyola. Thankfully, I didn't know that I was actually meeting Carl Brans (somehow I missed his profile on the Loyola physics faculty listings) because that would have completely overwhelmed me. It was only halfway through the interview that I realized that I was talking to someone who knew a lot more gravity than I did. Since then, we have become very good friends, his good nature, his humility, and his commitment to rigor is something that I cherish and I am inspired by. So, here is to Carl for showing the path that many others like me could follow. Cheers!

5.1.2 Towards Consistent Theories of Gravity

There have been numerous attempts to formulate a quantum theory of gravity [1–3], such as string theory (ST) [4], Loop Quantum Gravity (LQG) [5], Causal Set [6], and asymptotic safety [7]. In many of these approaches gravitational interactions yield non-local operators where the interactions are spread out over a space-time region. For instance, strings and branes are non-local objects, nonlocality also emerges in string field theory [8], non-commutative geometry [9], p-adic strings [10], zeta strings [11], and strings quantized on a random lattice [12, 13], for a review, see [14]. A key feature of all these stringy models is the presence of an *infinite series of higher-derivative* terms incorporating the non-locality in the form of an *exponential kinetic* correction [15–17], or equivalently modifying the graviton propagator by an *entire function* [18–21]. Similar infinite-derivative modifications have also been argued to arise in the asymptotic safety approach to quantum gravity [22].[1]

Only very recently, the concrete criteria for any covariant gravitational theory (including infinite-derivative theories) to be free from ghosts and tachyons around the Minkowski vacuum was obtained in Refs. [25, 26]. The class of action considered were assumed to be free from torsion, have a well defined Newtonian limit and to be parity conserving. It was also shown in [25, 26], that one could construct

---

[1]Finite higher derivative theories suffer from Ostrogradsky instabilities, see Ref. [23]. However, the Ostrogradsky argument relies on having a highest "momentum" associated with the highest derivative in the theory, in which the energy comes as a linear term, as opposed to quadratic. In a classical theory this would lead to instability and in a quantum theory, this would yield ghosts or extra poles in the propagator. A classic example is Stelle's 4th derivative theory of gravity [24], which has been argued to be UV finite, but contains massive spin-2 ghost, therefore shows vacuum instabilities.

theories that have no extra poles in the propagators, so that there are no new degrees of freedom, ghosts or otherwise. The only dynamically relevant degrees of freedom are the massless gravitons; the graviton propagator, however, can be modified by a multiplicative *entire function*. In particular, one can choose the entire function to correspond to be a gaussian, which would suppress the ultraviolet (UV) modes *possibly* making the theory asymptotically free [27], see also [18–21] for similar arguments in slightly different models. We should point out that at a classical level it has already been shown that in these infinite derivative gravity (IDG) models one can find cosmological solutions bereft of singularities [17, 25, 28, 29], as well as non-singular static, spherically symmetric metrics [29–31].[2] These classical results corroborates the idea that IDG theories may provide us with an asymptotically safe/free theory of gravity, for a review on these models, see [32].

The aim here is to go beyond the analysis around the Minkowski vacuum. We would like to find a robust algorithm to construct the most general action of gravity that is "consistent" around constant curvature maximally symmetric space-times, viz. de Sitter (dS) and anti-de Sitter (AdS) and Minkowski. The analysis in [25] essentially gave us constraints that quadratic curvature terms (such as R^2, $R\Box R$, etc.) must satisfy in order for the theory to be free from instabilities around the Minkowski space-time. However, the higher curvature terms remained completely arbitrary. If however, we believe that the "ultimate" theory of gravity must be consistent on any background, then requiring that it be so should provide constraints on the higher curvature terms. The ultimate hope is that this may provide us with new insights on how to construct a consistent and finite quantum theory of gravity. Looking at dS/AdS backgrounds is a first step towards this process where we will start with a gravitational action that is covariant, parity preserving, torsion-free and possesses a well defined Newtonian limit. Our goal will be to study the quadratic fluctuations around dS and AdS backgrounds. In this article we will argue that for this purpose it is sufficient to study the fluctuations around an "equivalent" action which has terms that are at most quadratic in curvatures. This is a crucial simplification which makes it possible to study the dynamics of linearized fluctuations around dS/AdS/Minkowski backgrounds for a very general class of covariant gravity theories.

To briefly outline our analysis, we note that in 4 space-time dimensions, a priori, there are a total 10 independent degrees of freedom in the metric, out of which two degrees of freedom are associated with a massless spin-2 field (tensor mode), two more degrees of freedom with a massless spin-1 field (vector mode), and two spin-0 fields (scalar modes), along with 4 gauge degrees of freedom. In a companion paper, using a covariant formalism we were able to show that in the equivalent action (and this really means for any action by our previous argument) the dangerous ghost-like vector mode and one of the scalar modes are absent from the theory, as one might expect from Bianchi identities. Further, following a rather elaborate calculation, in [37] we were also able to decompose the equivalent quadratic curvature action

---

[2]The action of Ref. [25] also provides the UV complete Starobinsky inflation [33–35]. Also, it was noted that the gravitational entropy for this action, for a static spherically symmetric background, gets no contribution from the quadratic curvature part [36].

into the remaining propagating degrees of freedom, the spin-2 gravitation and the spin-0 Brans-Dicke scalar[3] and obtain the conditions under which the tensor and the scalar mode can be made ghost and tachyon free in dS/AdS. While we recommend the readers to our companion paper [37] for all the details of the derivations leading up to the consistency conditions, in this article we will briefly outline the important results. Indeed, our results will match the Minkowski space-time analysis of Ref. [29], when we let the cosmological constant vanish.

Let us now begin our discussion by obtaining the most general form of the gravitational action that is relevant for studying the classical and quantum properties of the fluctuations around dS/AdS backgrounds.

5.2 Higher Derivative Actions on (anti-)de Sitter Space-Times

5.2.1 Obtaining the General Form of the Covariant Derivative Structure

Our aim in this section is to arrive at the most general form of the gravitational action that is relevant for studying classical and quantum properties of the fluctuations around constant curvature backgrounds. While this was already investigated for the Minkowski space-time in 4 dimensions in Ref. [25], here we generalize the analysis to include dS/AdS backgrounds. Now, for investigating theoretical and observational consistencies of gravitational models, often it is sufficient to consider quadratic fluctuations around relevant background metrics, i.e., only keep $\mathcal{O}(h^2)$ terms in the action, where $h_{\mu\nu}$ corresponds to fluctuations around the background metric, $\bar{g}_{\mu\nu}$:

$$g_{\mu\nu} = \bar{g}_{\mu\nu} + h_{\mu\nu}. \tag{5.1}$$

In this article we will restrict ourselves to constant curvature, maximally symmetric space-times, i.e., $\bar{g}_{\mu\nu}$ is dS/AdS or the Minkowski metric. Keeping this in mind, let us first identify the most general form of a covariant action that we need to consider if we are only interested in keeping the $\mathcal{O}(h^2)$ terms in the action. Conversely, this will tell us how to obtain the $\mathcal{O}(h^2)$ action starting from any arbitrary covariant metric theory of gravity. Our arguments will closely resemble what was discussed for Minkowski space-times in [25, 26] (see [36] for its generalization to any dimensions), but they will become more intricate for dS/AdS backgrounds.

As was first noted in [25], any covariant action with a well defined Minkowski limit can be written as

---

[3] Although, Brans and Dicke formulated their theory by adding a new nonminimally coupled scalar field, as is well known, this scalar degree of freedom can be incorporated within the metric degrees of freedom by replacing $R \rightarrow F(R)$ in the gravitational action [38]. This is the approach that naturally emerges in our analysis.

$$S = \int d^4x \sqrt{-g} \left[\mathcal{P}_0 + \sum_i \mathcal{P}_i \prod_I (\hat{\mathcal{O}}_{iI} \mathcal{Q}_{iI}) \right] \qquad (5.2)$$

where $\mathcal{P}$, $\mathcal{Q}$'s are functions of the Riemann and the metric tensor, while the differential operators $\hat{\mathcal{O}}$'s are made up solely from covariant derivatives, and contains at least one of them. Essentially, any action which admits a Taylor series expansion in covariant derivatives is included in our discussion. However, nonlocal operators such as $\Box^{-1}$ (see for instance [39]) falls outside the purview of our analysis.

First of all, it is easy to see that even if the $\mathcal{Q}$'s are complicated functions of the Riemann tensor to begin with, one can always use simple rules of calculus to break up $\hat{\mathcal{O}}_I \mathcal{Q}_I$ into a sum of terms where each term is of the form $\prod_J (\hat{\mathcal{O}}_J \mathcal{R}_J)$, where $\mathcal{R}_J$'s now represent just the Riemann tensors. We note in passing that if a metric contraction is present inside the $\mathcal{Q}$'s they can be moved to $\mathcal{P}$ as the metric is annihilated by the covariant derivatives, $\nabla_\mu g_{\nu\rho} = 0$. In other words, without loss of any generality, we can write our action in the form

$$S = \int d^4x \sqrt{-g} \left[\mathcal{P}_0 + \sum_i \mathcal{P}_i \prod_I (\hat{\mathcal{O}}_{iI} \mathcal{R}_{iI}) \right] . \qquad (5.3)$$

Purposely, we have not specified the index structure of the differential operators and the curvature tensors. The most useful property of the maximally symmetric constant curvature space-times is that the Riemann tensor can be completely expressed in terms of the metric and therefore the covariant derivatives annihilate the Riemann tensor and any functions thereof. Mathematically,

$$\hat{\mathcal{O}} \bar{\mathcal{R}} = \hat{\mathcal{O}} \bar{\mathcal{P}} = 0 \qquad (5.4)$$

This, in turn, implies that at most we need to consider terms which contain two $\hat{\mathcal{O}}$'s: If one has a term like $\hat{\mathcal{O}} \mathcal{R}$, then if both $\hat{\mathcal{O}}$ and $\mathcal{R}$ take on the background curvature values term must vanish. This implies that we need to vary at least one of them, and since we are only interested in quadratic variations, at most we can accommodate two such variations. The relevant action then reduces to

$$S = \int d^4x \sqrt{-g} \left[\mathcal{P}_0 + \sum_i \mathcal{P}_{1i} (\hat{\mathcal{O}}_{1i} \mathcal{R}_{1i})(\hat{\mathcal{O}}_{2i} \mathcal{R}_{2i}) + \sum_i \mathcal{P}_{2i} (\hat{\mathcal{O}}_{3i} \mathcal{R}_{3i}) \right] \qquad (5.5)$$

Let us simplify the second term. First consider the situation that $\mathcal{P}_1$ is just a constant. In this case, applying repeated integration by parts one can convert the term into the form of the last term. So, $\mathcal{P}_1$ must contain Riemann tensors. In this case, schematically:

$$\int d^4x \sqrt{-g} \mathcal{P}_{1i} (\hat{\mathcal{O}}_{1i} \mathcal{R}_{1i})(\hat{\mathcal{O}}_{2i} \mathcal{R}_{2i})$$

$$= - \int d^4x \sqrt{-g} \left(\frac{\hat{O}_{1i}}{\nabla} \mathcal{R}_{1i} \right) \left[(\nabla \mathcal{P}_{1i})(\hat{O}_{2i} \mathcal{R}_{2i}) + \mathcal{P}_{1i} (\nabla \hat{O}_{2i} \mathcal{R}_{2i}) \right].$$

The first term is a product of three operators (as long as $\hat{O}_{i1}$ contains more than one derivatives) and hence do not contribute to the quadratic fluctuations, and one can continue to integrate by parts the "second" terms to keep reducing the number of derivatives from $\mathcal{O}_{i1}$. This process can continue till we are left with only a single covariant derivative in $\hat{O}_{i1}$. Thus, the relevant action reduces to the form

$$S = \int d^4x \sqrt{-g} \left[\mathcal{P} + \sum [\mathcal{P}(\nabla \mathcal{R})(\hat{O} \mathcal{R}) + \mathcal{P} \hat{O} \mathcal{R}] \right]. \tag{5.6}$$

We have suppressed the indices, but remind the readers that $\mathcal{P}$'s are just made up of the metric and Riemann tensors, while $\hat{O}$'s are made up of covariant derivatives.

It is convenient to make a last rearrangement. Since $\mathcal{P}$, $\mathcal{R}$ contains an even number of indices, the $\hat{O}$ appearing in the third term must contain at least two covariant derivatives. Integrating by parts it is then trivial to see that this term can always be recast as the second. Thus our relevant action is of the form

$$S = \int d^4x \sqrt{-g} \left[\mathcal{P}_0 + \sum_{i=1} \mathcal{P}_i (\nabla \mathcal{R})(\hat{O}_i \mathcal{R}) \right]. \tag{5.7}$$

In other words, given any arbitrary higher derivative action which possesses a well defined Minkowski limit, $\mathcal{R} \to 0$, we can always obtain an action of the form (5.7) plus additional terms which do not contribute to the quadratic action involving $h_{\mu\nu}$.

5.2.2 Constant Curvature Background Solutions

Before proceeding any further, we need to determine the vacuum solution around which we want to perturb our action. This, in particular will also tell us whether (5.7) provides us with an dS/AdS or Minkowski solution. For this question we need to look at linear variations of the action. However, since all the terms except the first contain covariant derivatives acting on two curvatures, and covariant derivatives annihilate the background curvatures, linear variations of these terms must vanish. Thus, we are only left to consider the linear variation of the first term, i.e., $\delta(\int d^4x \sqrt{-g} \mathcal{P}_0(\mathcal{R}))$. This has already been discussed in previous literature, but for completeness, below we provide a discussion and the main result.

Firstly, it becomes useful from this point onwards to consider $\mathcal{P}_0$ as a function of the scalar curvature, the traceless Ricci tensor (we will refer to this as the TR tensor from here on)

$$S_{\mu\nu} = R_{\mu\nu} - \frac{1}{4}Rg_{\mu\nu},$$

and the Weyl tensor

$$C^{\mu}_{\alpha\nu\beta} = R^{\mu}_{\alpha\nu\beta} - \frac{1}{2}(\delta^{\mu}_{\nu}R_{\alpha\beta} - \delta^{\mu}_{\beta}R_{\alpha\nu} + R^{\mu}_{\nu}g_{\alpha\beta} - R^{\mu}_{\beta}g_{\alpha\nu}) + \frac{R}{6}(\delta^{\mu}_{\nu}g_{\alpha\beta} - \delta^{\mu}_{\beta}g_{\alpha\nu}),$$

as the latter two are traceless and vanish on dS/AdS/Minkowski space-times.

The key point is that since the Lagrangian is a scalar quantity, in all the scalar polynomials that appear in the Lagrangian there cannot be any term that contains a single TR or Weyl tensor, there has to be at least two TR tensors, or two Weyl, or one Weyl plus one TR tensor. Otherwise, their indices have to be contracted with the metric tensor which makes them vanish. This means that while taking a single variation of any such scalar polynomial, there will always remain another TR or Weyl tensor which then has to take on the background value and hence must vanish. To conclude, we only need to worry about the function

$$\mathcal{P}_R(R) = \mathcal{P}_0(R, S = 0, C = 0), \tag{5.8}$$

and the variation of the action (5.7) reduces to

$$\delta S = \int d^4x \sqrt{-\bar{g}} \left[\frac{h}{2}\mathcal{P}_R(R) - \frac{h}{4}\mathcal{P}'_R(R)R \right], \tag{5.9}$$

where we have dropped some total derivatives. Thus the background curvature, $\bar{R}$, is determined by the equation

$$2\mathcal{P}_R(\bar{R}) - \bar{R}\mathcal{P}'_R(\bar{R}) = 0. \tag{5.10}$$

5.2.3 Classification Based on Quadratic Curvature Action

We are now going to perform a final simplification or rather a classification: For a given action of the form (5.7), we will attempt to find an action which has a much simpler form, but which nevertheless gives the same quadratic (in $h_{\mu\nu}$) action as that of the original action. It will become evident that several different actions of the form (5.7) will have the same simple *equivalent* action. Also, if a particular action admits several background curvatures, i.e., (5.10) has more than one solution, then it will have different equivalent actions depending upon the background about which one wants to find the quadratic action.

Having made these clarifications, let us proceed. We first observe that while obtaining the quadratic action for the fluctuations, $\delta\sqrt{-g}$ or $\delta\mathcal{P}_i$, cannot contribute in the variation of the second term in (5.7), else the covariant derivatives will annihilate the background Riemann tensors. Thus, the quadratic variation must be given by

$$\delta S = \int d^4x \left[\delta(\sqrt{-g}\mathcal{P}_0) + \sum_{i=1} \sqrt{-g}\mathcal{P}_i(\bar{\mathcal{R}})\delta(\nabla\mathcal{R})\delta(\hat{\mathcal{O}}_i\mathcal{R}) \right]. \qquad (5.11)$$

Now, the background Riemann tensor can be written completely in terms of the metric, $\mathcal{P}_i(\bar{\mathcal{R}}) = \tilde{\mathcal{P}}_i(\bar{g})$. Then the terms involving $\mathcal{P}_i$'s can be simplified as follows:

$$\int d^4x\sqrt{-\bar{g}}\mathcal{P}_i(\bar{\mathcal{R}})\delta(\nabla\mathcal{R})\delta(\hat{\mathcal{O}}\mathcal{R}) = \int d^4x\sqrt{-\bar{g}}\delta(\nabla\mathcal{R})\delta(\tilde{\mathcal{P}}_i(\bar{g}_{\mu\nu})\hat{\mathcal{O}}\mathcal{R})$$

$$\approx \int d^4x\sqrt{-\bar{g}}\delta(\nabla\mathcal{R})\delta(\tilde{\mathcal{P}}_i(g_{\mu\nu})\hat{\mathcal{O}}\mathcal{R}) = \int d^4x\sqrt{-\bar{g}}\delta(\nabla\mathcal{R})\delta(\tilde{\mathcal{O}}\mathcal{R}) \qquad (5.12)$$

What we have shown here is that the quadratic variation of any action of the form (5.7) is exactly the same as the variation coming from an *equivalent* action of the form

$$S = \int d^4x\sqrt{-g}\left[\mathcal{P} + \sum(\nabla\mathcal{R})(\tilde{\mathcal{O}}\mathcal{R})\right], \qquad (5.13)$$

where $\tilde{\mathcal{O}}$ can be obtained from $\hat{\mathcal{O}}$ according to the prescription above. Therefore, for the purpose of understanding the linearized fluctuation dynamics, we need only to consider actions of the form (5.13).

Now, these actions were precisely the type of actions that were considered in [25, 26], and the Bianchi identities along with the commutativity of the covariant derivatives (we are considering a torsionless theory) enable one to recast it in the following rather simple form:

$$S = \int d^4x\sqrt{-g}[\mathcal{P}_0(\mathcal{R}) + R\mathcal{F}_1(\Box)R + S_{\mu\nu}\mathcal{F}_2(\Box)S^{\mu\nu} + C_{\mu\nu\lambda\sigma}\mathcal{F}_3(\Box)C^{\mu\nu\lambda\sigma}].$$

$$(5.14)$$

where the $\mathcal{F}_i$'s are of the form

$$\mathcal{F}_i(\Box) = \sum_{n=1}^{\infty} c_{i,n}\Box^n$$

We note that although we continue to use the same symbol, $\mathcal{P}_0$, this term can actually change as one goes over from (5.13) to (5.14). More details including illustrative examples will be provided in the companion paper [37].

To complete the reduction, let us focus on the variation of the $\mathcal{P}_0(R)$ piece, see [40, 41] for similar discussions and conclusions. Once more, since both the Weyl and TR tensors vanish on Minkowski/dS/AdS, we can at most have two of those. Moreover, we also can't have terms containing a single symmetric or Weyl tensor, since their indices have to necessarily be contracted which makes them vanish, and by the same token a mixed term with one symmetric and one Weyl also cannot be non-vanishing. In other words, the only relevant part of $\mathcal{P}_0$ in action (5.14) that survives is of the form:

$$\mathcal{P}_0 = \mathcal{P}_R(R) + \mathcal{P}_S(R) S_{\mu\nu} S^{\mu\nu} + \mathcal{P}_C(R) C_{\mu\nu\rho\sigma} C^{\mu\nu\rho\sigma} , \qquad (5.15)$$

where

$$\mathcal{P}_S(R) = \frac{1}{2} \left(\frac{\partial^2 \mathcal{P}_0}{\partial S_{\mu\nu} \partial S^{\mu\nu}} \right)_{S=C=0} \quad \text{and} \quad \mathcal{P}_C(R) = \frac{1}{2} \left(\frac{\partial^2 \mathcal{P}_0}{\partial C_{\mu\nu\rho\sigma} \partial C^{\mu\nu\rho\sigma}} \right)_{S=C=0} \qquad (5.16)$$

Finally, for the S- and C-terms the quadratic variations must originate from S and C tensors, so the R can take on the background value, $\bar{R}$. It is also obvious that the $\mathcal{P}_R(R)$ reduces around the dS/AdS/Minkowski background to

$$\mathcal{P}_R \rightarrow \frac{M_P^2}{2} R + c_{1,0} R^2 - \Lambda ,$$

where the parameters of the equivalent action are given by

$$M_P^2 = \frac{4}{\bar{R}} [\mathcal{P}_R(\bar{R}) - \frac{1}{2} \bar{R}^2 \mathcal{P}_R''(\bar{R})], \ c_{1,0} = \frac{1}{2} \mathcal{P}_R''(\bar{R}), \ \Lambda = \mathcal{P}_R(\bar{R}) - \frac{1}{2} \bar{R}^2 \mathcal{P}_R''(\bar{R}) = \frac{M_P^2 \bar{R}}{4} . \qquad (5.17)$$

The last inequality was indeed expected in accordance with (5.10).

Thus, the equivalent action involving the non-derivative terms are given by

$$S = \int d^4 x \sqrt{-g} \left[\frac{M_P^2}{2} R + c_{1,0} R^2 + c_{2,0} S_{\mu\nu} S^{\mu\nu} + c_{3,0} C_{\mu\nu\lambda\sigma} C^{\mu\nu\lambda\sigma} - \Lambda \right] , \qquad (5.18)$$

where

$$c_{2,0} = \mathcal{P}_S(\bar{R}) , \quad c_{3,0} = \mathcal{P}_C(\bar{R}) , \qquad (5.19)$$

and the other coefficients are given by (5.17).

To summarize, we have shown that in order to investigate quadratic fluctuations around dS/AdS/Minkowski space-times in a generic gravitational theory, all we need to focus our attention on are actions of the form:

$$S = \int d^4 x \sqrt{-g} \left[\frac{M_P^2}{2} R - \Lambda + R \mathcal{F}_1(\Box) R + S_{\mu\nu} \mathcal{F}_2(\Box) S^{\mu\nu} + C_{\mu\nu\lambda\sigma} \mathcal{F}_3(\Box) C^{\mu\nu\lambda\sigma} \right] , \qquad (5.20)$$

where we have now redefined the $\mathcal{F}$'s to include the constant terms:

$$\mathcal{F}_i(\Box) = \sum_{n=0}^{\infty} c_{i,n} \Box^n .$$ (5.21)

We point out that typically one expects the higher derivative terms to become important at some scale $M \leq M_p$ which can be made explicit by rescaling the $c_{i,n}$'s and redefining $\mathcal{F}_i(\Box) \to \mathcal{F}_i(\Box/M^2)$. This is especially useful for constructing phenomenological models and will be discussed in [37], here though we will work with (5.21).

In the process of arriving at the equivalent action (5.20) we have also provided the algorithm on how to obtain the coefficients, $c_{i,n}$'s, starting from a generic covariant action that is regular as $\mathcal{R} \to 0$. Thus given any action of the form (5.7) (and indeed (5.2)) we can determine an action of the form (5.20) that is identical to the general action up to quadratic order in fluctuations around dS/AdS/Minkowski background. It is worth emphasising that all the coefficients c_i's depend on the background curvature parameter $\bar{R}$, which is determined according to (5.10) from the original action.

Finally, we note that the Gauss Bonnet scalar being a topological invariant in four dimensions allows us to set one of the coefficients among $c_{1,0}, c_{2,0}, c_{3,0}$ to zero, if we want to. This completes the derivation of the equivalent quadratic action in terms of the curvature tensors. In the next section we will provide the perturbative structure of action (5.20) and the conditions for having a ghost and tachyon free spectrum around the dS/AdS space-times.

5.3 Quadratic Fluctuations Around dS/AdS/Minkowski Background

5.3.1 Action and Field Equations

The goal of this subsection is to obtain the $\mathcal{O}(h^2)$ action starting from the equivalent action (5.20) in a form that is suitable to address issues of stability and consistency. For this purpose, it becomes imperative that we not only find an expression for the $\mathcal{O}(h^2)$ Lagrangian, but also that we decouple the Lagrangian into separate sectors containing the different physical degrees of freedom of the metric, and present it in a form where we can read off the corresponding propagators. So, we will have to decompose the metric tensor into its 10 degrees of freedom:

$$h_{\mu\nu} = h_{\mu\nu}^{\perp} + \nabla_{(\mu} A_{\nu)}^{\perp} + (\nabla_\mu \nabla_\nu - \frac{1}{4} g_{\mu\nu} \Box) B + \frac{1}{4} g_{\mu\nu} h ,$$ (5.22)

where $h_{\mu\nu}^{\perp}$ represents the transverse traceless massless spin-two graviton,

$$\nabla^{\mu} h_{\mu\nu}^{\perp} = g^{\mu\nu} h_{\mu\nu}^{\perp} = 0 \,, \tag{5.23}$$

containing 5 degrees of freedom, $A_{\mu}^{\perp}$ is the transverse vector,

$$\nabla^{\mu} A_{\mu}^{\perp} = 0 \,, \tag{5.24}$$

accounting for three degrees of freedom, and the two scalars B and h make up the remaining two degrees of freedom. We should mention that in all the calculations that follow we will be using the $-+++$ signature for the metric.

Our next step is to substitute (5.22) into (5.20) and simplify the Lagrangian to the point where we obtain decoupled actions for the different modes. It turns out that such a simplifying and decoupling process provides an extra-ordinary algebraic and technical challenge the details of which we provide in our companion paper [37]. Here we present the main physical arguments and results. As noted earlier, a-priori, the metric represents a massless spin-2, a massless spin-1, and two scalar fields; three gauge degrees reduce the spin two field to the two spin-two helicity states, while another gauge freedom can be used to eliminate the time like component of the vector to again leave us with the two spin-one helicity states.

Now, it is expected, and has been explicitly verified around the Minkowski background [25, 26], that only the spin-2 graviton and one of the scalar fields should survive. Indeed, one finds that when one substitutes the decomposed metric (5.22) into the action (5.20), all the terms involving the vector field, $A_{\mu}^{\perp}$, automatically drops out. Also, only one combination of the two scalar fields,

$$\phi \equiv \Box B - h \,, \tag{5.25}$$

survive.

After a tour-de-force calculation, we obtain a radically simplified action:

$$S = S_0 + S_2 + \mathcal{O}(h^3) \,, \tag{5.26}$$

where

$$S_2 \equiv \frac{1}{2} \int dx^4 \sqrt{-\bar{g}} \, \tilde{h}^{\perp \mu\nu} \left(\Box - \frac{\bar{R}}{6} \right) \left[1 + \frac{4}{M_p^2} c_{1,0} \bar{R} + \frac{2}{M_p^2} \left\{ \left(\Box - \frac{\bar{R}}{6} \right) \mathcal{F}_2(\Box) \right. \right.$$
$$\left. \left. + 2 \left(\Box - \frac{\bar{R}}{3} \right) \mathcal{F}_3 \left(\Box + \frac{\bar{R}}{3} \right) \right\} \right] \tilde{h}_{\mu\nu}^{\perp} \,, \tag{5.27}$$

and

$$
S_0 \equiv -\frac{1}{2} \int dx^4 \sqrt{-\bar{g}} \widetilde{\phi} \left(\Box + \frac{\bar{R}}{3} \right) \left[1 + \frac{4}{M_p^2} c_{1,0} \bar{R} - \frac{2}{M_p^2} \left\{ 6 \left(\Box + \frac{\bar{R}}{3} \right) \mathcal{F}_1(\Box) \right. \right.
$$
$$
\left. \left. + \frac{1}{2} \Box \mathcal{F}_2 \left(\Box + \frac{2}{3} \bar{R} \right) \right\} \right] \widetilde{\phi} . \tag{5.28}
$$

Here, we have introduced canonical fields

$$
\widetilde{h}_{\mu\nu}^{\perp} = \frac{1}{2} M_p h_{\mu\nu}^{\perp} \text{ and } \widetilde{\phi} = \sqrt{\frac{3}{32}} M_p \phi . \tag{5.29}
$$

It is now straight forward to obtain the field equations

$$
\left(\Box - \frac{\bar{R}}{6} \right) \left[1 + \frac{4}{M_p^2} c_{1,0} \bar{R} + \frac{2}{M_p^2} \left\{ \left(\Box - \frac{\bar{R}}{6} \right) \mathcal{F}_2(\Box) \right. \right.
$$
$$
\left. \left. + 2 \left(\Box - \frac{\bar{R}}{3} \right) \mathcal{F}_3 \left(\Box + \frac{\bar{R}}{3} \right) \right\} \right] \widetilde{h}_{\mu\nu}^{\perp} = \kappa \tau_{\mu\nu}
$$
$$
- \left(\Box + \frac{\bar{R}}{3} \right) \left[1 + \frac{4}{M_p^2} c_{1,0} \bar{R} \right.
$$
$$
\left. - \frac{2}{M_p^2} \left\{ 6 \left(\Box + \frac{\bar{R}}{3} \right) \mathcal{F}_1(\Box) + \frac{1}{2} \Box \mathcal{F}_2 \left(\Box + \frac{2}{3} \bar{R} \right) \right\} \right] \widetilde{\phi} = \kappa \tau
$$

$$\tag{5.30}$$

where $\tau_{\mu\nu}$, τ represents the appropriate stress-energy sources for the gravitational fields. We have performed several checks of the above result in [37].

5.3.2   Consistency Conditions

The condition for the theory not to have any ghost/tachyon-like states around the Minkowski space-time was obtained in [25, 26] by looking at the propagators. Although, essentially the propagators are the inverses of the field equation operators, obtaining its precise form is somewhat of a technical exercise on dS/AdS space-times, see for instance [42, 43] for a discussion. For us, all we need to care about is the number and nature of the zeroes in the field equation operators for the tensor and scalar modes respectively:

$$\mathcal{T}(\bar{R}, \Box) \equiv \left(\Box - \frac{\bar{R}}{6}\right)\left[1 + \frac{4\bar{R}}{M_p^2}c_{1,0} + \frac{2}{M_p^2}\left\{\left(\Box - \frac{\bar{R}}{6}\right)\mathcal{F}_2(\Box)\right.\right.$$
$$\left.\left. + 2\left(\Box - \frac{\bar{R}}{3}\right)\mathcal{F}_3\left(\Box + \frac{\bar{R}}{3}\right)\right\}\right],$$

$$\mathcal{S}(\bar{R}, \Box) \equiv -\left(\Box + \frac{\bar{R}}{3}\right)\left[1 + \frac{4\bar{R}}{M_p^2}c_{1,0} - \frac{2}{M_p^2}\left\{2(3\Box + \bar{R})\mathcal{F}_1(\Box)\right.\right.$$
$$\left.\left. + \frac{1}{2}\Box\mathcal{F}_2\left(\Box + \frac{2}{3}\bar{R}\right)\right\}\right]. \tag{5.31}$$

To see this, let us first look at the GR operators:

$$\mathcal{T}_{\mathrm{GR}}(\bar{R}, \Box) \equiv \left(\Box - \frac{\bar{R}}{6}\right),$$
$$\mathcal{S}_{\mathrm{GR}}(\bar{R}, \Box) \equiv -\left(\Box + \frac{\bar{R}}{3}\right). \tag{5.32}$$

As is evident, the function, $\mathcal{T}_{\mathrm{GR}}$, has a zero at $\Box = \bar{R}/6$, corresponding to a pole in the propagator that is known to represent the massless graviton state, the "artificial" mass is simply an artifact of the non-zero curvature of dS/AdS. $\mathcal{S}_{\mathrm{GR}}$ also possesses a zero at $\Box = -\bar{R}/3$, and a corresponding pole in the propagator. As in the Minkowski case, the ghost-like scalar state (note the negative sign in front of the $\mathcal{S}_{\mathrm{GR}}$ operator) is again needed to cancel the unphysical longitudinal degrees of freedom in the graviton field.

Let us now focus on the general $\mathcal{T}(\bar{R}, \Box)$, $\mathcal{S}(\bar{R}, \Box)$ functions. Firstly, we recognize the presence of the zeroes representing the graviton and scalar modes that are present in normal GR. Secondly, just as in the Minkowski case, to ensure that we do not introduce a Weyl ghost in the tensorial mode, we must impose that there are no extra zeroes in $\mathcal{T}(\bar{R}, \Box)$, or equivalently,

$$a(\bar{R}, \Box) \equiv 1 + \frac{4\bar{R}}{M_p^2}c_{1,0} - \frac{2}{M_p^2}\left[\left(\Box - \frac{\bar{R}}{6}\right)\mathcal{F}_2(\Box) + 2\left(\Box - \frac{\bar{R}}{3}\right)\mathcal{F}_3\left(\Box + \frac{\bar{R}}{3}\right)\right]$$
$$\tag{5.33}$$

should not have any zeroes. Finally, again as in the Minkowski case, the scalar function, $\mathcal{S}(\bar{R}, \Box)$ can have one extra zero, as that would correspond to a pole in the propagator which will have the correct residue sign. Indeed this zero corresponds to the Brans-Dicke scalar degree of freedom. Thus, the function,

$$b(\bar{R}, \Box) \equiv 1 + \frac{4\bar{R}}{M_p^2}c_{1,0} + \frac{2}{M_p^2}\left[2(3\Box + \bar{R})\mathcal{F}_1(\Box) + \frac{1}{2}\Box\mathcal{F}_2\left(\Box + \frac{2}{3}\bar{R}\right)\right],$$
$$\tag{5.34}$$

can at least have a single zero. If $b(\bar{R}, \Box)$ does contain a zero, then one has to ensure that the resulting scalar degree of freedom is not tachyonic:

$$\text{If } b(\bar{R}, m^2) = 0 \text{ then } m^2 > -\frac{\bar{R}}{3} . \tag{5.35}$$

Several comments are now in order:

- The conditions that we obtained obviously reduces to the conditions that were previously enumerated for the Minkowski case in [25, 26] when $\Lambda \to 0$.
- It is appropriate to point out a particular special case where $\mathcal{F}_2 = \mathcal{F}_3 = 0$ and $\mathcal{F}_1 = c_{1,0}$, a constant. In this case the tensor mode does not get any correction from it's GR counterpart, but the scalar propagator picks up an extra pole. This is indeed the Brans-Dicke scalar mode that appears in the Starobinsky inflationary model [44].
- It should be apparent that both the scalar and tensor propagators depend on the background curvature, and thus if a particular model of gravity admits more than one constant curvature background, it is possible that the theory is consistent on one background and not the other. To view it differently, requiring that a theory of gravity be consistent around all possible backgrounds may be a powerful way to narrow down the list of acceptable theories of gravity.

5.4 Discussion

To summarize, here we have provided a formalism on how to find a quadratic (in curvatures) order action of gravity that is equivalent to any given covariant gravitational action as far as linearized fluctuations are concerned around constant curvature maximally symmetric space-times. As elaborated in [37], while perturbing the quadratic curvature action around dS/AdS/Minkowski metrics we found that only the spin-2 massless graviton and possibly a spin-0 Brans-Dicke scalar can propagate in these backgrounds. We also enumerated the conditions under which the theory can be made perturbatively stable, i.e. the conditions for a given theory to be free from ghosts and tachyons. Our results match the limits of Minkowski space-time [44] for quadratic curvature gravity with infinite derivatives, as well as the limit of pure Einstein-Hilbert action on dS/AdS backgrounds.

While our analysis can be applied to obtain viable cosmological models involving inflationary or bouncing cosmology as well as the modified gravity models motivated by the cosmic speed-up problem, it also provides encouraging signs for efforts in constructing a more fundamental gravity model which is bereft of the UV problems of GR. Classically, for the IDG theories, the next big step would be to be able to compute perturbations around cosmological and spherically symmetric solutions, because that would help us analyse a wide array of phenomenological applications that have made GR such a success. On the quantum front, while toy models have provided us with some encouraging results regarding finiteness of higher loops in infinite derivative theories [27], see also [13, 45, 46], whether there is any chance that the higher loops can be made finite in IDG theories remains an intriguing question for future!

Acknowledgments We would like to thank Spyridon Talaganis for discussions. TB would like to thank Carl for his insightful comments on the general subject matter of IDG theories. AM is supported by the STFC grant ST/J000418/1. AK is supported by the FCT Portugal fellowship SFRH/BPD/105212/2014.

References

1. M.J.G. Veltman, Quantum theory of gravitation. Conf. Proc. C **7507281**, 265 (1975)
2. B.S. DeWitt, Quantum theory of gravity. 1. The canonical theory. Phys. Rev. **160**, 1113 (1967). B.S. DeWitt, Quantum theory of gravity. 2. The manifestly covariant theory. Phys. Rev. **162**, 1195 (1967). B.S. DeWitt, Quantum theory of gravity. 3. Applications of the covariant theory. Phys. Rev. **162**, 1239 (1967)
3. B.S. DeWitt, G. Esposito, An introduction to quantum gravity. Int. J. Geom. Meth. Mod. Phys. **5**, 101 (2008), arXiv:0711.2445 [hep-th]
4. J. Polchinski, *String Theory: Superstring Theory and Beyond*, vol. 2 (Cambridge, UK: Univ. Pr 1998) , p. 531
5. A. Ashtekar, Introduction to loop quantum gravity and cosmology. Lect. Notes Phys. **863**, 31 (2013)
6. J. Henson, in *The Causal Set Approach to Quantum Gravity*, ed. by D. Oriti, Approaches to Quantum Gravity, pp. 393–413, arXiv:gr-qc/0601121 (for a review)
7. S. Weinberg, in *Ultraviolet Divergences in Quantum Theories of Gravitation*. eds. by S.W. Hawking (Cambridge Univ. (UK)); W. Israel (Alberta Univ., Edmonton (Canada). Theoretical Physics Inst.), pp. 790–831; ISBN 0 521 22285 0; 1979; pp. 790–831; University Press; Cambridge
8. E. Witten, Noncommutative geometry and string field theory. Nucl. Phys. B **268**, 253 (1986)
9. A. Smailagic, E. Spallucci, Lorentz invariance, unitarity in UV-finite of QFT on noncommutative spacetime. J. Phys. A **37**, 1 (2004) [*Erratum-ibid. A***37**, 7169 (2004)] arXiv:hep-th/0406174
10. P.G.O. Freund, M. Olson, Nonarchimedean strings. Phys. Lett. B **199**, 186 (1987). P.G.O. Freund, E. Witten, Adelic string amplitudes. Phys. Lett. B **199**, 191 (1987). L. Brekke, P.G.O. Freund, M. Olson, E. Witten, Nonarchimedean string dynamics. Nucl. Phys. B **302**, 365 (1988). P.H. Frampton, Y. Okada, Effective scalar field theory of P^-adic string. Phys. Rev. D **37**, 3077 (1988)
11. B. Dragovich, Zeta strings, arXiv:hep-th/0703008
12. M.R. Douglas, S.H. Shenker, Strings in less than one-dimension. Nucl. Phys. B **335**, 635 (1990). D.J. Gross, A.A. Migdal, Nonperturbative solution of the ising model on a random surface. Phys. Rev. Lett. **64**, 717 (1990). E. Brezin, V.A. Kazakov, Exactly solvable field theories of closed strings. Phys. Lett. B **236**, 144 (1990). D. Ghoshal, p-adic string theories provide lattice discretization to the ordinary string worldsheet. Phys. Rev. Lett. **97**, 151601 (2006)
13. T. Biswas, M. Grisaru, W. Siegel, Linear regge trajectories from worldsheet lattice parton field theory. Nucl. Phys. B **708**, 317 (2005), arXiv:hep-th/0409089
14. W. Siegel, Introduction to string field theory, arXiv:hep-th/0107094
15. W. Siegel, Stringy gravity at short distances, arXiv:hep-th/0309093
16. A.A. Tseytlin, On singularities of spherically symmetric backgrounds in string theory. Phys. Lett. B **363**, 223 (1995), arXiv:hep-th/9509050

17. T. Biswas, A. Mazumdar, W. Siegel, Bouncing universes in string-inspired gravity. JCAP **0603**, 009 (2006), arXiv:hep-th/0508194
18. E. Tomboulis, Renormalizability and asymptotic freedom in quantum gravity. Phys. Lett. B **97**, 77 (1980). E.T. Tomboulis, in *Renormalization and Asymptotic Freedom in Quantum Gravity*, ed. by S.M. Christensen. Quantum Theory of Gravity, pp. 251–266 and Preprint - TOMBOULIS, E.T. (REC.MAR.83) p. 27. E.T. Tomboulis, Superrenormalizable gauge and gravitational theories, arXiv:hep-th/9702146. E. T. Tomboulis, arXiv:1507.00981 [hep-th]
19. L. Modesto, Super-renormalizable quantum gravity, Phys. Rev. D **86**, 044005 (2012), arXiv:1107.2403 [hep-th]
20. A.O. Barvinsky, Y.V. Gusev, New representation of the nonlocal ghost-free gravity theory, arXiv:1209.3062 [hep-th]. A.O. Barvinsky, aspects of nonlocality in quantum field theory, quantum gravity and cosmology, arXiv:1209.3062 [hep-th]
21. J.W. Moffat, Ultraviolet complete quantum gravity. Eur. Phys. J. Plus **126**, 43 (2011), arXiv:1008.2482 [gr-qc]
22. K. Krasnov, Renormalizable non-metric quantum gravity? arXiv:hep-th/0611182. K. Krasnov, Non-metric gravity i: field equations. Class. Quant. Grav. **25**, 025001 (2008), arXiv:gr-qc/0703002
23. D.A. Eliezer, R.P. Woodard, Nucl. Phys. B **325**, 389 (1989)
24. K.S. Stelle, Renormalization of higher derivative quantum gravity. Phys. Rev. D **16**, 953 (1977)
25. T. Biswas, E. Gerwick, T. Koivisto, A. Mazumdar, Towards singularity and ghost free theories of gravity. Phys. Rev. Lett. **108**, 031101 (2012), arXiv:1110.5249 [gr-qc]
26. T. Biswas, T. Koivisto, A. Mazumdar, Nonlocal theories of gravity: the flat space propagator, arXiv:1302.0532 [gr-qc]
27. S. Talaganis, T. Biswas, A. Mazumdar, Class. Quant. Grav. **32**(21), 215017 (2015). doi:10.1088/0264-9381/32/21/215017, arXiv:1412.3467 [hep-th]
28. T. Biswas, T. Koivisto, A. Mazumdar, Towards a resolution of the cosmological singularity in non-local higher derivative theories of gravity. JCAP **1011**, 008 (2010), arXiv:1005.0590 [hep-th]
29. T. Biswas, A.S. Koshelev, A. Mazumdar, S.Y. Vernov, Stable bounce and inflation in non-local higher derivative cosmology. JCAP **1208**, 024 (2012), arXiv:1206.6374 [astro-ph.CO]
30. V.P. Frolov, A. Zelnikov, T. de Paula Netto, JHEP **1506**, 107 (2015). doi:10.1007/JHEP06(2015)107, arXiv:1504.00412 [hep-th]. V.P. Frolov, Phys. Rev. Lett. **115**(5), 051102 (2015). doi:10.1103/PhysRevLett.115.051102, arXiv:1505.00492 [hep-th]. V.P. Frolov, A. Zelnikov, arXiv:1509.03336 [hep-th]
31. T. Biswas, A. Conroy, A.S. Koshelev, A. Mazumdar, Class. Quant. Grav. **31**, 015022 (2014) [Class. Quant. Grav. **31**, 159501 (2014)]. doi:10.1088/0264-9381/31/1/015022, 10.1088/0264-9381/31/15/159501, arXiv:1308.2319 [hep-th]
32. T. Biswas, S. Talaganis, Mod. Phys. Lett. A **30**, no. 03n04, 1540009 (2015). doi:10.1142/S021773231540009X, arXiv:1412.4256 [gr-qc]
33. T. Biswas, A. Mazumdar, Class. Quant. Grav. **31**, 025019 (2014). doi:10.1088/0264-9381/31/2/025019, arXiv:1304.3648 [hep-th]
34. D. Chialva, A. Mazumdar, Mod. Phys. Lett. A **30**, no. 03n04, 1540008 (2015). doi:10.1142/S0217732315400088, arXiv:1405.0513 [hep-th]
35. B. Craps, T. De Jonckheere, A.S. Koshelev, Cosmological perturbations in non-local higher-derivative gravity, arXiv:1407.4982 [hep-th]
36. A. Conroy, A. Mazumdar, A. Teimouri, Phys. Rev. Lett. **114**(20), 201101 (2015). doi:10.1103/PhysRevLett.114.201101, arXiv:1503.05568 [hep-th]. A. Conroy, A. Mazumdar, S. Talaganis, A. Teimouri, arXiv:1509.01247 [hep-th]
37. T. Biswas, A.S. Koshelev, A. Mazumdar, Analysis of stability of gravitational theories around (anti-)deSitter backgrounds, (in preparation)
38. G. Magnano, L.M. Sokolowski, Phys. Rev. D **50**, 5039 (1994). doi:10.1103/PhysRevD.50.5039, arXiv:gr-qc/9312008
39. R.P. Woodard, Nonlocal Models of Cosmic Acceleration. Found. Phys. (2014) **44**(2), 213–233 (2014). doi:10.1007/s10701-014-9780-6 e-Print: arXiv:1401.0254 [astro-ph.CO]

40. T. Chiba, JCAP **0503**, 008 (2005), arXiv:gr-qc/0502070
41. A. Nunez, S. Solganik, Phys. Lett. B **608**, 189–193 (2005), arXiv:hep-th/0411102
42. B. Allen, Phys. Rev. D **34**, 3670 (1986). doi:10.1103/PhysRevD.34.3670. P.J. Mora, N.C. Tsamis, R.P. Woodard, J. Math. Phys. **53**, 122502 (2012). doi:10.1063/1.4764882, arXiv:1205.4468 [gr-qc]
43. E. D'Hoker, D.Z. Freedman, S.D. Mathur, A. Matusis, L. Rastelli, Nucl. Phys. B **562**, 330 (1999). doi:10.1016/S0550-3213(99)00524-6, arXiv:hep-th/9902042
44. A.A. Starobinsky, Phys. Lett. B **91**, 99 (1980). doi:10.1016/0370-2693(80)90670-X
45. T. Biswas, J.A.R. Cembranos, J.I. Kapusta, JHEP **1010**, 048 (2010), arXiv:1005.0430 [hep-th]
46. T. Biswas, J. Kapusta, A. Reddy, JHEP **1212**, 008 (2012), arXiv:1201.1580 [hep-th]

Chapter 6
Rotating Boson Stars

Eckehard W. Mielke

Abstract Recently, experimental evidence has been accumulated that fundamental scalar fields, like the Higgs boson, exist in Nature. The gravitational collapse of such a boson cloud would lead to a *boson star* (BS) as a new type of a compact object. Similarly as for white dwarfs and neutron stars (NSs), there exist a limiting mass, the Kaup limit, below which a BS is *stable* against complete gravitational collapse to a black hole (BH). Depending the self-interaction of the basic scalars, one can distinguish mini-, axi-dilaton, soliton, charged, oscillating and rotating BSs. Their compactness normally prevents a Newtonian approximation, however, modifications of general relativity (GR), as in the case of Jordan-Brans-Dicke theory, would provide them with *gravitational memory*. Balance between the quantum pressure due to Heisenberg's uncertainty principle and gravity permits the existence of a completely stable branch of spherically symmetric configurations. Moreover, as a coherent state, like the vortices of Bose-Einstein condensates, it allows for rotating solutions with *quantized angular momentum*. In this review, we concentrate on the fascinating possibility of weakening the BH uniqueness theorem for rotating configurations and soliton-type collisions of excited BSs. (Dedicated to Carl Brans' 80th birthday, the author's professor at Princeton in the fall of 1973, then lecturing on complex relativity).

6.1 Introduction: From Geons to Boson Stars

Scalar fields are the basic states in Wigner's classification of irreducible unitary representations of the Poincaré group of relativity. They are postulated as variable gravitational 'constant' $1/G$ in Jordan-Brans-Dicke (JBD) theory [2], in the Brout-Englert-Higgs mechanism [16] of spontaneous symmetry breaking (SSB) and, as well, in inflationary cosmology. Such proposals of the late 50's arose from the

E.W. Mielke (✉)
Departamento de Física, Universidad Autónoma Metropolitana Iztapalapa,
Apartado Postal 55-534, C.P. 09340 Mexico, D.F., Mexico
e-mail: ekke@xanum.uam.mx

© Springer International Publishing Switzerland 2016
T. Asselmeyer-Maluga (ed.), *At the Frontier of Spacetime*,
Fundamental Theories of Physics 183, DOI 10.1007/978-3-319-31299-6_6

fundamental question, why objects have mass, i.e. in a broad sense [17, 27, 33], is GR founded (Fig. 6.1) on a Machian type principle?

The Higgs particle, as an excitation of SSB, is a necessary ingredient of the standard model in order to understand mass generation, and may also decay into dark matter particles. Thus the recent discovery [32] of the Higgs boson of mass $m_h \cong 125$ GeV/c^2 at the Large Electron Positron (LEP) collider at CERN gives this strand of investigation a fresh impetus.

Mathematically, a *boson star* (BS) is a fascinating soliton-like object built out of a huge number N of scalar particles in a coherent state with a self-generated gravitational confinement.

It started in 1955 with Wheeler's geons being unstable localized solutions of the coupled Einstein-Maxwell equations, like a ball lightning (in German: Kugelblitz) [6]. Then, the electromagnetic field was replaced by some tentative scalar field in the pioneering works [10, 18, 35] of Feinblum and McKinley 1968, Kaup 1968, as well as Ruffini and Bonazzola in 1969.

In a nutshell, BSs represent soliton-type regular configurations composed of a complex, massive scalar field which are confined by their self-generate gravitational field. They are localized objects with a finite mass M and particle number N. Its associated $U(1)$ symmetry gives rise to a conserved current and conserved charge

Fig. 6.1 Carl Brans together with his colleagues and the author (*middle*) at the entrance of LIGO, Louisiana, December 7, 2001

$Q = eN$ in terms [34] of the particle number N. The term *mini-BS* was coined [21] by T.D. Lee in 1989 and, in the form of mini-MACHOS [30], has been resurrected as candidates for dark matter (DM).

6.2 Self-Gravitating Scalar Field

The Lagrangian density of gravitationally coupled complex scalar field Φ reads

$$\mathcal{L}_{BS} = \frac{\sqrt{|g|}}{2\kappa} \left\{ R + \kappa \left[g^{\mu\nu} (\partial_\mu \Phi^*)(\partial_\nu \Phi) - U(|\Phi|^2) \right] \right\}. \tag{6.1}$$

Here $\kappa = 8\pi G$ is the gravitational constant in natural units, g the determinant of the metric $g_{\mu\nu}$, and $R := g^{\mu\nu} R_{\mu\nu}$ the curvature scalar with Tolman's sign convention.

Using the principle of variation, one finds the *coupled Einstein-Klein-Gordon equations*

$$G_{\mu\nu} := R_{\mu\nu} - \frac{1}{2} g_{\mu\nu} R = -\kappa T_{\mu\nu}(\Phi), \tag{6.2}$$

$$\left(\Box + \frac{dU}{d|\Phi|^2} \right) \Phi = 0, \tag{6.3}$$

where

$$T_{\mu\nu}(\Phi) = \frac{1}{2} \left[(\partial_\mu \Phi^*)(\partial_\nu \Phi) + (\partial_\mu \Phi)(\partial_\nu \Phi^*) \right] - g_{\mu\nu} \mathcal{L}(\Phi)/\sqrt{|g|} \tag{6.4}$$

is the *stress-energy tensor* and $\Box := (1/\sqrt{|g|}) \, \partial_\mu (\sqrt{|g|} g^{\mu\nu} \partial_\nu)$ the generally covariant d'Alembertian.

6.2.1 Maximal Mass of Stable BS

Particular choices for the self-interacting potential are

• Mini-BS with merely a mass term

$$U(|\Phi|) = m^2 \Phi^* \Phi = m^2 |\Phi|^2 \longrightarrow M_{crit} \propto \frac{M_{Pl}^2}{m} \quad \text{(Kaup limit)}, \tag{6.5}$$

where $M_{Pl} = \sqrt{\hbar c/G}$ is the Planck mass. Stable configurations exist only below the Kaup limit $M_{crit} \leq 0.633 M M_{Pl}^2/m$. This critical mass corresponds to an effective radius of $R_{BS} = 2.47 R_S = 2.47 \times 2GM/c^2$, and total particle number $N_{crit} \sim (M_{Pl}/m)^3$. For 1 GeV constituents, a mini-BS has $M_{BS} \sim 10^{-19} M_\odot$, i.e.

a tiny fraction of a solar mass and only $R_{BS} \sim 1$ femto m. Therefore, it may not be realistic DM candidate.

- A repulsive quartic interaction [7]

$$U_{CWS}(|\Phi|) = m^2|\Phi|^2 + \lambda|\Phi|^4 \longrightarrow M_{crit} \propto \frac{M_{Pl}^3}{m^2} \quad \text{(Chandrasekhar type limit)} \tag{6.6}$$

rescales [30] its critical mass till $M_{crit} \approx 0.1\sqrt{\lambda}\left(GeV/mc^2\right)^2$. For $\lambda = 1$, and $m_p \sim 1 GeV/c^2$ to about one solar mass.

- A soliton type interaction

$$U(|\Phi|) = \lambda|\Phi|^2\left(|\Phi|^4 - a|\Phi|^2 + b\right) \longrightarrow M_{crit} \propto \frac{M_{Pl}^4}{m^3} \tag{6.7}$$

similarly to a Φ^6 interaction [28].

- Another Φ^6 potential is

$$U_{LE}(\phi) = \frac{m^2}{2}\phi^2\left(1 - \chi\phi^4\right), \tag{6.8}$$

inspired by the Lane-Emden (LE) equation of astrophysics.

6.2.2 Conserved Noether Charge

Due to the $U(1)$ symmetry of a complex scalar field, the global phase invariance $\phi \to \phi e^{i\alpha}$ leads to Noether's current density

$$j^\mu := \frac{1}{2}\sqrt{|g|}\, g^{\mu\nu}[\Phi^*\partial_\nu\Phi - \Phi\partial_\nu\Phi^*] \tag{6.9}$$

which is *locally* conserved, i.e., $\partial_\mu j^\mu = 0$. In order to obtain a finite boson number

$$N := \int j^0 dv \tag{6.10}$$

or charge $Q = eN$, the field should have only a time-dependence in the phase of the scalar field

$$\Phi(t, r) = P(r)\, e^{-i\omega t} \quad \text{spherically symmetric BS} \tag{6.11}$$

$$\Phi(t, r, \theta, \varphi) = P(r, \theta)\, e^{-i\omega t - ia\varphi} \quad \text{rotating BS with quantized rotation} \tag{6.12}$$

The normalized energy $\omega = E/\hbar$ determines the number of bosons bound in the star, whereas a is the azimuthal quantum number. BSs do not possess a sharp radius: Instead, an exponential fall-off of the scalar field prevails, similarly as the density of the Sun modeled by the Lane-Emden equation.

Only an effective radius can be defined for BS, but the asymptotic Schwarzschild metric gives a general lower bound on the radius of any type of object, namely $R > M/M_{Pl}^2$ in natural units.

6.2.3 Spherically Symmetric Solitons

The stationarity Ansatz

$$\Phi(r, t) = P(r)e^{-i\omega t} \tag{6.13}$$

(and its complex conjugate) describe a spherically symmetric bound state, when subjected to the spherical symmetry line-element

$$ds^2 = e^{\nu(r)}dt^2 - e^{\lambda(r)}\left[dr^2 + r^2\left(d\theta^2 + \sin^2\theta d\varphi^2\right)\right]. \tag{6.14}$$

This isotropic metric is static and the functions $\nu = \nu(r)$ and $\lambda = \lambda(r)$ depend only on the Schwarzschild type radial coordinate r. Thus there arises a system of three coupled nonlinear equations for the radial parts of the scalar and the strong gravitational tensor field.

The Klein-Gordon equation reduces to the *radial Schrödinger equation*

$$\left[\partial_{r*^2} - V_{\text{eff}}(r^*) + \omega^2 - m^2\right] P = 0 \tag{6.15}$$

for the radial function $P(r) := \Phi e^{i\omega t}$. The curved spacetime enters essentially via an *effective gravitational* potential

$$V_{\text{eff}}(r^*) = \frac{e^\nu dU}{(d|\Phi|^2)} + \frac{e^\nu l(l+1)}{r^2} + \frac{(\nu' - \lambda')e^{\nu-\lambda}}{2r} \tag{6.16}$$

when written in terms of the tortoise coordinate $r^* := \int \exp[(\lambda - \nu)/2]\, dr$.

Then, it can be easily realized that localized solutions decrease asymptotically as

$$P(r) \sim (1/r)\exp\left(-\sqrt{m^2 - \omega^2}\, r\right) \tag{6.17}$$

in a Schwarzschild-type asymptotic background. As first shown numerically by Kaup [18], cf. [35] for the real scalar field case, metric and curvature associated with a BS remain *completely regular*. Via an analytic shooting argument, it has been mathematically proven [1] that *globally regular* mini-BSs exists.

In order to facilitate a visualization of a spherical BS, in Fig. 6.2 the gravitational back-reaction is neglected in order to obtain a sequence of exact Lane-Emden solitons or 'lumps'.

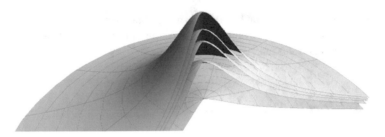

Fig. 6.2 Lame-Emden soliton as a non-gravitational model of a BS

6.2.4 Scalar Matter as an Anisotropic 'fluid'

The canonical stress-energy tensor of a BS, unlike a classical fluid, is in general *anisotropic* as already noticed by Kaup. For a spherically symmetric configuration, it becomes diagonal, i.e.

$$T_\mu{}^\nu(\Phi) = \text{diag}\,(\rho, -p_r, -p_\perp, -p_\perp) \tag{6.18}$$

with

$$\rho = \frac{1}{2}(\omega^2 P^2 e^{-\nu} + P'^2 e^{-\lambda} + U)$$

and

$$p_r = \rho - U,$$

$$p_\perp = p_r - P'^2 e^{-\lambda}.$$

In contrast to NSs, where the ideal fluid approximation demands the isotropy of the pressure, for spherically symmetric BSs there are different stresses p_r and $p_\perp$ in radial or tangential directions, respectively. The notion of *fractional anisotropy* $a_f := (p_r - p_\perp)/p_r = P'^2 e^{-\lambda}/(\rho - U)$ depends essentially on the self-interaction U.

The contracted Bianchi identity $\nabla^\mu G_\mu{}^\nu \equiv 0$ is, in GR, equivalent to *generalization of the Tolman-Oppenheimer-Volkoff equation*

$$\frac{d}{dr} p_r = -\nu'\left(\rho + p_r - \frac{2}{r}(p_r - p_\perp)\right) \tag{6.19}$$

of 'hydrostatic' equilibrium for an anisotropic fluid.

For a spherically symmetric BS in its ground state two different layers of the scalar matter are separated by $p_\perp(R_c) = 0$, i.e. a zero of the tangential pressure. Near the center, $p_\perp$ is positive and, after passing through zero at the *core radius* R_c, it stays negative until radial infinity. The core radius R_c is still *inside* the BS and contains most of the scalar matter. Hence, all three stresses are positive inside the BS core; the boundary layer contains a matter distribution with $p_r > 0$ and $p_\perp < 0$.

6.3 Boson Stars in Jordan-Brans-Dicke Theory

In the so-called scalar-tensor (ST) theories, a real time-dependent scalar field replaces the inverse Newton's gravitational constant G, whose coupling strength to the metric is given by a function $\varpi(\phi_{BD})$. In the simplest scenario, the Jordan-Brans-Dicke (JBD) theory [2], ϖ is a constant. GR is attained in the limit $1/\varpi \to 0$. To ensure that the weak-field limit of this theory agrees with current observations of pulsars [11], ϖ must exceed 25,000. This is within a factor of 1.7 of the precision of the Cassini experiment in the solar system, i.e. experiments taking place in the current cosmic epoch. Scalar tensor theories have regained some popularity through inflationary scenarios, and because a JBD model with $\varpi = -1$ is the low-energy limit of superstring theory.

Here, we exhibit models which contain, beside the complex scalar field of the BS matter, the real scalar field of a JBD type theory.

The Lagrangian for our system of ST gravity coupled to a self-interacting, complex scalar field in the (physical) Jordan frame is

$$
\mathcal{L}_{BD} = \frac{\sqrt{|\tilde{g}|}}{2} \left[\phi_{BD}\tilde{R} - \frac{\varpi(\phi_{BD})}{\phi_{BD}} \tilde{g}^{\mu\nu} \partial_\mu \phi_{BD} \partial_\nu \phi_{BD} + \tilde{V}(\phi_{BD}) + \tilde{g}^{\mu\nu} \partial_\mu \Phi^* \partial_\nu \Phi - U_{CSW} \right].
$$

(6.20)

The gravitational scalar is ϕ_{BD} and $\varpi(\phi_{BD})$ is the Jordan frame coupling of ϕ_{BD} to the matter. In more general theories, the real scalar ϕ_{BD} possesses even a potential $\tilde{V}$. The complex scalar Φ has mass m and is self-interacting[1] through the potential term $U_{CSW} = U_{CSW}(|\Phi|)$.

There is an alternative representation of Lagrangian (6.20) in the so-called Einstein frame. The transition to this frame is effected by the conformal change

$$
\tilde{g}_{\mu\nu} = e^{2W(\varphi_E)} g_{\mu\nu},
$$

(6.21)

where

$$
\phi_{BD}^{-1} = \kappa e^{2W(\varphi_E)}
$$

(6.22)

and $W(\varphi_E)$ is *Wagoner transformation*[2] from ϕ_{BD} to the gravitational scalar φ_E in the Einstein frame. The relationship between $\varpi(\phi_{BD})$ and $W(\varphi_E)$ is obtained by requiring

$$
\left(\frac{\partial W}{\partial \varphi_E} \right)^2 = \frac{1}{2\varpi + 3}.
$$

(6.23)

---

[1] One motivation is that extended inflation models based on BD theory explain the completion of the phase transition in a more natural manner, without fine-tuning.

[2] It is also related to bifurcations [36] of effective higher order curvature Lagrangians.

Using the potential $V(\varphi_E) = e^{4W(\varphi_E)} \tilde{V}[\phi_{BD}(\varphi_E)]$, we find the Lagrangian in the Einstein frame

$$\mathcal{L}_{BD} = \frac{\sqrt{|g|}}{2\kappa}\left[R - 2g^{\mu\nu}\partial_\mu\varphi_E\partial_\nu\varphi_E + V(\varphi_E)\right] \qquad (6.24)$$
$$+ \frac{\sqrt{|g|}}{2}e^{2W(\varphi_E)}\left[g^{\mu\nu}\partial_\mu\Phi^*\partial_\nu\Phi - e^{2a(\varphi_E)}U_{CSW}\right].$$

Mathematically, the transition from the Jordan to the Einstein frame is a conformal change of metric, cf. [24], and a field redefinition or 'renormalization' of the scalar field, the afore mentioned Wagoner transformation. The question which "frame" describes the true, physical metric that measures the distance between spacetime points is a subtle one which depends on the coupling to matter (see the careful analysis of Brans [3]). However, the Einstein frame is, at times, more useful for calculations of BSs leading to results close enough to GR.

As in GR, exited states of BS in general are not stable. They form BH if they cannot lose enough energy to go to the ground state.

6.3.1 Gravitational Memory

The evolution of conventional nuclear burning stars follows the Hertzsprung-Russell diagram of the luminosity as a function of the temperature. There is the possibility that a BS could go through a continually evolution as well, if the gravitational attraction changes during the evolution of the Universe.

In JBD or scalar tensor theory, a real scalar field regulates the strength of the gravitational force, and suppose that at the time of BS formation, the gravitational strength was different from the one prevailing today.

Inside the BS, the strength of the gravitational constant at formation time may still be conserved. According to this *gravitational memory effect*, the BS is book-keeping the evolution of G. Because the BD scalar field has also radial dependence and may change its value at infinity, there should be a repercussion on the BS. But if this change is much slower than the cosmological evolution of G, the star is practically static. If, instead, the BD field adapts quickly inside as changes occur at infinity, the BS evolves as well.

Thus, the BS mass increases with increasing $\phi_{BD}(\infty) = 1/G$, i.e. with time. One can understand that an increasing BD field pumps energy into the BS and increases the mass. Thereby, the *evolution* of a BS in JBD theory resembles the evolution of conventional stars (cf. Ref. [39] for more details).

6.4 Rotating Boson Stars

Since BSs resemble *gravitational atoms*,[3] with the usual wave function

$$\Phi \propto \frac{1}{r\sqrt{4\pi}} R(r)\, P_l^{|a|}(\cos\theta)\, P(r)\, e^{-ia\varphi - i\omega t}, \qquad (6.25)$$

it is rather natural to expect that the angular momentum J is quantized by a, the analog of the *magnetic* or *azimuthal quantum number*. In fact, it is characterized by the principal quantum number $n \geq l + 1$, the *angular momentum* quantum number l and the *azimuthal* quantum number $|a| \leq l$. Moreover, it can be shown via catastrophe theory [20] that the *ground state*, the 1s state with $n = 1$ and $l = 0$, is spherically symmetric and a *stable* configuration for a mass below the Kaup limit.

In spite of previous disclaims, *rapidly rotating* BSs, [29, 37] exist in the Einstein-KG system: An isotropic stationary *axisymmetric* line element

$$ds^2 = f(r,\theta)dt^2 - 2k(r,\theta)dtd\varphi - l(r,\theta)d\varphi^2 - e^{\mu(r,\theta)}\left(dr^2 + r^2 d\theta^2\right) \quad (6.26)$$

is used, where f, k, l, μ are functions to be determined numerically. In order to find rotating BSs, the stationary Ansatz for the scalar field in such a gravitational background is

$$\Phi(r,\theta,\varphi,t) := P(r,\theta)\exp[-i(a\varphi + \omega t)], \qquad (6.27)$$

where the dependence on the azimuthal angle φ occurs only in the phase. Uniqueness of the scalar field under a complete rotation $\Phi(\varphi) = \Phi(\varphi + 2\pi)$ requires the azimuthal quantum number $a = 0, \pm 1, \pm 2, \ldots$ to be quantized, like the rigid rotator in quantum mechanics (Figs. 6.3, 6.4 and 6.5).

Fig. 6.3 H-atom in an eigenstate of $n = 4$, $l = 3$, and azimuthal quantum number $a = 1$

[3]In Ref. [28], the BS is composed from several complex scalars which are in the same ground state of a 't Hooft-Polyakov type *monopole configuration*. Complications due to a possible dependence on the azimuthal angle φ are there avoided by averaging the energy-momentum tensor, leaving merely an angular momentum term in the field equations.

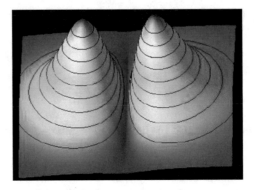

Fig. 6.4 Isocontour plot of the scalar field modulus for rotating BS with $a = 1$ in the a meridian plane, cf. [13]

Fig. 6.5 Surfaces of constant scalar energy density of a *rotating BS*. Inside prevails a toroidal topology

After integrating the time component j^0 over the whole space, we find the particle number N:

$$N = 4\pi \int_0^{\pi/2} \int_0^\infty (\omega l - ak)\frac{P^2}{\sqrt{fl + k^2}}e^\mu r dr d\theta . \qquad (6.28)$$

For asymptotically flat spacetimes in GR, Komar [25] has shown that conserved quantities for solutions with a Killing vector field ξ^α are generated by $K := \int \xi^\alpha n_\beta (T_\alpha{}^\beta - \frac{1}{2}\delta_\alpha^\beta T_\mu{}^\mu) dV$, where $n_\beta = \delta_\beta^0$ is a unit vector in the timelike direction.

The choice $\xi^\alpha = 2n_\beta$ leads to Tolman's expression for the total mass M, whereas the rotational Killing vector $\xi^\alpha = \delta_3^\alpha$ interconnects the canonical angular momentum current with the $U(1)$ Noether current (6.9) via $T_3{}^0 \sqrt{|g|} = a j^0$, essentially due to $i\partial_\varphi \Phi = a\Phi$. Consequently, the *total angular momentum*

$$J = \int T_3{}^0 \sqrt{|g|} d^3x = aN \qquad (6.29)$$

of a rotating BS is exactly proportional to the particle number N and quantized.

Instead of the ring singularity of the Kerr metric, it exhibits an effective 'mass torus' [38] built from coherent scalar fields. Consequently, the suggestion to fill in the Kerr metric, in view of its ring singularity, with a regular toroidal rather than a spherical source, is realized to some extent. In a sense, it becomes the field-theoretical pendant of rotating NSs [23].

There exist sequences of rotating mini-BSs: The energy density ρ vanishes identically at the axis of rotation due to 'centrifugal forces', i.e., vacuum predominates in the interior; cf. the situation of the rotating BS in $(2+1)$-dimensions. Nevertheless, its critical mass will increase above the Kaup limit.

If sub-millisecond pulsars would be detected, today's realistic equation of states (EOSs) for NSs had to be subjected to a major revision. Central cores built from strange matter or fermion condensates would be need for denser stars, sustaining an even faster rotation. The core of a pulsar would resemble a rotating BS whose physical parameters may indirectly affect the observable quantities at the surface.

6.5 Sagitarius Sgr A^*, a Compact Radio Source in Our Galaxy

The center of the Milky way is a radio source at $R_0 = 8.33 \pm 0.35$ kpc distance and, due to dust, it is difficult to observe even the Kepler orbits (Fig. 6.6) of nearby stars. Accordingly, the central region has an estimated mass of $M = (4.31 \pm 0.736) \times 10^6$ $M_\odot$ and is believed to harbor a BH.

Fig. 6.6 Orbit of the star S2 around the galactic center Sgr A^*

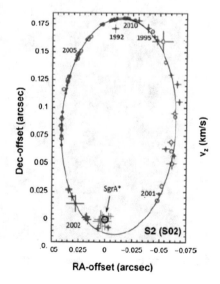

In contrast, the center of our nearby sister galaxy M31 can be observed much better via the Hubble space telescope, cf. Fig. 6.7.

6.5.1 Kerr Black Hole Versus a Rotating BS?

Since the central region of Sgr A^* appears to be rather compact, it is widely believed that it harbors a supermassive BH. However, the Kerr metric has difficulties when matching it consistently to a interior metric generated from a rotating realistic matter source. Thus, there are distinguished doubts [19] whether "...the [gravitational] field outside such a fast rotating collapsed object can be Kerr." Only asymptotically, it may represent the spacetime of an astrophysical rotating object.

The uniqueness of BH has been assessed by John Wheeler's famous phrase: "Black holes have no Hair". Accordingly, generalizations of spherically symmetric BHs with a bound state of scalar fields do not exist.

More recently however, the exterior of rotating BS may provide scalar "hair" to Kerr BHs, since a metrical Ansatz with the two Killing vectors $\xi^\alpha = 2n_\beta$ and $\xi^\alpha = \delta_3^\alpha$ is feasible [14, 15]. These are, however, not Killing vectors of the full solution: The scalar field depends on both only through a phase. This type of 'hair' turns out to generate an intermediate state between rotating BSs and Kerr BHs with a scalar

Fig. 6.7 Core of the galaxy Andromeda (M31)

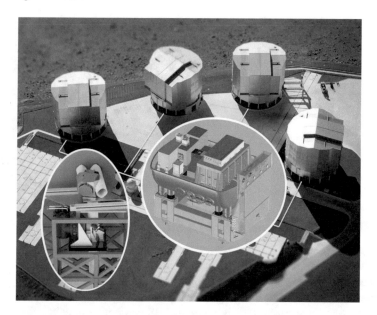

Fig. 6.8 GRAVITY: optical interferometer involving four 8 m telescopes at VLT (Cerro Paranal, Chile)

halo. Contrary to the case of the Kerr spacetime, two invariants [41] built from the Simon-Mars tensor are non-zero for a rotating BS or NS.

For our Galaxy, a Kerr BH or a rotating BS, acting as a 'black hole mimicker', are marginally compatible [42] with the observations of Sgr A^* so far. Although rather unlikely, should we face the bizarre possibility of Sgr A^* harboring a BS made of some kind of DM anchored to its extended halo? In the near future, BHs may well be distinguished observationally from BS or other more exotic configurations.

The GRAVITY instrument [9] is an optical interferometer (Fig. 6.8) in the near-infrared with an astrometric precision of 10 microarcseconds (10μ as) and phase-referenced imaging with 4 milliarcsecond resolution. The most prominent goal is to observe highly relativistic motions of matter around the center of our galaxy. It should provide the ultimate empirical test whether or not the Galactic center harbors a BH of four million solar masses. Thereby, GRAVITY may even be able to verify or exclude alternative models of GR in the presently unexplored strong field limit.

A more ambitious future project is to study images (Fig. 6.9) of BH shadows [8] via the Event Horizon Telescope (Sub)mm VLBI all over the world, which is expected to have an angular resolution of 1μ as!

Also for DM halos in galaxies, gravitational lensing [40] provides an excellent window of observations. In fact, one of the main predictions of GR is the deflection of light by the Sun. Up to a factor of 2, this was suggested by Soldner already in 1801 in the context of Newton's theory.

Fig. 6.9 Accretion disk
around a Schwarzschild BH
with $\Theta_{\mathrm{Sgr}A^*} = 53\mu$ as

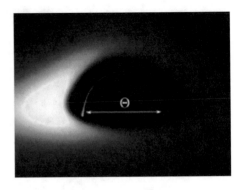

6.6 Boson-Anti-Boson Star in a Saturn Type Configuration

Since the scalar constituents of a BS are unknown, QCD axions are another possible
ingredient. Such *axion stars* (ASs) [31] may collide with NSs and thereby produce
Gamma ray bursts without invoking an ad hoc 'beaming mechanism'.

A more radical scenario [26] is a *toroidal* BS configuration containing an antimatter ring around a centrally condensed core, thus resembling partially a *positron-
ium* atom. Such *toroidal structures* have also been considered before for NSs.

When considering 'excited scalar geons' [28], the next excited state, the 1p state,
is *axisymmetric* with quantized angular momentum $J = aN$, where N is the particle
number of the BS. As we have learned before, the first rotating state with $a = 1$ has
the form of a *mass torus* and is marginally stable against gravitational radiation. The
lowest BS mode which owns a quadrupole moment is the 3d state with $n = 3$ and
$l = 2$. Since this 'gravitational soliton' allows $\Delta J = 2$ transitions to its ground state,
it may rapidly decay by radiating, in addition to scalar modes, gravitational waves.

Let us now consider a "Saturn type configuration", i.e. spherically symmetric BS
surrounded by a toroidal boson star configuration in a marginal stable configuration,
cf. Fig. 6.4 of Ref. [12]. For the argument's simplicity, we assume that mass and
particle number add up linearly and the torus is the only responsible for the total
angular momentum $J_{\mathrm{tot}} = J_{32}$.

Since a BS is built from a complex scalar field, such an object could form not only
from matter, but alternatively, from both, matter and bosonic antimatter. Accordingly,
two possibilities with respect to the total particle number emerge:

$N_{\mathrm{tot}} = N_{10} + N_{32}$ for a boson-boson star (BBS), or
$N_{\mathrm{tot}} = N_{10} - N_{32}$ for an boson-antiboson star (BAS),

where the sub-indices in N_{nl} etc. refer to the principal and angular momentum quantum numbers n and l, respectively.

Through a merger of the torus with the core, such a bizarre BAS would annihilate
all its scalar particles.

6.7 Head-On Collisions of BS

Head-on collisions of massive galaxy clusters, like those occurring in the so-called *bullet cluster* are a challenge for the cold DM paradigm. Instead, one may suspect that the halo is composed of a gas of 'axion mini clusters' or mini-ASs. In fact, axion like scalar fields and the Lane-Emden truncation of their periodic potential has been analyzed [4, 5] as a model of DM halos.

So far, for the LE potential (6.8), an exact auto-Bäcklund transformation for generating multi-solitons has yet not been found. Instead a approximate mapping [4, 5] to a kink-kink pair may serve us as a guide: At large separations from the central interaction region, cf. Figure 6.10, the two soliton solution clearly decouples asymptotically into a (non-interacting) kink–kink pair (Fig. 6.10).

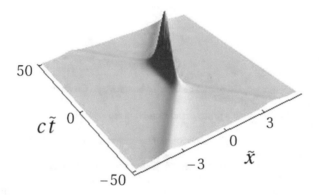

Fig. 6.10 Minkowski diagram of a kink-kink collision in 2D monitored via the absolute value of its spatial derivative

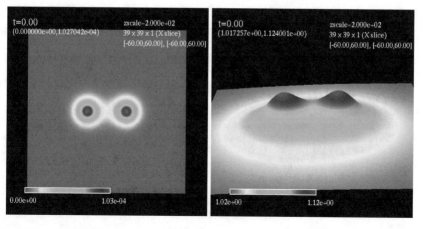

Fig. 6.11 Animation of the scalar amplitude $|\Phi|^2$ and metric component g_{xx} in the $z = 0$ plane

On the other hand, for a superposition

$$\Phi \simeq \Phi_1 + \Phi_2 \tag{6.30}$$

of two single BSs with the same mass at spatial infinity, it is necessary to perform a 3 + 1 decomposition of spacetime. Then the Einstein equations turn into an evolutionary system. In the harmonic or de Donder gauge $\partial^\mu h_{\mu\nu} = \partial_\nu h$, the Einstein equations reduce to a wave equation $\Box g_{ab} = \cdots$ nonlinear in the metric. In order to ensure stability of the numerical code, the next step is to convert this second order system into a first order one, where a 3rd order Runge-Kutta method can be applied to integrate in time. Moreover, an adaptive mesh refinement in space and time is employed, cf. Ref. [22].

As a result, the collision of BSs will be partially inelastic as in the case [5] of solitons (Fig. 6.11).

References

1. P. Bizoń, A. Wasserman, On existence of mini-boson stars. Commun. Math. Phys. **215**, 357 (2000)
2. C. Brans, R.H. Dicke, Mach's principle and a relativistic theory of gravitation. Phys. Rev. **124**, 925 (1961)
3. C. Brans, Non-linear Lagrangians and the significance of the metric. Class. Quantum Grav. **5**, L197 (1988)
4. D. Castañeda Valle, E.W. Mielke, Solitonic axion condensates modeling dark matter halos. Ann. Phys. **336**, 245 (2013)
5. D. Castañeda Valle, E.W. Mielke, Increased infall velocities in galaxy clusters from solitonic collisions? Phys. Rev. D **89**, 043504, 1–5 (2014)
6. J. Cen, P. Yuan, S. Xue, Observation of the optical and spectral characteristics of ball lightning. Phys. Rev. Lett. **112**, 035001 (2014)
7. M. Colpi, S.L. Shapiro, I. Wasserman, Boson stars: gravitational equilibria of self-interacting scalar fields. Phys. Rev. Lett. **57**, 2485 (1986)
8. P.V.P. Cunha, C.A.R. Herdeiro, E. Radu, H.F. Runarsson, Shadows of Kerr black holes with scalar hair. Phys. Rev. Lett. **115**, 211102 (2015)
9. F. Eisenhauer et al., GRAVITY: observing the universe in motion. The Messenger **143**, 16 (2011)
10. D.A. Feinblum, W.A. McKinley, Stable states of a scalar particle in its own gravitational field. Phys. Rev. **168**, 1445 (1968)
11. P.C.C. Freire, N. Wex, G. Esposito-Farèse, J.P.W. Verbiest, M. Bailes, B.A. Jacoby, M. Kramer, I.d H. Stairs, J. Antoniadis, G.H. Janssen, The relativistic pulsar–white dwarf binary PSR J1738+0333 II. The most stringent test of scalar–tensor gravity. Mon. Not. R. Astron. Soc. **423**, 3328–3343 (2012)
12. R. Ferrell, M. Gleiser, Gravitational atoms: Gravitational radiation from excited boson stars. Phys. Rev. D **40**, 2524 (1989)
13. P. Grandclément, C. Somé, E. Gourgoulhon, Models of rotating boson stars and geodesics around them: new type of orbits. Phys. Rev. D. **90**, 024068 (2014)
14. C.A.R. Herdeiro, E. Radu, A new spin on black hole hair. Int. J. Mod. Phys. D **23**, 1442014 (2014)
15. C.A.R. Herdeiro, E. Radu, Asymptotically flat black holes with scalar hair: a review. Int. J. Mod. Phys. D **24**, 1542014 (2015)

16. P. Higgs, Evading the Goldstone Theorem (Nobel Lecture, 8 December 2013). Int. J. Mod. Phys. A **30**, 1530021 (2015)
17. D.I. Kaiser, When fields collide. Sci. Am. **296N6**, 41–47 (2007)
18. D.J. Kaup, Klein-Gordon geon. Phys. Rev. **172**, 1331 (1968)
19. R.P. Kerr, Rotating black holes and the Kerr metric. in *Relativistic Astrophysics*, 5th Sino-Italian Workshop, eds. by D.–S. Lee, W. Lee, S.–S. Xue, AIP Conference Proceedings, vol 1059, 9–12 (2008)
20. F.V. Kusmartsev, E.W. Mielke, F.E. Schunck, Gravitational stability of boson stars. Phys. Rev. D **43**, 3895 (1991)
21. T. D. Lee, Y. Pang, Stability of miniboson stars. Nucl. Phys. B. **315**, 477 (1989)
22. S.L. Liebling, C. Palenzuela, Dynamical boson stars. Living Rev. Relativ. **15**, 6 (2012)
23. V.S. Manko, E.W. Mielke, J.D. Sanabria-Gómez, Exact solution for the exterior field of a rotating neutron star. Phys. Rev. D **61**, R081501 (2000)
24. E.W. Mielke, Conformal changes of metrics and the initial-value problem of general relativity. Gen. Rel. Grav. **8**, 321 (1977)
25. E.W. Mielke, Affine generalization of the Komar complex of general relativity, Phys. Rev. **D63**, 044018 (2001) 1 (2001)
26. E.W. Mielke, Gamma rays from boson anti-boson star mergers, in *Proceedings of the ICGA-5, Gravitation and Cosmology Supplements*, vol.8, Supplement II N2 (ICGA-5 Proceedings), pp. 111–113 (2002)
27. E.W. Mielke, Weak equivalence principle from a spontaneously broken gauge theory of gravity. Phys. Lett. B **702**, 187–190 (2011)
28. E.W. Mielke, R. Scherzer, Geon-type solutions of the nonlinear Heisenberg-Klein-Gordon equation. Phys. Rev. D **24**, 2111 (1981)
29. E.W. Mielke, F.E. Schunck, Rotating boson stars. in *Gravity, Particles and Space-Time*, eds. by P. Pronin, G. Sardanashvily, (World Scientific: Singapore 1996), pp. 391–420
30. E.W. Mielke, F.E. Schunck, Boson stars: alternatives to primordial black holes? Nucl. Phys. B **564**, 185 (2000)
31. E.W. Mielke, F.E. Schunck, Are axidilaton stars massive compact halo objects? Gen. Rel. Grav. **33**, 805 (2001)
32. J.L. Miller, The Higgs particle, or something much like it, has been spotted. Phys. Today **65**(9), 12 (2012)
33. P. Parsons, *3-Minute Einstein: Digesting His Life, Theories and Influence in 3-minute Morsels*, Foreword by John Gribbin (Apple Press, 2011)
34. D. Pugliese, H. Quevedo, J.A. Rueda, R. Ruffini, On charged boson stars. Phys. Rev. D **88**, 024053 (2013)
35. R. Ruffini, S. Bonazzola, Systems of self-gravitating particles in general relativity and the concept of an equation of state. Phys. Rev. **187**, 1767 (1969)
36. F.E. Schunck, F.V. Kusmartsev, E.W. Mielke, Dark matter problem and effective curvature Lagrangians. Gen. Rel. Grav. **37**(8), 1427–1433 (2005)
37. F.E. Schunck, E.W. Mielke, Rotating boson stars, in Proceedings of the Bad Honnef Workshop *Relativity and Scientific Computing: Computer Algebra, Numerics, Visualization*, eds. by F.W. Hehl, R.A . Puntigam, H. Ruder, (Springer-Verlag, Berlin, 1996), pp 8–11; 138–151
38. F.E. Schunck, E.W. Mielke, Rotating boson star as an effective mass torus in general relativity. Phys. Lett. A **249**, 389 (1998)
39. F.E. Schunck, E.W. Mielke, TOPICAL REVIEW: general relativistic boson stars. Class. Quantum Gravity **20**, R301–R356 (2003)
40. F.E. Schunck, B. Fuchs, E.W. Mielke, Scalar field haloes as gravitational lenses. Mon. Not. R. Astron. Soc. **369**, 485–491 (2006)
41. C. Somé, P. Grandclément, E. Gourgoulhon, A characterization of 3+1 spacetimes via the Simon-Mars tensor, arXiv:1412.6542 (2014)
42. Y.-F. Yuan, R. Narayan, M.J. Rees, Constraining alternate models of black holes: type I X-ray bursts on accreting fermion-fermion and boson-fermion stars. Astrophys. J. **606**, 1112 (2004)

Chapter 7
The Lambda-CDM Model Is Not an Universal Attractor of the Brans–Dicke Cosmology

Israel Quiros

Abstract By means of the tools of the dynamical systems theory it is shown that the general relativity de Sitter solution is an attractor of the Jordan frame (dilatonic) Brans–Dicke theory only for the exponential potential $U(\varphi) \propto \exp\varphi$, which corresponds to the quadratic potential $V(\phi) \propto \phi^2$ in terms of the original Brans–Dicke field $\phi = \exp\varphi$, or for potentials which approach to $\exp\varphi$ at the stable point. I find bounds on the Brans–Dicke coupling constant ω_{BD}, which are consistent with well-known results.

7.1 Introduction

The Brans–Dicke (BD) theory of gravity [1] represents the simplest modification of general relativity (GR) by the addition of a new (scalar) gravitational degree of freedom ϕ in addition to the $10°$ which are associated with the metric tensor $g_{\mu\nu}$. This theory has been cornerstone for a better understanding of several other modifications of general relativity such as the $f(R)$ theories of gravity [2]. In contrast to Einstein's GR, the BD theory is not a fully geometrical theory of gravity since, while one of the propagators of the gravitational field: the metric tensor, defines the metric properties of the spacetime, the scalar field ϕ modifies the local strength of the gravitational interactions through the effective gravitational coupling $G_{\mathrm{eff}} \propto \phi^{-1}$.

Many aspects of BD theory have been well-explored in the past (see the textbooks [3, 4]), while other aspects have been cleared up just recently. Thanks to the chameleon effect [5], for instance, it was just recently understood that the lower experimental bounds on the BD coupling parameter ω_{BD}, which were set up through experiments in the solar system, might not apply in the large cosmological scales if consider BD theory with a potential. According to the chameleon effect, the effective mass of the scalar field m_ϕ computed in the Einstein's frame, depends on the background energy density of the environment: In the large cosmological scales where

I. Quiros (✉)

Dpto. Ingeniería Civil, División de Ingeniería, Universidad de Guanajuato, CP 36000 Leon, Guanajuato, Mexico

e-mail: iquiros@fisica.ugto.mx

© Springer International Publishing Switzerland 2016

T. Asselmeyer-Maluga (ed.), *At the Frontier of Spacetime*,

Fundamental Theories of Physics 183, DOI 10.1007/978-3-319-31299-6_7

133

the background energy density is very small (of the order of the critical density), the effective mass is also very small, so that the scalar field degree of freedom has impact in the cosmological dynamics. Meanwhile, in the solar system, where the averaged energy density of the environment is huge compared with the one in the cosmological scale, the effective mass is large, so that the Yukawa component of the gravitational interaction associated with the scalar field $\propto e^{-m_\phi r}/r$, is short-ranged, leading to an effective screening of the scalar field degree of freedom in the solar system.

The Brans–Dicke theory has found very interesting applications specially in cosmology [4], where it has been explored as a possible explanation of the present stage of the accelerated expansion of the universe [6]. The problem with this is that, but for some anomalies in the power spectrum of the cosmic microwave background [7], at the present stage of the cosmic evolution any cosmological model has to approach to the so called concordance or ΛCDM model [8]. The mathematical basis for the latter is the GR (Einstein–Hilbert) action plus a matter action piece:

$$S_{\Lambda CDM} = \frac{1}{16\pi G} \int d^4x \sqrt{|g|} \, (R - 2\Lambda) + \int d^4x \sqrt{|g|} \mathcal{L}_{CDM}, \qquad (7.1)$$

where $\mathcal{L}_{CDM}$ is the Lagrangian density of (pressureless) cold dark matter (CDM). On the other hand, it has been known for decades, that GR can be recovered from the BD theory only in the limit when the BD coupling constant $\omega_{BD} \to \infty$.

In the references [9–11], by means of the tools of the dynamical systems theory, it was (apparently) shown that the Jordan frame (JF) Brans–Dicke theory leads naturally to the ΛCDM model since, as the authors showed in [9], the GR–de Sitter solution is an attractor of JF–BD theory, independent on the choice of the self-interaction potential for the BD scalar field.[1] The interesting thing is that the bounds on ω_{BD} found in [9] ($\omega_{BD} \approx -3/2$), and in [10] ($\omega_{BD} \approx -1$), are far from the solar system–based experimental bound $\omega_{BD} > 40000$ [15].

Although the chameleon effect could (in principle) explain such a discrepancy between the bounds on ω_{BD} based in solar system experimentation (see, however, Ref. [16]), and those based in cosmological considerations, nevertheless, the bounds estimated on the basis of cosmological arguments: $\omega_{BD} > 120$ in [17], and $10 < \omega_{BD} < 10^7$ in [18], neither are consistent with the ones found in the references [9] and [10]. Besides, we stress that the conclusion on the existence of the GR–de Sitter (stable) critical point independent on the assumed potential in [9–11], is misleading. As a matter of fact, in [11] the authors seem to recognize that the de Sitter

---

[1] For prior works where the de Sitter solutions are investigated within the frame of the scalar-tensor theories, see Ref. [12], where de Sitter exact and intermediate inflationary solutions are found for FRW models with appropriate choice of the coupling function $\omega_{BD}(\varphi)$. In [13] it is shown that intermediate "almost de Sitter" solutions might arise also when CDM is included. Other, more resent works on this subject, are also found [14].

equilibrium configuration arises only for the quadratic monomial $V(\phi) \propto \phi^2$, and for the lineal $V(\phi) \propto \phi$ potentials. Since the estimates of [9–11] are based on the analysis of linearized solutions which are, in fact, very small perturbations around the stable GR–de Sitter critical point (hence are highly dependent on the choice of the initial conditions), we suspect that these estimates could be physically meaningless. Actually, linear solutions around general relativity (plus a cosmological constant), which is obtained in the formal limit $\omega_{BD} \to \infty$ of Brans–Dicke theory, can not be reliable sources of bounds on the parameters of the BD theory.

In this paper we shall apply the tools of the theory of the dynamical systems in order to uncover the dynamics of cosmological models which are based in the BD theory of gravity, for several self-interaction potentials, in a convenient phase space. Unlike [9–11], here we shall explore specific potentials other than (but also including) the quadratic and the lineal monomials: $V(\phi) \propto \phi^2$ and $V(\phi) \propto \phi$, respectively. For a better understanding of our analysis we shall study the vacuum BD theory and the BD theory with matter, separately (see Sects. 7.4 and 7.5, respectively). In the Sect. 7.6, we shall show that it is not enough that the de Sitter solution be a critical point of the dynamical system, in order for the ΛCDM model to be an attractor of the BD theory with a potential. It is a necessary condition for the latter that the GR–de Sitter solution, i.e., the de Sitter point which leads to $\phi = \phi_0 = const.$, to be a stable critical point instead. It will be shown that, as a matter of fact, only for the quadratic monomial potential $V(\phi) \propto \phi^2$, or for potentials that approach to ϕ^2 at the stable point, the ΛCDM model is an attractor of the BD cosmology.

For simplicity of mathematical handling we shall use the dilatonic field variable φ instead of the standard BD field ϕ. These variables are related by Eq. (7.4). At the end of the contribution the reader can find an appendix with concrete criticism on the procedure and on the results discussed in Ref. [9].

7.2 Basic Setup

Here we assume the Brans–Dicke theory [1] with the potential, to dictate the dynamics of gravity and matter. In the Jordan frame it is depicted by the following action:

$$S_{JF}^{\phi} = \int d^4x \sqrt{|g|} \left\{ \phi R - \frac{\omega_{BD}}{\phi} (\partial \phi)^2 - 2V + 2\mathcal{L}_m \right\}, \tag{7.2}$$

where $(\partial \phi)^2 \equiv g^{\mu\nu} \partial_\mu \phi \partial_\nu \phi$, $V = V(\phi)$ is the scalar field self-interaction potential, ω_{BD} is the BD coupling parameter, and $\mathcal{L}_m = \mathcal{L}_m(\chi, \partial\chi, g_{\mu\nu})$ is the Lagrangian density of the matter degrees of freedom, collectively denoted by χ. Unless the contrary is specified, the natural units $8\pi G = 1/M_{PL}^2 = c = 1$, are adopted. The field equations which are derived from (7.2) are the following:

$$G_{\mu\nu} = \frac{\omega_{\text{BD}}}{\phi^2}\left[\partial_\mu\phi\partial_\nu\phi - \frac{1}{2}g_{\mu\nu}\left(\partial\phi\right)^2\right] - g_{\mu\nu}\frac{V}{\phi} + \frac{1}{\phi}\left(\nabla_\mu\partial_\nu\phi - g_{\mu\nu}\nabla^2\phi\right) + \frac{1}{\phi}T^{(m)}_{\mu\nu},$$

$$\nabla^2\phi = \frac{2}{3 + 2\omega_{\text{BD}}}\left(\phi\partial_\phi V - 2V + \frac{1}{2}T^{(m)}\right),\tag{7.3}$$

where $G_{\mu\nu} = R_{\mu\nu} - g_{\mu\nu}R/2$, is the Einstein's tensor, $\nabla^2 = g^{\mu\nu}\nabla_\mu\nabla_\nu$, is the D'Alembertian operator, and $T^{(m)}_{\mu\nu} = -(2/\sqrt{|g|})\partial\left(\sqrt{|g|}\,\mathcal{L}_m\right)/\partial g^{\mu\nu}$, is the conserved stress-energy tensor of the matter degrees of freedom $\nabla^\mu T^{(m)}_{\mu\nu} = 0$.

It is also convenient to rescale the BD scalar field and, consequently, the self-interaction potential:

$$\phi = e^\varphi,\ V(\phi) = e^\varphi\,U(\varphi),\tag{7.4}$$

so that, the action (7.2) is transformed into the string frame BD action [19]:

$$S^\varphi_{\text{SF}} = \int d^4x\sqrt{|g|}e^\varphi\left\{R - \omega_{\text{BD}}(\partial\varphi)^2 - 2U + 2e^{-\varphi}\mathcal{L}_m\right\}.\tag{7.5}$$

The following motion equations are obtained from (7.5):

$$G_{\mu\nu} = (\omega_{\text{BD}} + 1)\left[\partial_\mu\varphi\partial_\nu\varphi - \frac{1}{2}g_{\mu\nu}(\partial\varphi)^2\right] - g_{\mu\nu}\left[\frac{1}{2}(\partial\varphi)^2 + U(\varphi)\right]$$

$$+ \nabla_\mu\partial_\nu\varphi - g_{\mu\nu}\nabla^2\varphi + e^{-\varphi}T^{(m)}_{\mu\nu},$$

$$\nabla^2\varphi + (\partial\varphi)^2 = \frac{2\left[\partial_\varphi U - U + \frac{e^{-\varphi}}{2}T^{(m)}\right]}{3 + 2\omega_{\text{BD}}},\tag{7.6}$$

where $\nabla^2 \equiv g^{\mu\nu}\nabla_\mu\partial_\nu$, $G_{\mu\nu} = R_{\mu\nu} - g_{\mu\nu}R/2$, and, as before, $T^{(m)}_{\mu\nu}$ is the (conserved) stress-energy tensor of the matter degrees of freedom: $\nabla^\mu T^{(m)}_{\mu\nu} = 0$.

In this contribution we shall consider Friedmann-Robertson-Walker (FRW) space-times with flat spatial sections for which the line-element takes the simple form:

$$ds^2 = -dt^2 + a^2(t)\delta_{ij}dx^idx^j,\ i, j = 1, 2, 3.$$

We assume the matter content of the Universe in the form of a cosmological perfect fluid, which is characterized by the following state equation $p_m = w_m\rho_m$, relating the barotropic pressure p_m and the energy density ρ_m of the fluid, where w_m is the so called equation of state (EOS) parameter. Under these assumptions the cosmological equations (7.6) are written as it follows:

$$3H^2 = \frac{\omega_{\text{BD}}}{2}\dot\varphi^2 - 3H\dot\varphi + U + e^{-\varphi}\rho_m,$$

$$\dot H = -\frac{\omega_{\text{BD}}}{2}\dot\varphi^2 + 2H\dot\varphi + \frac{\partial_\varphi U - U}{3 + 2\omega_{\text{BD}}} - \frac{2 + \omega_{\text{BD}}(1 + w_m)}{3 + 2\omega_{\text{BD}}}e^{-\varphi}\rho_m,$$

$$\ddot{\varphi} + 3H\dot{\varphi} + \dot{\varphi}^2 = 2\frac{U - \partial_\varphi U}{3 + 2\omega_{BD}} + \frac{1 - 3w_m}{3 + 2\omega_{BD}} e^{-\varphi} \rho_m,$$

$$\dot{\rho}_m + 3H(w_m + 1)\rho_m = 0, \tag{7.7}$$

where $H \equiv \dot{a}/a$ is the Hubble parameter.

Due to the complexity of the system of non-linear second-order differential equations (7.7), it is a very difficult (and perhaps unsuccessful) task to find exact solutions. Yet, even when an analytic solution can be found it will not be unique but just one in a large set of them. This is in addition to the problem of the stability of given solutions. In this case the dynamical systems tools come to our rescue. These very simple tools give us the possibility to correlate such important concepts in the phase space like past and future attractors (also saddle equilibrium points), limit cycles, heteroclinic orbits, etc., with generic behavior of the dynamical system derived from the set of equations (7.7), without the need to analytically solve them. A very compact and basic introduction to the application of the dynamical systems in cosmological settings with scalar fields can be found in the references [20–25].

7.3 Dynamical Systems

As it is for any other physical system, the possible states of a cosmological model may be also correlated with points in an equivalent state space or phase space. However, unlike in the classical mechanics case, where the phase space is spanned by the generalized coordinates and their conjugate momenta, in a cosmological context the choice of the phase space variables is not a trivial issue. This leads to a certain degree of uncertainty in the choice of an appropriate set of variables of the phase space. There are, however, certain—not written—rules one follows when choosing these variables: (i) these should be dimensionless variables, and (ii) whenever possible, these should be bounded. The latter requirement is necessary to have a bounded phase space where all of the existing equilibrium points are "visible", i.e., none of then goes to infinity. Unfortunately it is not always possible to find such bounded variables.

In general, when one deals with BD cosmological models it is customary to choose the following variables [9–11]:

$$x \equiv \frac{\dot{\varphi}}{\sqrt{6}H} = \frac{\varphi'}{\sqrt{6}}, \quad y \equiv \frac{\sqrt{U}}{\sqrt{3}H}, \quad \xi \equiv 1 - \frac{\partial_\varphi U}{U}, \tag{7.8}$$

where the tilde means derivative with respect to the variable $\tau \equiv \ln a$—the number of e-foldings. As a matter of fact x and y in Eq. (7.8), are the same variables which are usually considered in similar dynamical systems studies of FRW cosmology, within the frame of Einstein's general relativity with a scalar field matter source [20]. In terms of the above variables the Friedmann constraint in Eq. (7.7) can be written as

$$\Omega_m^{\text{eff}} \equiv \frac{e^{-\varphi}\rho_m}{3H^2} = 1 + \sqrt{6}x - \omega_{\text{BD}}\,x^2 - y^2 \geq 0. \tag{7.9}$$

Notice that one might define a dimensionless potential energy density and an "effective kinetic" energy density

$$\Omega_U = \frac{U}{3H^2} = y^2, \quad \Omega_K^{\text{eff}} = x\left(\omega_{\text{BD}}x - \sqrt{6}\right), \tag{7.10}$$

respectively, so that the Friedmann constraint can be re-written in the following compact form: $\Omega_K^{\text{eff}} + \Omega_U + \Omega_m^{\text{eff}} = 1$.

The definition for the dimensionless effective kinetic energy density Ω_K^{eff} has not the same meaning as in GR with a scalar field. It may be a negative quantity without challenging the known laws of physics. Besides, since there is not restriction on the sign of Ω_K^{eff}, then, it might happen that $\Omega_U = U/3H^2 > 1$. This is due to the fact that the dilaton field in the BD theory is not a standard matter field but it is a part of the gravitational field itself. This effective (dimensionless) kinetic energy density vanishes whenever: $x = \sqrt{6}/\omega_{\text{BD}} \Rightarrow \dot{\varphi} = 6H/\omega_{\text{BD}} \Rightarrow \varphi = 6\ln a/\omega_{\text{BD}}$, or if: $x = 0 \Rightarrow \dot{\varphi} = 0 \Rightarrow \varphi = const.$, which, provided that the matter fluid is cold dark matter, corresponds to the GR–de Sitter universe, i.e., to the ΛCDM model.

The following are useful equations which relate $\dot{H}/H^2$ and $\ddot{\varphi}/H^2$ with the phase space variables x, y and ξ:

$$\frac{\dot{H}}{H^2} = 2\sqrt{6}x - 3\omega_{\text{BD}}x^2 - \frac{3y^2\xi}{3 + 2\omega_{\text{BD}}} - \frac{2 + \omega_{\text{BD}}(1 + w_m)}{3 + 2\omega_{\text{BD}}}3\Omega_m^{\text{eff}},$$

$$\frac{\ddot{\varphi}}{H^2} = -3\sqrt{6}x - 6x^2 + \frac{6y^2\xi}{3 + 2\omega_{\text{BD}}} + \frac{1 - 3w_m}{3 + 2\omega_{\text{BD}}}3\Omega_m^{\text{eff}}. \tag{7.11}$$

Our goal will be to write the resulting system of cosmological equations (7.7), in the form of a system of autonomous ordinary differential equations (ODE-s) in terms of the variables x, y, ξ, of some phase space. We have:

$$x' = \frac{\ddot{\varphi}}{\sqrt{6}H^2} - x\frac{\dot{H}}{H^2}, \quad y' = y\left[\frac{\sqrt{6}}{2}(1 - \xi)x - \frac{\dot{H}}{H^2}\right],$$

$$\xi' = -\sqrt{6}x(1 - \xi)^2(\Gamma - 1), \quad \Gamma \equiv \frac{U\partial_\varphi^2 U}{(\partial_\varphi U)^2}, \tag{7.12}$$

or, after substituting Eqs. (7.11) into (7.12), we obtain the following autonomous system of ODE-s:

$$x' = -3x \left(1 + \sqrt{6}x - \omega_{\text{BD}}x^2\right) + \frac{x + \sqrt{2/3}}{3 + 2\omega_{\text{BD}}} 3y^2\xi$$

$$+ \frac{\frac{1-3w_m}{\sqrt{6}} + [2 + \omega_{\text{BD}}(1 + w_m)]\, x}{3 + 2\omega_{\text{BD}}} 3\Omega_m^{\text{eff}},$$

$$y' = y \left[3x \left(\omega_{\text{BD}}x - \frac{\xi + 3}{\sqrt{6}}\right) + \frac{3y^2\xi}{3 + 2\omega_{\text{BD}}} + \frac{2 + \omega_{\text{BD}}(1 + w_m)}{3 + 2\omega_{\text{BD}}} 3\Omega_m^{\text{eff}}\right],$$

$$\xi' = -\sqrt{6}x\, (1 - \xi)^2\, (\Gamma - 1), \tag{7.13}$$

where Ω_m^{eff} is given by Eq. (7.9), and it is assumed that $\Gamma = U\partial_\varphi^2 U/(\partial_\varphi U)^2$ can be written as a function of ξ [25]: $\Gamma = \Gamma(\xi)$. Hence, the properties of the dynamical system (7.13) are highly dependent on the specific functional form of the potential $U = U(\varphi)$.

7.4 Vacuum Brans-Dicke Cosmology

A significant simplification of the dynamical equations is achieved when matter degrees of freedom are not considered. In this case, since $\Omega_m^{\text{eff}} = 0 \Rightarrow y^2 = 1 + \sqrt{6}x - \omega_{\text{BD}}\,x^2$, then the system of ODE-s (7.13) simplifies to a plane-autonomous system of ODE-s:

$$x' = \left(-3x + 3\frac{x + \sqrt{2/3}}{3 + 2\omega_{\text{BD}}}\xi\right)\left(1 + \sqrt{6}x - \omega_{\text{BD}}x^2\right),$$

$$\xi' = -\sqrt{6}x\, (1 - \xi)^2\, (\Gamma - 1). \tag{7.14}$$

In the present case one has $\Omega_K^{\text{eff}} + \Omega_U = 1$, where

$$\Omega_U = \frac{U}{3H^2} = y^2 = 1 + \sqrt{6}x - \omega_{\text{BD}}x^2, \quad \Omega_K^{\text{eff}} = x\left(\omega_{\text{BD}}x - \sqrt{6}\right). \tag{7.15}$$

We recall that the definition of the effective (dimensionless) kinetic energy density Ω_K^{eff}, has not the same meaning as in GR with scalar field matter, and it may be, even, a negative quantity. In this paper we consider non-negative self-interaction potentials $U(\varphi) \geq 0$, so that the dimensionless potential energy density $\Omega_U = y^2$, is restricted to be always non-negative: $\Omega_U = 1 + \sqrt{6}x - \omega_{\text{BD}}x^2 \geq 0$. Otherwise, $y^2 < 0$, and the phase-plane would be a complex plane. Besides, we shall be interested in expanding cosmological solutions exclusively ($H \geq 0$), so that $y \geq 0$. Because of this the variable x is bounded to take values within the following interval:

$$\alpha_- \leq x \leq \alpha_+, \quad \alpha_\pm = \sqrt{3/2}(1 \pm \sqrt{1 + 2\omega_{\text{BD}}/3})/\omega_{\text{BD}}. \tag{7.16}$$

This means that the phase space for the vacuum Brans–Dicke theory Ψ_{vac} can be defined as: $\Psi_{vac} = \{(x, \xi) : \alpha_- \leq x \leq \alpha_+\}$, where the bounds on the variable ξ—if any—are set by the concrete form of the self-interaction potential (see below).

Another useful quantity is the deceleration parameter

$$q = -1 - \frac{\dot{H}}{H^2} = -1 - 2\sqrt{6}x + 3\omega_{BD}x^2 + \frac{3(1 + \sqrt{6}x - \omega_{BD}x^2)\xi}{3 + 2\omega_{BD}}. \quad (7.17)$$

Seemingly, in accordance with the results of [9–11], without the specification of the function $\Gamma(\xi)$, there are found four dilatonic equilibrium points $P_i : (x_i, \xi_i)$, in the phase space corresponding to the dynamical system (7.14). The first one is the GR–de Sitter phase: $(0, 0) \Rightarrow x = 0 \Rightarrow \varphi = \varphi_0$, and $y^2 = 1 \Rightarrow 3H^2 = U = const.$, which corresponds to accelerated expansion $q = -1$. Given that, the eigenvalues of the linearization matrix around this point depend on the concrete form of the function $\Gamma(\xi)$,

$$\lambda_{1,2} = -\frac{3}{2}\left(1 \pm \sqrt{1 + \frac{8(1 - \Gamma)}{3(3 + 2\omega_{BD})}}\right),$$

at first sight it appears that nothing can be said about the stability of this solution until the functional form of the self-interaction potential is specified. Notice, however, that since $\xi = 0$ at this equilibrium point, this means that $U(\varphi) \propto e^\varphi$, i.e., the function Γ is completely specified: $\Gamma = 1$. As a matter of fact, the eigenvalues of the linearization matrix around $(0, 0)$ are: $\lambda_1 = -3$, $\lambda_2 = 0$. This means that $(0, 0)$ is a non-hyperbolic point.

We found, also, another de Sitter solution: $q = -1 \Rightarrow \dot{H} = 0$, which is associated with scaling of the effective kinetic and potential energies of the dilaton:

$$P : \left(\frac{1}{\sqrt{6}(1 + \omega_{BD})}, 1\right) \Rightarrow \frac{\Omega_K^{eff}}{\Omega_U} = -\frac{6 + 5\omega_{BD}}{12 + 17\omega_{BD} + 6\omega_{BD}^2},$$

$$\lambda_1 = -\frac{4 + 3\omega_{BD}}{1 + \omega_{BD}}, \quad \lambda_2 = 0, \quad (7.18)$$

where, as before, λ_1 and λ_2 are the eigenvalues of the linearization matrix around the critical point. We call this as BD–de Sitter critical point to differentiate it from the GR–de Sitter point.

In order to make clear what the difference is between both de Sitter solutions, let us note that the Friedmann constraint (7.9), evaluated at the BD–de Sitter point above, can be written as

$$e^{-\varphi}\rho_m = 3H_0^2 + \frac{6 + 5\omega_{BD}}{6(1 + \omega_{BD})^2} 3H_0^2 - U_0,$$

i.e., $e^{-\varphi}\rho_m = const.$ This means that the weakening/strengthening of the effective gravitational coupling $(G_{eff} \propto e^{-\varphi})$ is accompanied by a compensating growing/decreasing property of the energy density of matter $\rho_m \propto e^\varphi$, which leads to an exponential rate o expansion $a(t) \propto e^{H_0 t}$. This is to be contrasted with the GR–de Sitter solution: $3H_0^2 = U_0 \Rightarrow a(t) \propto e^{\sqrt{U_0/3}\,t}$, which is obtained only for vacuum, $\rho_{vac} = U_0; \rho_m = 0$.

The effective stiff-dilaton critical points ($\Omega_K^{eff} = 1$):

$$P_\pm : (\alpha_\pm, 1) \Rightarrow q_\pm = 2 + \sqrt{6}\,\alpha_\pm, \quad \lambda_1^\pm = 6\left(1 + \sqrt{2/3}\,\alpha_\pm\right), \quad \lambda_2 = 0, \quad (7.19)$$

are also found, where the $\alpha_\pm$ are defined in Eq. (7.16).

Before we said that, seemingly (in accordance with the results of the references [9–11]), the obtained critical points are quite independent of the form of the function Γ. Notice, however, that this is not true at all. For the GR–de Sitter point, for instance, $\xi = 0$, which means that $\xi = 1 - \partial_\varphi U/U = 0 \Rightarrow U \propto \exp\varphi$, forcing $\Gamma = 1$. For the remaining equilibrium points, $\xi = 1 \Rightarrow U = const$, and $\Gamma =$ undefined. This means that the equilibrium points listed above exist only for specific self–interaction potentials, but not for arbitrary potentials. Hence, contrary to the related statements in [9–11], the above results are not as general as they seem to be.

Given that the critical points obtained before were all non-hyperbolic, resulting in a lack of information on the corresponding asymptotic properties, in the following subsections we shall focus in the exponential potential: $U(\varphi) \propto \exp(k\varphi) \Rightarrow \xi = 1 - k$, which includes the particular case when $k = 1 \Rightarrow \xi = 0 \Rightarrow U(\varphi) = M^2 \exp\varphi \Rightarrow \Gamma = 1$, and the cosmological constant case $k = 0 \Rightarrow \xi = 1 \Rightarrow U = M^2$, with the hope to get more precise information on the stability properties of the corresponding equilibrium configurations.[2] These particular cases: $\xi = 0$, and $\xi = 1$, correspond to the four critical points obtained above. For completeness we shall consider also other potentials than the exponential.

7.4.1 Exponential Potential

Let us investigate the vacuum FRW–BD cosmology driven by the exponential potential $U(\varphi) \propto \exp(k\varphi)$. In this case, since $\xi = 1 - k$, is a constant, the plane-autonomous system of ODE-s (7.14), simplifies to a single autonomous ODE:

$$x' = -\left(\frac{(k + 2 + 2\omega_{BD}) x - \sqrt{\frac{2}{3}}(1 - k)}{1 + 2\omega_{BD}/3}\right)\left(1 + \sqrt{6}x - \omega_{BD}x^2\right). \quad (7.20)$$

---

[2] When the critical point under scrutiny is a non-hyperbolic point the linear analysis is not enough to get useful information on the stability of the point. In this case other tools, such as the center manifold theorem are to be invoked [26, 27].

The critical points of the latter dynamical system are ($\eta = k + 2 + 2\omega_{BD}$):

$$x_1 = \sqrt{2/3}\,(1 - k)/\eta, \ x_\pm = \alpha_\pm, \tag{7.21}$$

where the $\alpha_\pm$ are given by Eq. (7.16). Notice that, since $x_i \neq 0$ (but for $k = 1$, in which case $x_1 = 0$ and $q = -1$), there are not critical points associated with constant $\varphi = \varphi_0$. This means that the de Sitter phase with $\dot{\varphi} = 0$ ($\varphi = const$), $U(\varphi) = const.$, i.e., the one which occurs in GR and which stands at the heart of the ΛCDM model, does not arise in the general case when $k \neq 1$.

Hence, only in the particular case of the exponential potential with $k = 1$ ($\xi = 0$), which corresponds to the quadratic potential in terms of the original BD variables: $V(\phi) = M^2\phi^2$, the GR-de Sitter phase is a critical point of the dynamical system (7.20). In this case the critical points are (see Eq. (7.21)): $x_1 = 0$, $x_\pm = \alpha_\pm$. Worth noticing that $x_1 = 0$ corresponds to the GR–de Sitter solution $3H^2 = M^2 \exp\varphi_0$, meanwhile, the $x_\pm = \alpha_\pm$, correspond to the stiff-fluid (kinetic energy) dominated phase: $\Omega_K^{eff} = 1$. While in the former case the deceleration parameter $q = -1 - \dot{H}/H^2 = -1$, in the latter case it is found to be $q = 2 + \sqrt{6}\,\alpha_+ > 0$.

For small (linear) perturbations $\epsilon = \epsilon(\tau)$ around the critical points: $x = x_i + \epsilon$, $\epsilon \ll 1$, one has that, around the de Sitter solution: $\epsilon' = -3\epsilon \Rightarrow \epsilon(\tau) \propto \exp(-3\tau)$, so that it is an attractor solution. Meanwhile, around the stiff-matter solutions: $\epsilon_\pm(\tau) \propto \exp\left[3\left(2 + \sqrt{6}\,\alpha_\pm\right)\tau\right]$, so that, if assume non-negative $\omega_{BD} \geq 0$, the points $x_\pm$ are always past attractors (unstable equilibrium points) since $2 + \sqrt{6}\,\alpha_- > 0$. For negative $\omega_{BD} < 0$, these points are both past attractors whenever $\omega_{BD} < -3/2$. In this latter case, for $-3/2 < \omega_{BD} < 0$, the point x_+ is a past attractor, while the point x_- is a future attractor instead.

7.4.2 Constant Potential $U(\varphi) = M^2$

The constant potential is a particular case of the exponential when $k = 0$ ($\xi = 1$ $\Rightarrow U = const$). In this case the autonomous ODE (7.20) simplifies:

$$x' = \left[\frac{\sqrt{2/3} - 2(1 + \omega_{BD})x}{3 + 2\omega_{BD}}\right]\left(1 + \sqrt{6}x - \omega_{BD}x^2\right). \tag{7.22}$$

The critical points correspond to the following values of the independent variable x:

$$x_1 = 1/\sqrt{6}(1 + \omega_{BD}), \ x_\pm = \alpha_\pm. \tag{7.23}$$

Since, in this case,

$$\frac{\dot{H}}{H^2} = -\frac{3 - \sqrt{6}\omega_{BD}x}{3 + 2\omega_{BD}}\left[1 - \sqrt{6}(1 + \omega_{BD})x\right] \Rightarrow \left.\frac{\dot{H}}{H^2}\right|_{x_1} = 0 \Rightarrow H = H_0,$$

$$\tag{7.24}$$

the point x_1 corresponds to BD–de Sitter expansion ($q = -1$). At x_1 the effective kinetic and potential energies of the dilaton scale as

$$\frac{\Omega_K^{\text{eff}}}{\Omega_U} = -\frac{6 + 5\omega_{\text{BD}}}{12 + 17\omega_{\text{BD}} + 6\omega_{\text{BD}}^2},$$

where, as mentioned before, the minus sign is not problematic since Ω_K^{eff} is not the kinetic energy of an actual matter field. As already shown—see the paragraph starting below Eq. (7.18) and ending above Eq. (7.19)—this point does not correspond to a ΛCDM phase of the cosmic evolution, since, unlike in the GR case, in the BD theory the effective gravitational coupling $G_{\text{eff}} \propto e^{-\varphi}$ is not a constant and, besides, the de Sitter solution $H = H_0$ is obtained in the presence of ordinary matter with energy density $\rho_m \propto G_{\text{eff}}^{-1}$.

Given that under a small perturbation ($\epsilon \ll 1$) around x_1: $\epsilon(\tau) \propto \exp\left(-\frac{4+3\omega_{\text{BD}}}{1+\omega_{\text{BD}}}\tau\right)$, this is a stable equilibrium point (future attractor) if the BD parameter $\omega_{\text{BD}} \geq 0$. In case it were a negative quantity, instead, x_1 were a future attractor whenever $\omega_{\text{BD}} < -4/3$ and $-1 < \omega_{\text{BD}} < 0$.

The critical points $x_{\pm}$ in Eq. (7.23), correspond to kinetic energy–dominated phases, i.e., to stiff-matter solutions $\Omega_K^{\text{eff}} = 1$, where $q = 2 + \sqrt{6}\,\alpha_+ > 0$, and, under a small perturbation $\epsilon' = \lambda_{\pm}\epsilon$, with $\lambda_{\pm} = 6\left(1 + \sqrt{2/3}\,\alpha_{\pm}\right)$, so that, assuming non-negative $\omega_{\text{BD}} \geq 0$, the points $x_{\pm}$ are always unstable (source critical points). In the case when $\omega_{\text{BD}} < 0$ is a negative quantity, the point x_- is unstable if $\omega_{\text{BD}} < -4/3$ (the critical point x_+ is always unstable).

7.4.3 Other Potentials Than the Exponential

The concrete form of the dynamical system (7.14) depends crucially on the function $\Gamma(\xi)$. For a combination of exponentials (M^2, N^2, k and m are free constant parameters):

$$U(\varphi) = M^2 e^{k\varphi} + N^2 e^{m\varphi}, \tag{7.25}$$

which corresponds to the BD potential $V(\phi) = M^2\phi^{k+1} + N^2\phi^{m+1}$, for instance, one has:

$$x' = \left(-3x + 3\frac{x + \sqrt{2/3}}{3 + 2\omega_{\text{BD}}}\xi\right)\left(1 + \sqrt{6}x - \omega_{\text{BD}}x^2\right),$$
$$\xi' = -\sqrt{6}x\left[k + m - km - 1 - (k + m - 2)\xi - \xi^2\right]. \tag{7.26}$$

In this case (assuming that $m > k$), since

$$\xi = \frac{1 - k + (1 - m)\left(\frac{N}{M}\right)^2 e^{(m-k)\varphi}}{1 + \left(\frac{N}{M}\right)^2 e^{(m-k)\varphi}}, \tag{7.27}$$

as φ undergoes $-\infty < \varphi < \infty \Rightarrow 1 - m \le \xi \le 1 - k$. Hence, the phase space where to look for equilibrium points of the dynamical system (7.26), is the bounded compact region of the phase plane (x, ξ), given by

$$\Psi_{\text{vac}}^{\text{c,exp}} = \{(x, \xi) : \alpha_- \le x \le \alpha_+,\ 1 - m \le \xi \le 1 - k\},$$

where, we recall, $\alpha_\pm = \sqrt{3/2}(1 \pm \sqrt{1 + 2\omega_{\text{BD}}/3})/\omega_{\text{BD}}$ (see Eq. (7.16)).

In the case of the cosh and sinh-like potentials,

$$U(\varphi) = M^2 \cosh^k(\mu\varphi), \quad V(\phi) = M^2\phi\left[\cosh(\ln \phi^\mu)\right]^k, \tag{7.28}$$

respectively, one has:

$$x' = \left(-3x + 3\frac{x + \sqrt{2/3}}{3 + 2\omega_{\text{BD}}}\xi\right)\left(1 + \sqrt{6}x - \omega_{\text{BD}}x^2\right),$$

$$\xi' = -\frac{\sqrt{6}}{k}x\left(k^2\mu^2 - 1 + 2\xi - \xi^2\right). \tag{7.29}$$

The difference between the cosh and the sinh-like potentials is in the phase space where to look for critical points of (7.29). For the cosh-like potentials one has that the phase space is the following bounded and compact region of the phase plane

$$\Psi_{\text{vac}}^{\cosh} = \{(x, \xi) : \alpha_- \le x \le \alpha_+,\ 1 - k\mu \le \xi \le 1 + k\mu\},$$

while, for the sinh-like potentials the phase space is the unbounded region $\Psi_{\text{vac}}^{\sinh} = \Psi_{\text{vac}}^{\sinh-} \cup \Psi_{\text{vac}}^{\sinh+}$, where $\Psi_{\text{vac}}^{\sinh-} = \{(x, \xi) : \alpha_- \le x \le \alpha_+,\ 1 + k\mu \le \xi < \infty\}$, $\Psi_{\text{vac}}^{\sinh+} = \{(x, \xi) : \alpha_- \le x \le \alpha_+,\ -\infty < \xi \le 1 - k\mu\}$.

A distinctive feature of the dynamical systems (7.26) and (7.29), is that the GR–de Sitter critical point with $x = \xi = 0$, $P_{\text{dS}} : (0, 0) \Rightarrow H = H_0$, $\varphi = \varphi_0$, is shared by all of them. However, as it will be shown in the following sections, this does not mean that for potentials of the kinds (7.25) and (7.28) with arbitrary free parameters, the ΛCDM model is an equilibrium point of the corresponding dynamical system. As a matter of fact, only for those arrangements of the free parameters which allow that the given potential approaches to the exponential $U \propto \exp\varphi$ as an asymptote, the ΛCDM model is an equilibrium configuration of the corresponding dynamical system.

7.5 Brans–Dicke Cosmology with Matter

In the former section we have investigated the dynamical properties of the vacuum Brans–Dicke cosmology in the phase space. Here we shall explore the case when the field equations are sourced by CDM, i.e., by pressureless dust with $w_m = 0$, and for exponential potentials only, since, in this latter case, $\xi = 1 - k$, is a constant. This means that the relevant phase space will be a region of the phase plane (x, y). For this case the autonomous system of ODE-s (7.13) results in the following plane-autonomous system:

$$x' = -3x \left(1 + \sqrt{6}x - \omega_{BD}x^2\right) + \frac{3(1-k)}{3 + 2\omega_{BD}} \left(x + \sqrt{2/3}\right) y^2$$
$$+ \frac{1 + \sqrt{6}(2 + \omega_{BD})\, x}{\sqrt{6}(3 + 2\omega_{BD})}\, 3\Omega_m^{eff},$$
$$y' = y \left[3x \left(\omega_{BD}x - \frac{4-k}{\sqrt{6}}\right) + \frac{3(1-k)}{3 + 2\omega_{BD}} y^2 + \frac{2 + \omega_{BD}}{3 + 2\omega_{BD}} 3\Omega_m^{eff}\right], \quad (7.30)$$

which has physically meaningful equilibrium configurations only within the phase plane: $\Psi_{mat} = \left\{(x, y) : \alpha_- \leq x \leq \alpha_+, 0 \leq y \leq \sqrt{1 + \sqrt{6}x - \omega_{BD}x^2}\right\}$, where we have considered the facts that $\Omega_m^{eff} \geq 0$ and $y \in R^+ \cup 0$. The critical points of this dynamical system are:

$$P_{stiff}^{\pm} : \left(\frac{1 \mp \sqrt{1 + 2\omega_{BD}/3}}{\sqrt{2/3}\omega_{BD}}, 0\right), \quad \Omega_m^{eff} = 0;$$

$$P_{sc} : \left(\frac{1}{\sqrt{6}(1 + \omega_{BD})}, 0\right), \quad \Omega_m^{eff} = \frac{12 + 17\omega_{BD} + 6\omega_{BD}^2}{6(1 + \omega_{BD})^2};$$

$$P_{sc}' : \left(-\frac{\sqrt{3/2}}{k+1}, \frac{\sqrt{k + 4 + 3\omega_{BD}}}{\sqrt{2}(k+1)}\right), \quad \Omega_m^{eff} = \frac{2k^2 - 3k - 8 - 6\omega_{BD}}{2(k+1)^2};$$

$$P_* : \left(-\frac{\sqrt{2/3}(k-1)}{\eta}, \frac{\beta}{\eta}\right), \quad \Omega_m^{eff} = \frac{3(2 - k - k^2)}{\eta^2} + \frac{(7 - 2k - 5k^2)\omega_{BD}}{2\eta^2},$$

$$(7.31)$$

where, in the last critical point we have defined the new parameters:

$$\beta = \sqrt{1 + 2\omega_{BD}/3}\sqrt{8 + 6\omega_{BD} - k(k - 2)}, \quad \eta = k + 2 + 2\omega_{BD}.$$

The equilibrium points $P_{stiff}^{\pm}$ represent stiff-fluid solutions, meanwhile the remaining points represent scaling between the energy density of the dilaton and the CDM.

Let us to focus into two of the above critical points: P_{sc}' and P_*. As it was for vacuum BD cosmology, the de Sitter critical point does not arise unless $k = 1$. In this latter case $(k = 1)$, for the last equilibrium point in Eq. (7.31), one gets:

$$P_* : (0, 1) , \ q = -1 \ (H = H_0), \ \Omega_m^{\text{eff}} = 0, \ \lambda_{1,2} = -3,$$

where λ_1 and λ_2 are the eigenvalues of the linearization matrix around $P_* : (0, 1)$. This means that, for the exponential potential $U(\varphi) \propto \exp \varphi$, the GR–de Sitter solution is an attractor of the dynamical system (7.30).

For the scaling point P_{sc}', the deceleration parameter is given by $q = (k - 2)/2(k + 1)$, so that, for $k = 0$, which corresponds to the constant potential $U = U_0$, the BD–de Sitter solution is obtained $q = -1 \Rightarrow a(t) \propto e^{H_0 t}, \ e^{-\varphi} \rho_m = const.$ However, since

$$\Omega_m = \frac{2k^2 - 3k - 8 - 6\omega_{\text{BD}}}{2(k + 1)^2},$$

at $k = 0$, $\Omega_m^{\text{eff}} = -(4 + 3\omega_{\text{BD}})$, is a negative quantity, unless the Brans–Dicke coupling parameter falls into the very narrow interval $-3/2 < \omega_{\text{BD}} \leq -4/3$. Hence, for $k = 0$, but for $-1.5 < \omega_{\text{BD}} \leq -1.33$, the point P_{sc}' does not actually belong in the phase space Ψ_{mat}.

7.6 (Non)emergence of the ΛCDM Phase from the Brans–Dicke Cosmology

This problem has been generously discussed before in the reference [9]. The conclusion on the emergence of the ΛCDM cosmology starting from the Brans–Dicke theory, seems to be supported by the existence of a de Sitter phase, which was claimed to be independent on the concrete form of the self-interaction potential of the dilaton field in [9, 10], and then, in Ref. [11] the same authors somewhat corrected their previous claim. In this section we shall address this problem and we will clearly show that, in general (but for the exponential potential $U(\varphi) \propto e^{\varphi}$), the ΛCDM model is not an attractor of the FRW–BD cosmology.[3]

Before we go any further, we want to make clear that the latter statement on the non-universality of the GR–de Sitter equilibrium point, does not forbids the possible existence of exact de Sitter solutions for several choices of the self-interaction potential (see, for instance, Ref. [14]). What the statement means is that, in case such solutions are found, these would not be generic solutions, but very particular (unstable) solutions instead, which are unable to represent any sensible cosmological scenario.

It will be useful to state that a de Sitter solution arises whenever $q = -1 \Rightarrow \dot{H} = 0 \Rightarrow H = H_0 \Rightarrow a(t) \propto e^{H_0 t}$. This condition can be achieved even if $x \neq 0$. However, only when $x = 0 \Rightarrow \dot{\varphi} = 0 \Rightarrow \varphi = \varphi_0$, the de Sitter solution can lead to the ΛCDM model, where by ΛCDM model we understand the FRW cosmology

---

[3]A detailed analysis of the procedure and of the misleading conclusions in [9] can be found in the appendix at the end of this contribution.

within the frame of Einstein's GR, with a cosmological constant Λ and cold dark matter as the sources of gravity. Actually, only if $\varphi = \varphi_0$, is a constant, the action (7.2)—up to a meaningless factor of $1/2$—is transformed into the Einstein-Hilbert action plus a matter source:

$$S = \frac{1}{8\pi G_N} \int d^4x \sqrt{|g|}\, \{R - 2U_0\} + 2 \int d^4x \sqrt{|g|} \mathcal{L}_m,$$

where $e^{\varphi_0} = 1/8\pi G_N$. When $\mathcal{L}_m$ is the Lagrangian of CDM, the latter action—compare with Eq. (7.1)—is the mathematical expression of what we call as the ΛCDM cosmological model. In the remaining part of this section we shall discuss on the (non)universality of the ΛCDM equilibrium point. For this purpose, in order to find useful clues, we shall explore first the simpler situation of vacuum BD cosmology and, then, the Brans–Dicke cosmology with CDM will be explored.

7.6.1 Vacuum FRW–BD Cosmology

In this simpler situation the de Sitter phase arises only if assume an exponential potential of the form $U(\varphi) \propto \exp \varphi \Rightarrow V(\phi) = M^2 \phi^2$, which means that $\xi = 0$ and $\Gamma = 1$, are both completely specified, or if $\xi = 1$, i.e., if $U(\varphi) = M^2 \Rightarrow V(\phi) = M^2 \phi$. As a matter of fact, as shown in Sect. 7.4, for exponential potentials of the general form: $U(\varphi) = M^2 \exp(k\varphi) \Rightarrow V(\phi) = M^2 \phi^{k+1}$, with $k \neq 1$ and $k \neq 0$, the de Sitter critical point does not exist. In other words, speaking in terms of the original BD variables: but for the quadratic and the lineal monomials, $V(\phi) \propto \phi^2$ and $V(\phi) \propto \phi$, respectively—also for those potentials which approach to either ϕ^2 or ϕ at the stable point of the potential—the de Sitter solution is not an equilibrium point of the corresponding dynamical system (see the worked examples in the appendix).

Now we want to show that, even when a de Sitter solution is a critical point of (7.22), the existence of a de Sitter equilibrium point in the vacuum BD cosmology, by itself, does not warrant that the ΛCDM model is approached. As an illustration of this statement, let us choose the vacuum FRW–BD cosmology driven by a constant potential (see Sect. 7.4.2). In this case one of the equilibrium points of the dynamical system (7.22): $x_1 = 1/\sqrt{6}(1 + \omega_{BD}) \neq 0$, corresponds to the de Sitter solution since $q = -1 \Rightarrow \dot{H}/H^2 = 0 \Rightarrow H = H_0$. The tricky situation here is that, although the de Sitter solution ($H = H_0$) is a critical point of the dynamical system (7.22), the ΛCDM model is not mimicked. Actually, at x_1,

$$x = \frac{\dot{\varphi}}{\sqrt{6}\,H} = \frac{1}{\sqrt{6}(1 + \omega_{BD})} \Rightarrow \dot{\varphi} = \frac{H_0}{1 + \omega_{BD}} \Rightarrow \varphi(t) = \frac{H_0\, t}{1 + \omega_{BD}} + \varphi_0,$$

i.e., the scalar field evolves linearly with the cosmic time t. This point corresponds to BD theory and not to GR since, while in the latter the Newton's constant G_N is a true constant, in the former the effective gravitational coupling (the one measured in Cavendish-like experiments) evolves with the cosmic time: $G_{\text{eff}} = e^{-\varphi}(4 + 2\omega_{\text{BD}})/(3 + 2\omega_{\text{BD}}) \Rightarrow \dot{G}_{\text{eff}}/G_{\text{eff}} = -H_0/(1 + \omega_{\text{BD}})$. Taking the Hubble time to be $t_0 = 13.817 \times 10^9$ yr (as, for instance, in [9]), i.e., the present value of the Hubble constant $H_0 = 7.24 \times 10^{-11}$ yr^{-1}, one gets

$$\dot{G}_{\text{eff}}/G_{\text{eff}} = -(1 + \omega_{\text{BD}})^{-1} 7.24 \times 10^{-11} \text{ yr}^{-1}. \qquad (7.32)$$

As a consequence of the above, if consider cosmological constraints on the variability of the gravitational constant [28], for instance the ones in [29], which uses WMAP-5yr data combined with SDSS power spectrum data: -1.75×10^{-12} yr$^{-1} < \dot{G}/G < 1.05 \times 10^{-12}$ yr^{-1}, or the ones derived in Ref. [30], where the dependence of the abundances of the D, ^{3}He, ^{4}He, and ^{7}Li upon the variation of G was analyzed: $|\dot{G}/G| < 9 \times 10^{-13}$ yr^{-1}, from Eq. (7.32) one obtains the following bounds on the value of the BD coupling constant: $\omega_{\text{BD}} > 40.37 \mid \omega_{\text{BD}} < -69.95$, and $\omega_{\text{BD}} > 79.44 \mid \omega_{\text{BD}} < -81.44$, respectively. These constraints contradict the results of [9, 10], and are more in the spirit of the estimates of [17, 31] (see, also, Ref. [18]).

7.6.2 Other Potentials

As seen in Sect. 7.4.3, for other potentials, such as the combination of exponentials (7.25), and the cosh and sinh-like potentials (7.28), the GR–de Sitter solution is a critical point of the corresponding dynamical system. However, do not get confused: the above statement is not true for any arrangement of the free constants.

Take, for instance, the combination of exponentials. The GR–de Sitter point $x = \xi = 0$ entails that (see Eq. (7.27)), either $k = m = 1 \Rightarrow \xi = 0$, or, for $m = 1$, arbitrary k, the point is asymptotically approached as $\varphi \to \infty$ if $k < 1$. In the former case ($k = m = 1$) the combination of exponentials $U(\varphi) = M^2 e^{k\varphi} + N^2 e^{m\varphi}$, coincides with the simple exponential $U(\varphi) = (M^2 + N^2) e^{\varphi}$, while in the latter case ($m = 1$, k arbitrary), assuming that $k < 1$, the above potential tends asymptotically ($\varphi \to \infty$) to the exponential $U(\varphi) \approx N^2 e^{\varphi}$.

For the cosh and sinh-like potentials one has:

$$U(\varphi) = M^2 \left(e^{\mu\varphi} \pm e^{-\mu\varphi} \right)^k, \qquad (7.33)$$

where the "+" sign is for the cosh potential, while the "−" sign is for the sinh potential, and the 2^{-k} has been absorbed in the constant factor M^2. On the other hand, one has the following relationships:

$$\xi = 1 - k\mu \frac{e^{\mu\varphi} - e^{-\mu\varphi}}{e^{\mu\varphi} + e^{-\mu\varphi}}, \quad \xi = 1 - k\mu \frac{e^{\mu\varphi} + e^{-\mu\varphi}}{e^{\mu\varphi} - e^{-\mu\varphi}},$$

where the left-hand equation is for the cosh-like potential, while the right-hand one is for the sinh-like potential. Since at the GR–de Sitter point: $x = \xi = 0$, then, from the above equations it follows that this critical point exists, for the cosh and sinh-like potentials, only if $k\mu = 1$, in which case, the mentioned potentials (7.33) asymptotically approach to the exponential as $\varphi \to \infty$: $U(\varphi) \approx M^2 e^{k\mu\varphi} = M^2 e^{\varphi}$.

Summarizing: Only for the exponential potential $U(\varphi) \propto \exp \varphi$, or for any other potential which, as $\varphi \to \infty$, tends asymptotically to the exponential $\exp \varphi$, the GR–de Sitter solution is an attractor of the dynamical system (7.14). This is easily visualized if realize that, by the definition of the variable ξ: $\xi = 1 - \partial_\varphi U/U$. Hence, if assume $\xi = 0$, which is a necessary condition for the existence of the GR–de Sitter point, then, necessarily: $\partial_\varphi U/U = 1 \Rightarrow U(\varphi) \propto e^{\varphi}$.

7.6.3 FRW–BD Cosmology with Matter

In the case when we consider a matter source for the Brans–Dicke equations of motion, in particular CDM, the existence of a de Sitter critical point with $x = 0$ $\Rightarrow \dot{\varphi} = 0$—which means that the effective gravitational coupling is a real constant that can be made to coincide with the Newton's constant—is to be associated with the ΛCDM model.

The autonomous system of ODE-s that can be obtained out of the cosmological FRW–BD equations of motion, when these are sourced by CDM, is the one in Eq. (7.30). The critical points of this dynamical system are given in Eq. (7.31). Notice that only one of them:

$$P_* : \left(-\frac{\sqrt{2/3}(k-1)}{k+2+2\omega_{BD}}, \frac{\beta}{k+2+2\omega_{BD}} \right),$$

where $\beta = \sqrt{1 + 2\omega_{BD}/3}\sqrt{8 + 6\omega_{BD} - k(k-2)}$, can be associated with GR–de Sitter expansion, i.e., with what we know as the ΛCDM model, in the special case when $k = 1$. In this latter case $P_* : (0, 1)$. Since we are considering exponential potentials of the form $U(\varphi) \propto \exp(k\varphi)$, then the GR–de Sitter equilibrium configuration is associated, exclusively, with the potential $\partial_\varphi U/U = k = 1 \Rightarrow U(\varphi) \propto e^{\varphi}$.

Although in Sect. 7.5 we have considered only exponential potentials in FRW–BD cosmology with background dust, it is clear that the result remains the same as for the vacuum case: Only for the exponential potential $U(\varphi) \propto \exp \varphi$, or for potentials that approach asymptotically to $\exp \varphi$, the GR–de Sitter solution is an equilibrium configuration of the corresponding dynamical system.

7.7 Discussion

Why do our results differ from those in Ref. [9, 10], even when the tools used are the same? To start with we shall concentrate, specifically, in the result related with what the authors of [9, 10] call as the asymptotic value of the scalar field mass at the de Sitter

point,[4] which is the value of the BD scalar field mass computed with the help of the
following known equation [9, 10, 32]: $m^2 = 2 \left[\phi \partial_\phi^2 V(\phi) - \partial_\phi V(\phi) \right] / (3 + 2\omega_{\mathrm{BD}})$,
or, in terms of the field variables φ and $U = U(\varphi)$ in Eq. (7.4), the mass squared of
the dilaton:

$$m^2 = 2 \left(\partial_\varphi^2 U - U \right) / (3 + 2\omega_{\mathrm{BD}}), \qquad (7.34)$$

evaluated at the GR–de Sitter equilibrium point.

According to [9, 10], the asymptotic value of the scalar field mass $m|_*$, at the
de Sitter point, is given by $m|_* \approx 1.84 \times 10^{-33}/\sqrt{3 + 2\omega_{\mathrm{BD}}}$ eV. Then the authors
constrain the BD coupling parameter ω_{BD} by contrasting the above value $m|_*$ with
known estimates signaling at $m|_* \sim 10^{-22}$ eV. The obtained bound $\omega_{\mathrm{BD}} \approx -3/2 +$
10^{-22}, is a singular value and, if matter is taken into account, is very problematic
since consistency of the BD motion equations require that only traceless matter can
be coupled to the BD scalar field if $\omega_{\mathrm{BD}} = -3/2$. The above bound on ω_{BD} is to be
contrasted with our result in the Sect. 7.6.1, or with our demonstrated result that the
GR–de Sitter solution can be attained only if

$$U \propto e^\varphi \ \Rightarrow \ \partial_\varphi^2 U - U = 0 \ \Rightarrow \ m^2 = 0,$$

i.e., if the dilaton is massless.

Given that the scalar field is necessarily massless at the GR–de Sitter point, which
would be the meaning of the tiny, yet non-vanishing, asymptotic value $m|_*$ computed
in [9]? In this regard notice that the computations in [9, 10] are based on the linearized
solutions (perturbations would be more precise) around the de Sitter point, which are
valid up to linear terms in the initial conditions. Besides, in order to obtain the bound
$\omega_{\mathrm{BD}} \approx -3/2 + 10^{-22}$ on the BD coupling parameter, the authors of [9, 10] assumed
what they called as "special initial conditions". Then, the mass of the BD scalar
field computed in the mentioned references is the mass of the field at the linearized
(perturbed) solutions around the de Sitter point, but not at the point itself where the
dilaton is actually massless as we have shown.

The next question would be: which is the actual meaning of the linearized solutions? The linearized solutions correspond to points in the phase space which are
very close to the stable equilibrium point—the de Sitter critical point in the present
case—so that the linear approximation takes place:

$$x(\tau) \approx x_c + \epsilon_x(\tau), \ y(\tau) \approx y_c + \epsilon_y(\tau), \qquad (7.35)$$

where x_c, y_c are the coordinates of the given equilibrium point, and the perturbations
$\epsilon_x \sim \epsilon_y \ll 1$, are very small. These solutions can be viewed as small deformations
of the stable GR–de Sitter solution. Just as an illustration, let us consider the FRW–
BD theory driven by the exponential potential $U(\varphi) \propto \exp \varphi$, in a background of

---

[4]A detailed analysis with specific criticism of the work in [9, 10] is included in the appendix.

CDM (see Sect. 7.5). The small perturbations around the de Sitter point $P : (0, 1)$, very quickly tend to vanish, restoring the system into the stable equilibrium state: $\epsilon_x \sim \epsilon_y \propto \exp(-3\tau)$, where we have taken into account that the eigenvalues of the linearization matrix at $(0, 1)$, coincide: $\lambda_1 = \lambda_2 = -3$. Then, the linearized solutions around the GR–de Sitter point look like[5]:

$$x(\tau) \approx A\, e^{-3\tau} \; \Rightarrow \; \varphi(a) \approx -\sqrt{2/3}A/a^3 + \varphi_0,$$
$$y(\tau) \approx 1 + B\, e^{-3\tau} \; \Rightarrow \; H^2(a) \approx M^2\, e^{\varphi_0 - \sqrt{2/3}A/a^3}/3(1 + B/a^3)^2.$$

These will eventually (perhaps very quickly) decay into the stable de Sitter solution: $x(\tau) = 0 \; \Rightarrow \; \varphi = \varphi_0, y(\tau) = 1 \; \Rightarrow \; H = H_0 = M\, e^{\varphi_0/2}/\sqrt{3}$.

We can say that the linearized solutions have a finite life-time in the sense that, within a finite amount of "time" τ, these decay into the stable solution. It is then clear that the mass of the dilaton computed at linearized solutions would be highly dependent on the assumed initial conditions, in contrast to the mass of the dilaton at the de Sitter point. Actually, while the mass of the field at perturbed (unstable) solutions, depends on the way the perturbations are generated, at the GR–de Sitter attractor, being a stable equilibrium configuration, the field is massless regardless of the initial conditions. Hence, making cosmological predictions on the base of perturbed solutions around equilibrium points, is meaningless due to the loss of predictability which is associated with the strong dependence on the initial conditions. The only useful information the dynamical systems theory allow to extract from the given cosmological dynamical system is encoded in the equilibrium points themselves, but not in the (linear) perturbations around them. The latter serve only as probes to test the stability of the given critical point.

In addition we have to say that estimates on the parameters of the BD theory, such as the BD coupling constant ω_{BD}, made on the basis of linearized solutions in the neighborhood of the GR–de Sitter solution, are not reliable since these linearized solutions may not depart too much from general relativity (plus a cosmological constant) which is obtained from the BD theory in the formal limit $\omega_{BD} \to \infty$. The same reasoning applies to the computation of other derived quantities such as the ratio $\dot{G}/G$ (see Sect. 7.6.1).

We find no reason to believe that we are living in one such perturbed solution and not in the equilibrium configuration itself. Besides, if one wants to avoid the cosmic coincidence problem, an equilibrium configuration which attracts the cosmic history into a GR–de Sitter stage, is all what one needs. Making definitive conclusions about the entire cosmic history based in computations made at a perturbed solution is potentially misleading.

We want to underline that the above discussion was based on the assumption that the computations made in [9] are correct. However, as I show in the appendix, these

---

[5]Here A and B are integration constants, which depend linearly on the initial conditions $x(0)$, $y(0)$, and on other free parameters such as ω_{BD}.

computations are incorrect in general. Hence, after all, it may be that the discrepancies of the present results and the results claimed in [9] (see also [10, 11]) is due to the incorrect procedure performed in those references.

7.8   Conclusion

In the present paper I have explored the asymptotic properties of FRW–BD cosmo-logical models (7.7), by means of the tools of the dynamical systems theory [9–11, 20–26, 33]. I have shown that, in spite of known results [9–11], the GR–de Sitter phase is not an universal attractor of the BD theory.[6] Only for the specific exponential potential $U(\varphi) \propto \exp \varphi$, which, in terms of the original BD field ϕ, amounts to the quadratic monomial $V(\phi) \propto \phi^2$, or for potentials which asymptotically approach to $\exp \varphi$ (ϕ^2), the GR–de Sitter phase is a stable critical point, i.e., a future attractor in the phase space. I have shown, also, that at the GR–de Sitter critical point the effective mass of the dilaton m^2 in Eq. (7.34), vanishes.

We have learned that physically meaningful conclusions can be based only on computations performed at the equilibrium configurations as, for instance, at the stable GR–de Sitter critical point. On the contrary, the results based on computations made at perturbed solutions are highly dependent on the initial conditions chosen and, hence, useless to make physically meaningful predictions.

In particular, the computations performed at the stable BD–de sitter critical point yield to bounds on the value of the BD coupling parameter $\omega_{BD} > 40.37 \mid \omega_{BD} < -69.95$, or $\omega_{BD} > 79.44 \mid \omega_{BD} < -81.44$, depending on the observational data assumed, which are consistent with the estimates of [17, 18, 31]. These results are to be contrasted with the ones in Ref. [9]: $\omega_{BD} = -3/2$, or in [10]: $\omega_{BD} \approx -1$, which were based on computations made at perturbed solutions around the GR–de Sitter attractor.

Acknowledgments   I want to thank T Asselmeyer-Maluga for inviting me to join this initiative to celebrate the 80th birthday of one of the greatest minds of the XX century, our dear fiend Carl H. Brans, to whom I am profoundly indebted. I want to acknowledge, also, J D Barrow, S D Odintsov and A Alho, for pointing to me several indispensable bibliographic references. Thanks are due to CONACyT of México for support of this research. I am grateful to SNI-CONACyT for continuous support of my research activity.

---

[6]This result has been independently confirmed in [27] by means of the center manifold theorem.

Appendix: Incorrectness of the Claim on the Universal Character of the Λ-CDM Attractor in the Brans-Dicke Cosmology

Here we want to show in details the source of the incorrect claim on the global character of the de Sitter attractor in the BD theory, exposed in Ref. [9] (see also [10, 11]). Here we assume the cosmological equations written in terms of the BD scalar field ϕ (see Eq. (7.3)), as well as the same definition of the phase space variables used in [9]:

$$
x := \frac{\dot{\phi}}{H\phi}, \quad y := \sqrt{\frac{V(\phi)}{3\phi}}\frac{1}{H}, \quad \lambda := -\phi\frac{\partial_\phi V(\phi)}{V(\phi)}. \tag{7.36}
$$

The obtained autonomous system of ODE is given by (Eq. (15) in Ref. [9]):

$$
x' = -x\left(3 + x + \frac{\dot{H}}{H^2}\right) + \frac{6y^2(2+\lambda)}{3 + 2\omega_{BD}} + \frac{3(1 - 3w_m)}{3 + 2\omega_{BD}}\left(1 + x - \frac{\omega_{BD}}{6}x^2 - y^2\right),
$$

$$
y' = -\frac{1}{2}y\left[x(1 + \lambda) + 2\frac{\dot{H}}{H^2}\right], \quad \lambda' = x\lambda[1 - \lambda(\Gamma - 1)], \tag{7.37}
$$

where w_m is the barotropic parameter of the matter fluid, the function $\Gamma = V\partial_\phi^2 V/ (\partial_\phi V)^2$, encodes the information of the potential $V = V(\phi)$, and it is assumed to be a function of the parameter λ: $\Gamma = \Gamma(\lambda)$. Notice that this function Γ does not coincide with the similar function defined in Eq. (7.12). Another useful expression is (Eq. (14) in [9]):

$$
\frac{\dot{H}}{H^2} = 2x\left(1 - \frac{\omega_{BD}}{4}x\right) - \frac{3y^2(2+\lambda)}{3 + 2\omega_{BD}} - \frac{3[2 + \omega_{BD}(1 + w_m)]}{3 + 2\omega_{BD}}\left(1 + x - \frac{\omega_{BD}}{6}x^2 - y^2\right). \tag{7.38}
$$

The de Sitter Point

The main claim of the authors of Ref. [9] on the global character of the de Sitter attractor is based on the argument that the general relativity de Sitter solution, which corresponds to the critical point $(x, y, \lambda) = (0, \pm1, -2)$ of the dynamical system (7.37), is obtained independent of the specific functional form of $\Gamma = \Gamma(\lambda)$. This is seemingly the case, since by substituting $x = 0$, $y = \pm1$ and $\lambda = -2$ in the autonomous system of ODE (7.37), one gets $x' = 0$, $y' = 0$ and $\lambda' = 0$ without making any assumptions on the functional form of $\Gamma(\lambda)$.

Notice, however, that it is mandatory that $\lambda = -2$ for the existence of the GR–de Sitter solution. Actually, as clearly seen from (7.38), when evaluated at $x = 0$, $y = \pm 1$ (the de Sitter condition) one gets:

$$\left.\frac{\dot{H}}{H}\right|_{(0,\pm 1,\lambda)} = -\frac{3(2+\lambda)}{3+2\omega_{\mathrm{BD}}} \Rightarrow \dot{H} = 0 \Leftrightarrow \lambda = -2.$$

Having established this fact (as properly done in [9]), the next step is to notice that, taking into account the definition of the variable λ in (7.36) (Eq. (12) of Ref. [9]), then, at the de Sitter attractor where $\lambda = -2$: $\partial_\phi V(\phi)/V(\phi) = 2/\phi \Rightarrow V(\phi) \propto \phi^2$. Then, the existence of the general relativity de Sitter attractor requires that $\lambda = -2$ and this, in turn, necessarily entails that $V(\phi) \propto \phi^2$. As a straightforward consequence of this, since $V(\phi) = V_0\,\phi^2$, then one invariably gets that $\Gamma = V\partial_\phi^2 V/(\partial_\phi V)^2 = 1/2$. This means that, unlike the claim in [9], at the de Sitter point ($\lambda = -2$), the function $\Gamma(\lambda)$ is completely specified: $\Gamma(-2) = 1/2$.

Summarizing this very simple and straightforward argument: the GR–de Sitter attractor necessarily entails $\lambda = -2 \Leftrightarrow V(\phi) \propto \phi^2 \Leftrightarrow \Gamma(-2) = 1/2$, and there is not other way around. This is precisely our main claim in the present paper: the GR–de Sitter attractor exists exclusively for the quadratic potential or for potentials that approach to the quadratic one at the stable point, i.e., it is mandatory that $\Gamma(-2) = 1/2$.

In order to illustrate our arguments, let us choose a pair of examples [9].

Example 1

Let us start by the potential

$$V(\phi) = V_0 \left(\phi^2 - v^2\right)^2. \tag{7.39}$$

The extrema of this potential are at $\phi^2 = v^2$ (minimums) and at $\phi = 0$ (maximum). Since

$$\lambda = -\phi\frac{\partial_\phi V}{V} = -\frac{4\phi^2}{\phi^2 - v^2} \Rightarrow \frac{\phi^2}{v^2} = \frac{\lambda}{\lambda + 4},$$

then $\Gamma = \Gamma(\lambda) = V\partial_\phi^2 V/(\partial_\phi V)^2 = \left(3 - v^2/\phi^2\right)/4 = (\lambda - 2)/2\lambda$. The relevant values are precisely the extrema $\phi^2 = v^2$, and $\phi = 0$. At these values we have $\lambda = \infty$ (undefined), and $\lambda = 0$, respectively. Besides $\Gamma(\infty) = 1/2$, and $\Gamma(0) = \infty$ (undefined).

The attractor solutions for this potential are at $\phi = \pm v$, i.e., at $\lambda = \infty$ (undefined). Notice that the value $\lambda = -2$, which is a necessary requirement for the existence of the de Sitter critical point, is correlated with an unphysical solution. Actually, at $\lambda = -2$, $\phi^2 = -v^2 \Rightarrow \phi = \pm iv$, so that the scalar field is pure imaginary and, since the effective gravitational coupling $G \propto \phi^{-1} = \pm iv^{-1}$ is unphysical, the de Sitter solution is unphysical for this potential. In consequence it is not a critical point (not even a point) in the corresponding phase space. This was expected since, for the potential (7.39), $\Gamma(-2) = 1$, i.e., $\Gamma(-2) \neq 1/2$.

Example 2

The other potential included in the stability analysis of the de Sitter point in [9] is the following:

$$V(\phi) = \frac{1}{2} m^2 \phi^2 + \frac{\alpha}{4} \phi^4. \tag{7.40}$$

From the particles physics perspective the physically interesting case is for negative $m^2 < 0$ (as for instance in the Higgs mechanism). This case ($m^2 < 0$) is another nice example of a potential that does not approach to the quadratic potential at the stable point and, hence, does not drive a stable de Sitter phase. For this potential

$$\lambda = -4 \left(\frac{m^2 + \alpha \phi^2}{2m^2 + \alpha \phi^2} \right) \Rightarrow \frac{\alpha \phi^2}{m^2} = -\frac{2(\lambda + 2)}{\lambda + 4},$$

$$\Gamma(\lambda) = \frac{\left(2 + \frac{\alpha \phi^2}{m^2} \right) \left(1 + 3 \frac{\alpha \phi^2}{m^2} \right)}{\left(2 + 4 \frac{\alpha \phi^2}{m^2} \right)^2} = -\frac{4(8 + 5\lambda)}{(8 + 6\lambda)^2}.$$

Since $m^2 < 0$, then $\alpha \phi^2 = -m^2$ are minimums, while $\phi = 0$ is a maximum. Hence, the attractor solutions in this case are at $\lambda = 0$, while at the maximum, which corresponds to an unstable equilibrium point, one has: $\lambda = -2$. I.e., the de Sitter solution in this case exists but it is an unstable point, not an attractor as assumed in [9]. Besides, for this potential $\Gamma(-2) = 1/2$ at $\phi = 0$. This is because near of the maximum $\phi \sim 0$, the potential behaves like the (negative of the) quadratic potential. In this case despite that $\Gamma(-2) = 1/2$, the quadratic potential is not approached at the stable point(s) $\phi^2 = -m^2/\alpha$. Hence, $\Gamma(-2) = 1/2$ is a necessary condition for the existence of the GR–de Sitter attractor, but it is not a sufficient condition. It is required, additionally, that the latter condition is satisfied at the stable point of the potential.

If, on the contrary, choose the less physically motivated case when in (7.40), $m^2 > 0$ is a positive quantity, then $V(\phi) = m^2 \phi^2/2 + \alpha \phi^4/4$ has only one extremum: a minimum at $\phi = 0$, where $\lambda = -2$ and $\Gamma(-2) = 1/2$. As explained above, this happens because in the neighborhood of the minimum at $\phi = 0$, this potential behaves like $\propto \phi^2$. In this case ($m^2 > 0$), the potential approaches to the quadratic one at the stable point and enters the category of potentials for which the ΛCDM model is an attractor of the FRW-BD theory, i.e., it serves as an illustration of our main result.

Stability Conditions for the de Sitter State

As previously shown, since the present method assumes that $\Gamma = \Gamma(\lambda)$, then for the existence of the de Sitter attractor, necessarily, $\Gamma(-2) = 1/2$ at the stable point of the potential: $\lambda = -2 \Leftrightarrow V(\phi) \propto \phi^2 \Leftrightarrow \Gamma(-2) = 1/2$.

Although the authors of [9] seem to perform a kind of general analysis of the stability conditions of the GR–de Sitter point by introducing the parameter

$$3\delta/8 = \partial\lambda'/\partial x|_* = \lambda_* [1 - \lambda_* (\Gamma(\lambda_*) - 1)] = \delta(\lambda_*),$$

as a matter of fact, as shown, at the de Sitter attractor $\lambda_* = -2$ and $\Gamma(-2) = 1/2$, so that necessarily: $\delta = \delta(-2) = -16[1 + 2(\Gamma(-2) - 1)]/3 = 0$. This is precisely the case which in [9] (first paragraph below Eq. (18)) is called as "degenerated". In more "dynamical systems oriented" terms, what happens in this case is that the GR–de Sitter solution is a non-hyperbolic critical point and the Hartman-Grobman theorem can not be applied, so that the standard tools of the linear analysis are useless and one is obliged to resource to other tools such as the center manifold theorem, etc., as done, for instance, in [27]. Hence the analysis of the stability in [9] after Eq. (18) is incorrect and its claimed generality through $\delta \neq 0$ is spurious. This is where our analysis in the present paper starts departing from the one in [9].

References

1. C. Brans, R.H. Dicke, Phys. Rev. **124**, 925–935 (1961)
2. Th.P. Sotiriou, V. Faraoni, Rev. Mod. Phys. **82**, 451–497 (2010), arXiv:0805.1726
3. Y. Fujii, K.-I. Maeda, *The Scalar-Tensor Theory of Gravitation* (Cambridge University Press, UK, 2003)
4. V. Faraoni, *Cosmology in Scalar-Tensor Gravity* (Kluwer Academic Publishers, The Netherlands, 2004)
5. J. Khoury, A. Weltman, Phys. Rev. Lett. **93**, 171104 (2004), arXiv:astro-ph/0309300; Phys. Rev. D **69**, 044026 (2004), arXiv:astro-ph/0309411
6. S. Sen, T.R. Seshadri, Int. J. Mod. Phys. D **12**, 445–460 (2003), arXiv:gr-qc/0007079; D.F. Torres, Phys. Rev. D **66**, 043522 (2002), arXiv:astro-ph/0204504; N. Banerjee, D. Pavon, Class. Quant. Grav. **18**, 593 (2001), arXiv:gr-qc/0012098
7. P.A.R. Ade et al., Planck Collaboration. Astron. Astrophys. **571**, A23 (2014), arXiv:1303.5083
8. P.J.E. Peebles, B. Ratra, Rev. Mod. Phys. **75**, 559–606 (2003), arXiv:astro-ph/0207347; V. Sahni, A.A. Starobinsky. Int. J. Mod. Phys. D **9**, 373–444 (2000), arXiv:astro-ph/9904398
9. O. Hrycyna, M. Szydlowski, Phys. Rev. D **88**(6), 064018 (2013), arXiv:1304.3300
10. O. Hrycyna, M. Kamionka, M. Szydlowski, Phys. Rev. D **90**(12), 124040 (2014), arXiv:1404.7112
11. O. Hrycyna, M. Szydlowski, JCAP **1312**, 016 (2013), arXiv:1310.1961
12. J.D. Barrow, K. Maeda, Nucl. Phys. B **341**, 294–308 (1990)
13. J.D. Barrow, Phys. Rev. D **51**, 2729–2732 (1995)
14. E. Elizalde, S. Nojiri, S.D. Odintsov, D. Saez-Gomez, V. Faraoni, Phys. Rev. D **77**, 106005 (2008), arXiv:0803.1311
15. C.M. Will, Living Rev. Rel. **9**, 3 (2006), arXiv:gr-qc/0510072; B. Bertotti, L. Iess, P. Tortora, Nature **425**, 374 (2003)

16. I. Quiros, R. García-Salcedo, T. Gonzalez, F.A. Horta-Rangel, Phys. Rev. D **92**(4), 044055 (2015), arXiv:1506.05420
17. V. Acquaviva, C. Baccigalupi, S.M. Leach, A.R. Liddle, F. Perrotta, Phys. Rev. D **71**, 104025 (2005), arXiv:astro-ph/0412052
18. R. Nagata, T. Chiba, N. Sugiyama, Phys. Rev. D **69**, 083512 (2004), arXiv:astro-ph/0311274
19. J.E. Lidsey, D. Wands, E.J. Copeland, Phys. Rept. **337**, 343–492 (2000), arXiv:hep-th/9909061
20. E.J. Copeland, A.R. Liddle, D. Wands, Phys. Rev. D **57**, 4686 (1998), arXiv:gr-qc/9711068
21. A.A. Coley, arXiv:gr-qc/9910074
22. E.J. Copeland, M. Sami, S. Tsujikawa, Int. J. Mod. Phys. D **15**, 1753–1936 (2006), arXiv:hep-th/0603057
23. L.A. Urena-Lopez, JCAP **1203**, 035 (2012), arXiv:1108.4712
24. C.G. Boehmer, N. Chan, arXiv:1409.5585
25. R. García-Salcedo, T. Gonzalez, F.A. Horta-Rangel, I. Quiros, D. Sanchez-Guzmán, Eur. J. Phys. **36**(2), 025008 (2015), arXiv:1501.04851
26. V.I. Arnold, Ordinary Differential Equations, translated from Russian and edited by R.A. Silverman (The MIT Press, Cambridge, Massachusets and London, England, 1973); D.K. Arrowsmith, C.M. Place, Introduction to Dynamical Systems (Cambridge University Press, UK, 1990); L. Perko, Differential Equations and Dynamical Systems (Springer, USA, 2001)
27. A. Cid, G. Leon, Y. Leyva, arXiv:1506.00186
28. J.-P. Uzan, Rev. Mod. Phys. **75**, 403 (2003), arXiv:hep-ph/0205340
29. F. Wu, X. Chen, Phys. Rev. D **82**, 083003 (2010), arXiv:0903.0385
30. F.S. Accetta, L.M. Krauss, P. Romanelli, Phys. Lett. B **248**, 146 (1990)
31. X. Chen, M. Kamionkowski, Phys. Rev. D **60**, 104036 (1999), arXiv:astro-ph/9905368
32. V. Faraoni, Class. Quant. Grav. **26**, 145014 (2009), arXiv:0906.1901
33. J. Wainwright, G.F.R. Ellis, Dynamical Systems in Cosmology (Cambridge University Press, Cambridge, 1997); A.A. Coley, Dynamical Systems and Cosmology (Kluwer Academic Publishers, Dordrecht Boston London, 2003)

Chapter 8
New Setting for Spontaneous Gauge Symmetry Breaking?

Roman Jackiw and So-Young Pi

Abstract Over half century ago Carl Brans participated in the construction of a viable deformation of the Einstein gravity theory. The suggestion involves expanding the tensor-based theory by a scalar field. But experimental support has not materialized. Nevertheless the model continues to generate interest and new research. The reasons for the current activity is described in this essay, which is dedicated to Carl Brans on his eightieth birthday.

Brans and Dicke (also P. Jordan) [1] proposed a tensor/scalar ($g_{\mu\nu}/\varphi$) generalization of Einstein's general relativistic tensor gravity model. In their generalization a scalar field φ is coupled to the Ricci scalar, and further dynamics is posited for φ. The dynamical equations follow from a generalized Einstein-Hilbert Lagrangian

$$I_\alpha = -\int \mathcal{L}_\alpha \qquad (8.1)$$

$$\mathcal{L}_\alpha = \sqrt{-g}\left[\frac{\alpha}{12}\,\varphi^2 R + \frac{1}{2}\,g^{\mu\nu}\,\partial_\mu\varphi\,\partial_\nu\varphi + \lambda\varphi^4\right]$$

The parameter α measures the strength of the $R - \varphi^2$ interaction and suggests a dynamical origin for the gravitational constant $G \propto 1/\varphi^2$. (A self coupling of strength λ may also be included, but it plays no role in our present discussion.)

While the model is attractive in that it presents a very explicit modification of the Einstein theory, it fails to agree with the experimental values for the classic solar system tests of gravity theory. Nevertheless these days interest has revived in the Brans-Dicke model at $\alpha = 1 : \mathcal{L}_W = \mathcal{L}_{\alpha=1}$.

R. Jackiw (✉)
Center for Theoretical Physics, Department of Physics,
Massachusetts Institute of Technology, 77 Massachusetts Avenue,
Cambridge 02139, MA, USA
e-mail: jackiw@mit.edu

S.-Y. Pi
Department of Physics, Boston University, Boston, MA 02215, USA
e-mail: soyoung@bu.edu

© Springer International Publishing Switzerland 2016   159
T. Asselmeyer-Maluga (ed.), *At the Frontier of Spacetime*,
Fundamental Theories of Physics 183, DOI 10.1007/978-3-319-31299-6_8

$$\mathcal{L}_W = \sqrt{-g}\left\{\frac{1}{12}\,\varphi^2 R + \frac{1}{2}\,g^{\mu\nu}\,\partial_\mu\varphi\,\partial_\nu\varphi + \lambda\varphi^4\right\} \tag{8.2}$$

The $\alpha = 1$ model possesses Weyl invariance, i.e invariance against rescaling the dynamical variables by a local space-time transformation.

$$g^{\mu\nu} \rightarrow e^{2\theta}g^{\mu\nu} \tag{8.3a}$$

$$\varphi \rightarrow e^\theta\varphi \tag{8.3b}$$

Here θ is an arbitrary function on space time. The reasons for the contemporary interest in $\mathcal{L}_W$ are the following.

These days physicists are satisfied by the success that has been achieved in understanding and unifying all forces save gravity. This has been accomplished with the help of spontaneous breaking of local internal symmetries.

With the desire to include gravity in this framework, and in keeping with its presumed geometric nature, various people have suggested studying Weyl invariant dynamics, with the hope that scaling will help understand short distance phenomena. Additionally some are tantalized by the long-standing desire to extend conventional space-time symmetries to include local conformal (Weyl) symmetry [2].

$\mathcal{L}_W$ seems to bring closer the above goals: An operative Weyl symmetry appears to host a local gauge symmetry, which can be broken by choosing specific values for φ. Indeed $\varphi = 1$ renders $\mathcal{L}_W$ equal to the Einstein-Hilbert Lagrangian. In this framework Einstein theory is merely the "unitary gauge" version of $\mathcal{L}_W$.

Certainly such ideas are provocative and worthy of further examination and possible development. However, a critical viewpoint leads to the following questions and observations.

1. No gauge potential (connection) is present; in what sense does I_W define a "gauge theory"?
2. There is no dynamical/energetic reason for choosing the "unitary gauge" $\varphi = 1$. (In familiar spontaneous breaking, asymmetric solutions are selected by lowest energy considerations.)
3. By inverting the order of presentation, we recognize that φ is a spurion variable: upon replacing $g_{\mu\nu}$ in the Einstein-Hilbert action by $g_{\mu\nu}\varphi^2$, one arrives at the Weyl action [3].

$$I_{\text{Einstein-Hilbert}}\Big|_{g_{\mu\nu}\,\rightarrow\, g_{\mu\nu}\,\varphi^2} \rightarrow I_W$$

4. The Weyl symmetry current vanishes identically. The computation is performed according to Noether's first theorem (applicable when transformation parameters are constant) and her second theorem (applicable when transformation parameters depend on space-time coordinates). The former is a special case of the latter; both give the same result: no current. With no current, there is no charge and no symmetry generator.

The fact that the Weyl current vanishes cannot be attributed to the locality of the symmetry transformation parameter $\theta(x)$. An instructive example is electrodynamics, where $\delta A_\mu = \partial_\mu \theta$ and $\delta \Psi = -i\theta \Psi$ for a charged field Ψ. The current is non-vanishing and is identically conserved, i.e. it is a superpotential.

$$J^\mu = \partial_\nu (F^{\mu\nu} \theta) \tag{8.4}$$

(This is the Noether current for gauge symmetry, not the source current J^μ_{EM} that appears in the Maxwell equations.) While the dependence on an inhomogenous θ may make J^μ unphysical, the global limit produces a sensible result.

$$J^\mu = \partial_\nu (F^{\mu\nu}) \theta = J^\mu_{EM} \theta \tag{8.5}$$

In the Weyl case, setting the parameter θ to a constant leaves a global symmetry. Yet the current still vanishes.

Evidently the vanishing of the Weyl symmetry current, both local and *a forteriori* also global, reflects the particularly peculiar role of the Weyl "symmetry" in the examined models [4].

As yet we do not know how to assess the significance of the above observations for a physics program based on Weyl symmetry. Clearly it is interesting to explore the similarities to and differences from the analogous structures in conventional gauge theory. We conclude with two observations on the model.

By an alternate "gauge choice" we can set $\sqrt{-g}$ to a constant. Evidently, a unimodular scalar/tensor theory is "gauge equivalent" to the Einstein-Hilbert model.

The kinetic term for φ is not Weyl invariant and its non-invariance is compensated by the non-minimal interaction with R. Alternatively we may dispense with the non-minimal interaction and achieve invariance by introducing a gauge field W_μ, which transform as $W_\mu \to W_\mu - \partial_\mu \theta$. One verifies invariance of

$$\int d^4x \sqrt{-g} \left(\frac{1}{2} g^{\mu\nu} D_\mu \varphi \, D_\nu \varphi \right)$$
$$D_\mu \equiv \partial_\mu + W_\mu. \tag{8.6}$$

Expanding and integrating by parts shows that (8.6) is equivalent to (8.2) provided R is given by the formula [5, 6]

$$\frac{1}{12} R = D^\mu W_\mu + \frac{g^{\mu\nu}}{2} W_\mu W_\nu. \tag{8.7}$$

While these observations are provocative, they have not produced any useful insights. Indeed thus far the only established role for I_W is to generate the traceless new improved energy momentum tensor $\theta^{CCJ}_{\mu\nu}$ [7]

$$\theta_{\mu\nu}^{CCJ} = \frac{2}{\sqrt{-g}} \left. \frac{\delta I_W}{\delta g^{\mu\nu}} \right|_{g_{\mu\nu} \to \delta_{\mu\nu}} \tag{8.8}$$

Acknowledgments This research was supported in part by U.S. Department of Energy, Grant No. DE-SC0010025.

References

1. C. Brans, R.H. Dicke, Phys. Rev. **124**, 925 (1961)
2. G. 't Hooft, Local Conformal Symmetry: The Missing Symmetry Component for Space and Time. First prize in GRG Essay contest 2015. arXiv:1410.6675
3. J. Anderson, Phys. Rev. D **3**, 1689 (1971)
4. R. Jackiw, S.-Y. Pi, Phys. Rev. D **91**, 667501 (2015)
5. A. Iorio, L. O'Raifeartaigh, I. Sachs, C. Wiesendanger, Nucl. Phys. B **495**, 433 (1997)
6. R. Jackiw, Theor. Math. Phys. **148**, 941 (2000)
7. C. Callan, S. Coleman, R. Jackiw, Ann. Phys. **59**, 42 (1970)

Chapter 9
The Brans-Dicke Theory and Its Experimental Tests

Martin P. McHugh

Abstract Carl Brans submitted his doctoral dissertation to the Princeton committee in May of 1961. By November, the Brans-Dicke theory was disseminated widely with the publication of a 10-page paper in Physical Review. An extension of Einstein's general relativity, it generated great interest and was the subject of enormous effort to test its implications experimentally. We examine the history and impact of the experimental tests of this theory.

9.1 Introduction

Carl Brans is a theorist. This is clear to all who know him and his work. He can be self-effacing yet almost boastful in saying he is not interested in the practical and that any of his early tentative steps as a student in the laboratory were disastrous. He tells a story of when he was a lab assistant at Loyola University New Orleans (where he was an undergraduate, and later a faculty member) and he nearly killed himself with a high voltage cosmic ray detector. He tells another of an incident at Princeton when he was a graduate student taking an oral exam. Eugene Wigner had posed a problem (something to do with cosmic rays in the earth's magnetic field) and Carl was working through it at the board when Wigner suggested that some numerical estimates should be made. Carl informed him that he did not know one of the numbers needed—the radius of the earth in some suitable units—and Wigner was aghast. Understandably a bit uncomfortable, Carl explained that he had never felt the need to learn such things. The other examiners laughed, and one of them bailed Carl out. He passed both the oral and the written qualifying exam in his first year at Princeton. So it is interesting that Carl's name is forever tied to someone who is known as the consummate experimentalist—someone who reveled in designing and

M.P. McHugh (✉)
Loyola University, 6363 St. Charles Avenue, Box 92, New Orleans 70118, USA
e-mail: mmchugh@loyno.edu

© Springer International Publishing Switzerland 2016   163
T. Asselmeyer-Maluga (ed.), *At the Frontier of Spacetime*,
Fundamental Theories of Physics 183, DOI 10.1007/978-3-319-31299-6_9

building experiments—Robert Dicke. One must note however that it is Dicke who 'crosses over' for the association because it is the Brans-Dicke *theory* for which they both are famous. The amount of experimental effort behind tests of the Brans-Dicke theory is truly stupendous (and I became even more convinced of this while doing research for this short article). The direct and indirect spinoff work resulting from this effort could easily fill an entire volume. But I make no effort to be complete here, and will merely point out a couple of highlights that strike me as most interesting and relevant. I should also note that the subject has the obvious connection to my interest in the history of Robert Dicke's work.

So it is my tribute to Carl on the occasion of his 80th birthday, to marvel at the truly astounding body of experimental work that his theory has inspired (in spite of himself).

9.2 Background

Robert Dicke began working on gravitation about the time of a sabbatical leave he took from Princeton University beginning in the fall of 1954. During this year at Harvard, the motivation for his interest in gravity was, at least in part, his perception of a lack of strong experimental support for General Relativity (GR). He was also influenced by ideas that are sometimes called the 'Dirac large numbers'. These ideas suggested that the coincidence in the size of some large dimensionless physical quantities might point to possible underlying—but as yet unknown—connections. One of the hypotheses suggested a link between the strength of gravity and the age of the universe. If true it would mean that the strength of the gravitational interaction (embodied in Newton's G) would change with time. Other major influences on Dicke included the ideas of Mach. One version of Mach's principle can be summarized to say that the inertial forces do not arise from motion relative to absolute space, but rather due to distant matter in the universe [1]. These ideas (Mach and Dirac) formed the background for Dicke's discussions with Carl, who came to Princeton as a graduate student in 1957. The theory that Carl would formulate was born from these discussions.

For the experiments we will begin with the heart of the Brans-Dicke theory [2], the scalar field introduced in addition to the tensor of GR. In GR the strength of the gravitational interaction is determined, as it is in Newtonian theory, by the constant G. In Brans-Dicke the strength of gravity is tied to the scalar field. This leads to—as Carl likes to point out—the paradoxically named 'variable gravitational constant', or variable G. Geophysical considerations [3], tracking of satellite and planetary orbits [4, 5], etc. were all tests considered by Dicke and others to look for this varying G. An early example of a proposed direct test in response to Brans-Dicke was to look at the dynamics of massive stars in galactic clusters under the influence of a variable G [6].

9.3 The Three Classic Tests

The initial Brans-Dicke paper presented calculations for what are sometimes known as the three 'classic' experimental tests of GR. For the first, the gravitational redshift, there is no difference between Brans-Dicke and GR. In fact, it has been shown that the gravitational redshift does not form a test capable of distinguishing between relativistic gravitational theories [7].

For the second, the deflection of light in a gravitational field, there is a difference between Brans-Dicke and GR. To first order the deflection angle $\delta\theta$ differs from the GR value by a factor $\frac{3+2\omega}{4+2\omega}$ where ω is a dimensionless constant whose inverse gives the relative strength of the scalar part of the Brans-Dicke scalar-tensor theory. For large ω the importance of the scalar is small, and experimental results approach those of GR. For the deflection of light experimental results available in 1961, "The accuracy of the light deflection observations is too poor to set any useful limit to the size of ω", to quote Brans-Dicke [2].

For the third classic test, the perihelion rotation of the orbit of Mercury, an exact solution is found in Brans-Dicke, and the difference with GR is found to be a factor of $\frac{4+3\omega}{6+3\omega}$. Thus for $\omega = 1$, the extra perihelion shift due to relativistic effects would be roughly 20 % smaller than the 43″·arc/century of GR. We will return to these two significant experimental tests-the deflection of light and the orbit of Mercury—but first we will turn our attention to another foundational experiment, then to one that is not envisioned in that first Brans-Dicke paper but will be expanded upon later.

9.4 The Weak Equivalence Principle

The first gravitational experiment that Dicke had begun in the mid 1950's was the so-called Eötvös experiment. Baron Roland von Eötvös had performed by far the most precise experimental test of universal free fall—the phenomenon that the acceleration due to gravity is the same for all objects regardless of their composition. The experiment is naturally viewed as a test of the equivalence of gravitational and inertial mass, and embodies what came to be known as the 'weak equivalence principle'. Eötvös found that the gravitational and inertial mass of a test object were equal to within 3 parts in 10^9. This principle is so fundamental to the theory of gravity that Dicke deemed it of utmost importance to repeat, and if possible improve upon the results of Eötvös. He and his collaborators made several changes to the experimental technique, and although it took close to 10 years of effort before the final results were published, they found the weak equivalence principle to hold to within a part in 10^{11} [8]. The Brans-Dicke theory obeys the weak principle, so the experiment supports it as much as it supports GR. However, the 'strong principle of equivalence' does not hold for Brans-Dicke. The strong principle, to quote Brans-Dicke, states that all "the laws of physics, including numerical content (i.e. dimensionless physical constants), as observed locally in a freely falling laboratory, are independent of the location in time or space of the laboratory" [2]. In fact, the distinction between the 'strong' and 'weak' becomes apparent when clarifying the role of the principle of equivalence in GR and alternate theories of gravity.

So the experimental verification of universal free-fall supports the weak, but not the strong principle of equivalence—with one crucial caveat which we will explore next.

9.5 The Nordtvedt Effect

Universal free-fall has a wrinkle within the context of the Brans-Dicke theory. First recorded in lectures given in the picturesque town of Varenna on Lake Como in northern Italy during the summer of 1961 [9], Dicke considered the possibility that gravitational self-energy of an object could lead to anomalous free-fall in Brans-Dicke. In other words, massive bodies would fall at a different rate than less massive objects. In particular he considered looking for an effect in the orbit of the most massive planet in the solar system, Jupiter. He reasoned though, that the effect was smaller than the current orbital determinations could test. The issue thus languished for several years until a letter, dated February 20, 1967, came to Dicke from Kenneth Nordtvedt, then at Montana State University [10]. In the letter, he refers to an enclosed preprint that he claims shows that both Einstein's general relativity and the Brans-Dicke theory obey the equivalence principle (universal free-fall) for massive bodies. Dicke's brief reply expresses his interest, but is non-committal. He appears not to have been convinced because in an April 3 letter to scientist-astronaut Dr. F.C. Michel [11] Dicke writes, "I have been unable to prove the constancy of the gravitational acceleration when self gravitational energy is important". The story does not end there of course, but continues with a subsequent May 24, 1967 letter from Nordtvedt to Dicke [11], where he says that he (Nordtvedt) discovered a mistake in his earlier paper, and that there is, in fact, an anomalous acceleration for massive bodies in the Brans-Dicke theory. Nordtvedt goes on to publish a pair of articles [12, 13] which demonstrate the existence of the effect (in particular for the Brans-Dicke theory) and to propose new possible experimental tests. Even before the first two papers had appeared in print he submitted a third [14] which very significantly points out for the first time, that a lunar laser ranging experiment could be used to test this effect—soon known as the 'Nordtvedt effect'. With the anomalous contribution from the gravitational self-energy, the earth and the moon fall towards the sun at slightly different rates, and their mutual orbital motion is thus distorted. This distortion can be measured by accurate distance measurements between the earth and the moon.

Ranging measurements to artificial satellites [4] and to the moon [15] were one of the many experimental ideas that had be bandied about by the Dicke group with the idea of looking for a variation in Newton's gravitational constant G. Already in 1964 a letter from Dicke to Dr. Urner Liddel of NASA [16] indicated that NASA had endorsed the idea of putting a reflector on the moon. A full proposal for the experiment came in 1965 [17]. Coincident with the new insight of Nordtvedt for a direct test of Brans-Dicke, but also due to the desire to take some of the work load off of the astronauts for the first landing [18], the lunar laser ranging retro-reflector (LLLR) experiment was chosen towards the end of 1968 to be part of the Early Apollo Scientific Experiments Package (EASEP) to be placed on the moon by the Apollo 11 astronauts in July 1969. This part of the first experimental package was quick and easy to deploy. The LLLR array was simply placed on the surface at a sufficient distance from the lunar lander so that dust from the launch for the return trip would not cover the reflectors.

The first reflections from the array were detected back on earth by August 1, 1969, and additional reflector arrays were placed on the moon by Apollo 14, 15 and by the unmanned Soviet Luna 17 spacecraft [18]. As the data came in, the test of the Nordtvedt effect proceeded.

As late as early 1975 the laser ranging analysis team had a tentative (unpublished) result of an orbital distortion of about 1 meter consistent with a significant Nordtvedt effect implying an $\omega = 7.5$ for Brans-Dicke [19]. A new ephemeris model led to the discovery of a systematic error, and the effect was reduced to 9 cm—within the experimental uncertainties. By the end of 1975 when the results were published [20] a constraining limit on the Brans-Dicke theory of $\omega > 79$ could be inferred from the data. This result came at a time, as we will discuss further, when evidence from a number of tests began to come in to severely constrain the Brans-Dicke theory.

As part of the legacy of experiments inspired by Brans-Dicke, the lunar laser ranging has continued up to the present day, making it easily the longest-lived of the first experiments placed on the moon by Apollo astronauts. It has given a wealth of information including the limits on a changing G (its original goal) and limits on a violation of the equivalence principle [21].

9.6 The Deflection of Light

Indeed there is a long history of optical light deflection experiments. As stated before, at the time Brans-Dicke was published, the limits were insufficient to put constraints on the theory. To highlight the difficulties of such an experiment, a more modern version done by a team from the University of Texas in 1973 only measured the effect to an accuracy of 10% [22]. This would only give a constraint of $\omega > 4$. However, motivated by Brans-Dicke to do better, a number of groups doing deflection experiments using radio waves instead of visible light began to get results by the late 1960's. By 1975, radio interferometry measurements of quasar sources passing near the sun had restricted $\omega > 40$ [23]. The exquisite angular resolution of Very Long Baseline Interferometry (VLBI) has pushed that restriction to $\omega > 10,000$ by 2011 [24].

9.7 Solar Oblateness

In 1961, the strongest test of GR was the one that had been around the longest—the measured anomalous precession of the perihelion of Mercury. By the reckoning of the original Brans-Dicke paper, this result set a limit on ω right from the start—$\omega \geq 6$. It is interesting to note that this was based on an 8 % uncertainty in the experimental value of the anomalous precession, even though the Brans-Dicke paper quoted the value as $42.6'' \pm 0.9''$·arc/century—a 2 % uncertainty. The case to justify the extra 'wiggle room' on the uncertainty was made quite reasonably by Brans and Dicke by showing how astronomical values are routinely revised by amounts much greater that the previously reported uncertainties. The example of the mass of Saturn showed that a new value reported in 1960 differed from a previous determination by an amount roughly 16 times as large as the reported error of the older measurement. Systematic errors, being very difficult to estimate, were typically not taken into account, and strictly the formal probable error was reported. Therefore an error in the mass of Venus seemed a plausible culprit that would increase the uncertainty in Mercury's orbit. But only a year later, by the end of 1962, the results from NASA's Mariner II probe supported the accepted value for the mass of Venus. The wiggle room already appeared to be shrinking.

Sitting in an airport lounge in New York (then known as Idlewild but soon to become JFK) waiting for a flight to a quantum electronics symposium in Paris, Dicke began to think seriously about other sources of error in the perihelion shift calculation. The experimental effort that resulted was in many ways the most complex (and most convoluted) of any inspired by Brans-Dicke. The idea that hit Dicke that day was that a slight distortion of the mass distribution of the sun would contribute to the perihelion shift. If the mass distribution was not perfectly spherical as was assumed by Einstein and others in the calculations, but instead was slightly squashed or oblate, the amount of the perihelion shift that could be simply explained by Newtonian gravity—without invoking relativity—would change. Dicke would discover later that an astronomer named Simon Newcomb had contemplated an oblate sun to account for the perihelion anomaly long before the advent of general relativity [25]. But when Einstein's calculation so nicely accounted for the entire anomaly, the interest in a possible solar oblateness evaporated. Dicke would later write, "I have long believed that an experimentalist should not be unduly inhibited by theoretical untidiness" [26]. So Dicke revived the idea, and set out to measure the oblateness of the sun. He recruited the help of a couple of brilliant experimentalists who were young Princeton faculty members in his research group—Henry Hill and Mark Goldenberg.

Rai Weiss, a postdoc in the Dicke group at the time, tells an interesting story about the initial design of the apparatus for this experiment [27]. After a group meeting where some of the ideas were discussed, Dicke 'disappeared' for 3 or 4 weeks. When he returned he had a stack of machine blueprints outlining the whole apparatus. But more than that, and what struck Weiss as pure genius, Dicke had thought through in enormous detail many of the myriad sources of systematic error. He had included in the design, features that often came after months of working with an apparatus and discovering its weaknesses.

Even with this heroic effort, this was only a start, with long work ahead and major contributions to the experimental apparatus and design from Hill and Goldenberg.

In highly simplified terms, the basic idea for the experiment was to capture the image of the sun with a telescope equipped with a circular disk to block out nearly the entire surface of the sun. The remaining thin ring would then be examined to look for the excess light that would be produced by a bulge in the shape of the sun. A perfectly spherical sun would lead to a uniform ring, but a bulging sun would produce a non-uniform ring. After construction, the first preliminary measurements were made in the late summer of 1963 [28].

More hard work ensued, but by end of the summer of 1966 they had collected sufficient data, and in 1967 published initial results that they had observed an oblate sun with a fractional difference in the equatorial and polar radii of $(5.0 \pm 0.7) \times 10^{-5}$ [29]. This small bulge was enough to lead to a slight perihelion rotation for Mercury and reduce the presumably relativistic effect from 43″ to 39.6″ arc/century. This later value agrees with Brans-Dicke with $\omega \sim 6$. The Dicke-Goldenberg result was immediately attacked on several fronts, and Dicke spent a large amount of time answering the critics. The great difficulty of the experiment came from the great complexity of the sun. It was not until 1974 that a full account and complete data analysis of the experiment was presented [30].

Meanwhile, Henry Hill, who was a key member of the effort early on, had left his assistant professor job at Princeton for Wesleyan University in the fall of 1964. Dicke thought very highly of Hill; he wrote a recommendation letter (dated December 22, 1965) for Hill's promotion to full professor at Wesleyan stating Hill was "one of the country's very few really brilliant experimentalists" [31]. What Hill had been planning since moving to Wesleyan was a light deflection experiment. It was set up at the Sacramento Peak Observatory in Tucson Arizona, and the facility was called the Santa Catalina Laboratory for Experimental Relativity by Astrometry (SCLERA). A great weakness of the light deflection experiments was the fact that they could be done only during an eclipse, and thus they were done with temporary installations that were at the mercy of the local conditions over the brief time the measurements could be made. The initial goal of SCLERA was to measure the deflection of light for stars near the sun without benefit of an eclipse. It turned out, however, that the instrument was suited for a measurement of the solar oblateness. With the controversy over the Dicke-Goldenberg result raging, Hill turned his efforts toward this measurement. His result published in 1974 [32] gave a value approximately 5 times smaller than Dicke-Goldenberg for the solar oblateness, back in line with Einstein's theory. Within a short time of this publication, results for the other solar system tests (laser ranging, radio deflection, etc.) all came in consistent with GR, placing strict limits on Brans-Dicke. Therefore, the issue seemed to be resolved, with Hill's results confirming the conventional wisdom. However, given the subtle nature of the experiments, and the complexity of the solar structure, the issue was, in fact, far from being settled.

A key to understanding the results was understanding the brightness variations of the sun—pole to equator, and at the edge or 'limb'. How these brightness variations changed with time was also important. Through the measurements to look for solar oblateness, Henry Hill was among the first to observe a set of normal mode oscillations of the sun that led the way for observational helioseismology [33]. Now a large, and well-established field (see [34] and references therein for a review of the subject) the study of how the sun oscillates was in its infancy then, and the oblateness work helped move it forward. In the late 70's Dicke and collaborators built a new instrument for solar observations, and along with his students Jeffrey Kuhn and Kenneth Libbrecht, continued work and made significant contributions to the field of solar physics (see for example [35]). Dicke himself was author or coauthor on more than 50 publications on solar physics.

9.8   The Shapiro Delay

Irwin Shapiro proposed a 'fourth test' of relativistic gravitation not long after the publication of Brans-Dicke [36]. There is a predicted time delay of a light signal as it passes through a gravitational field. The time comparisons needed to test this effect require an out-and-back signal, and the initial proposal was to look at radar reflections off of the inner planets as the signal passed near the sun. First results published in 1968 [37] were not precise enough to place any restrictions on Brans-Dicke. The subsequent use of satellite probes allowed improved measurements, and by 1979 experiments using the Viking spacecraft gave a restriction of $\omega \geq 500$ [38].

9.9   Gravity Probe B

Yet another test of relativity, proposed by Leonard Schiff in 1960, was to observe a spinning gyroscope in orbit around the earth [39]. An expected precession of the angular momentum axis of the gyroscope is the result of two relativistic effects—a so-called geodetic or de Sitter term, and an additional frame-dragging or Lense-Thirring term due to the rotation of the earth 'dragging' the inertial frame around with it. After the Brans-Dicke publication, some authors touted this experiment as the best test of the theory [40]. The geodetic term would differ from GR by a factor $\frac{3\omega+4}{3\omega+6}$ and the frame-dragging term would differ by a factor of $\frac{2\omega+3}{2\omega+4}$. The experiment—known as Gravity Probe B—ended up having quite a long history. It launched in 2004 with final results published in 2011 [41]. Both the measured geodetic and frame-dragging terms agreed with the predictions of GR within the experimental uncertainties. The 90 % confidence limit determined by those uncertainties places restrictions on Brans-Dicke of $\omega \geq 118$.

9.10   Scalar Gravitational Waves

In Brans-Dicke theory gravitational radiation has additional terms beyond what is predicted in GR. There is experimental and observational work in this area that is of note for what some of those involved were inspired to go on to do later. One of

Dicke's students whose Ph.D. thesis was entitled *An astronomical and geophysical search for scalar gravitational waves* was W. Jason Morgan. He went on to be an eminent geophysicist who was one of the pioneers in the theory of plate tectonics. He won the National Medal of Science in 2002. Rai Weiss and Barry Block did an experiment to look for scalar gravitational waves by using a gravimeter to monitor the spherically symmetrical normal mode (or 'breathing mode') of the earth [42]. Weiss went on to invent the interferometric gravitational wave detector [43] and to be one of the founders of LIGO [44]—the large project based on that invention to realize the detection of gravitational radiation.

9.11   The Binary Pulsar

Early results from observations of the binary pulsar PSR 1913+16 considered gravitational radiation in the context of the Brans-Dicke theory [45]. By the late 1970's solar system test constraints were fairly strict, and dramatic differences between Brans-Dicke and GR predictions would only occur for orbiting stars with quite different gravitational binding energies. In that case, Brans-Dicke predicts an additional energy loss of the binary system due to dipole radiation, which does not occur in GR. For the binary pulsar however there was evidence for the companions to be quite similar in mass, so no further constraints on Brans-Dicke were possible.

9.12   The Cosmic Microwave Background

In the summer of 1964, casting about for cosmological tests, Dicke suggested looking for the remnant microwave background radiation of an early hot epoch in the universe. In a story told at length elsewhere (see for example [46]), members of Dicke's group, Dave Wilkinson and Peter Roll, began an experiment to look for the background while Jim Peebles began exploring the theoretical implications. They were famously 'scooped' in the discovery of the cosmic microwave background by Penzias and Wilson [47, 48], but an era of observational cosmology was launched, at least in part by Brans-Dicke.

9.13   Solar Neutrinos

There were other research efforts, although not directly conceived as tests of gravity, that nonetheless were influenced by Brans-Dicke. An early paper by John Bahcall on solar neutrinos [49] references the predictions of a solar model based on the variable G of Brans-Dicke. It was hoped that the neutrino detection experiments of Ray Davis might be able to shed light on these models. Of course the 'missing' neutrino problem that was revealed later agreed with none of the models. The standard solar models were later confirmed by helioseismology, which we have seen, had contributions from the work done on solar oblateness tests of Brans-Dicke. Confirming solar models showed that there was a real 'solar neutrino problem' which was later explained with the new physics of neutrino oscillations. Nobel prizes for these experiments followed for Raymond Davis Jr. and Masatoshi Koshiba in 2002 and for Takaaki Kajita and Arthur B. McDonald in 2015.

9.14 Conclusions

Carl's thesis project has been tested in locations ranging from mineshafts to the moon. Satellites, telescopes and the full might of NASA have been behind the massive body of work that tested the Brans-Dicke theory. And when we consider all the directions these experiments have taken, it is mind-boggling. For those interested in delving further into the many experimental tests of the Brans-Dicke theory, Clifford Will's book [50] or his online living review [51] are very good places to start.

References

1. R.H. Dicke, New research on old gravitation. Science **129**, 621 (1959)
2. C. Brans, R.H. Dicke, Mach's principle and a relativistic theory of gravitation. Phys. Rev. **124**, 925 (1961)
3. R.H. Dicke, Principle of equivalence and the weak interactions. Rev. Mod. Phys. **29**, 355 (1957)
4. R.H. Dicke, W.F. Hoffmann, R. Krotkov, Tracking and Orbit Requirements for Experiment to Detect Variations in Gravitational Constant, in *Proceedings of the Second International Space Science Symposium*, Florence, pp. 287–291, 10–14 April 1961
5. I.I. Shapiro, W.B. Smith, M.B. Ash, R.P. Ingalls, Gordon H. Pettengill, Gravitational Constant, Experimental bound on its time variation. Phys. Rev. Lett. **26**, 27 (1971)
6. A. Finzi, Test of possible variations of the gravitational constant by the observation of white dwarfs within galactic clusters. Phys. Rev. **128**, 2012 (1962)
7. R.H. Dicke, *The Theoretical Significance of Experimental Relativity* (Gordon and Breach, New York, 1964)
8. P.G. Roll, R. Krotkov, R.H. Dicke, The equivalence of inertial and passive gravitational mass. Ann. Phys. (N.Y.) **26**, 442 (1964)
9. R.H. Dicke, in *Mach's Principle and Equivalence*, ed. by C. Møller. Evidence for Gravitational Theories: Proceedings of Course 20 of the International School of Physics Enrico Fermi (Academic, New York, 1962), pp. 1-49
10. H. Robert, Dicke Papers, Box 4, Folder 5; Department of Rare Books and Special Collections, Princeton University Library
11. H. Robert, Dicke Papers, Box 5 Folder 1; Department of Rare Books and Special Collections, Princeton University Library
12. K. Nordtvedt Jr., The equivalence principle for massive bodies. I. Phenomenology, Phys. Rev. **169**, 1014 (1968)
13. K. Nordtvedt Jr., The equivalence principle for massive bodies. II. Theory, Phys. Rev. **169**, 1017 (1968)

14. K. Nordtvedt Jr., Testing relativity with laser ranging to the moon. Phys. Rev. **170**, 1186 (1968)
15. C.O. Alley, P.L. Bender, R.H. Dicke, J.E. Faller, P.A. Franken, H.H. Plotkin, D.T. Wilkinson, Optical radar using a corner reflector on the moon. J. Geophys. Res. **70**, 2267 (1965)
16. H. Robert, Dicke Papers, Box 19 Folder 17; Department of Rare Books and Special Collections, Princeton University Library
17. C.O. Alley et al., University of Maryland Proposal, 13 December 1965
18. P.L. Bender, D.G. Currie, R.H. Dicke, D.H. Eckhardt, J.E. Faller, W.M. Kaula, J.D. Mulholland, H.H. Plotkin, S.K. Poultney, E.C. Silverberg, D.T. Wilkinson, C.O. Alley, The lunar laser ranging experiment. Science **182**, 229 (1973)
19. H. Robert, Dicke Papers, Box 19 Folder 3; Department of Rare Books and Special Collections, Princeton University Library
20. J.G. Williams et al., New test of the equivalence principle from lunar laser ranging. Phys. Rev. Lett. **36**, 551 (1976)
21. S.M. Merkowitz, Tests of gravity using lunar laser ranging. Living Rev. Relativ. **13**, 7 (2010). http://www.livingreviews.org/lrr-2010-7 (cited on 23/11/2015)
22. Texas Mauritanian Eclipse Team, Gravitational deflection of light: solar eclipse of 30 June 1973 I. Description of procedures and final results. Astron. J. **81** (1976) 452
23. E.B. Fomalont, R.A. Sramek, A confirmation of Einstein's general theory of relativity by measuring the bending of microwave radiation in the gravitational field of the Sun. Ap. J. **199**, 749 (1975)
24. S.B. Lambert, C. Le Poncin-Lafitte, Improved determination of γ by VLBI. Astron. Astrophys. **529**, A70 (2011). doi:10.1051/0004-6361/201016370
25. S. Newcomb, The Elements of the Four Inner Planets and the Fundamental Constants of Astronomy, (p. 111) Supplement to the American Ephemeris and Nautical Almanac for 1897, Government Printing Office, Washington (1895)
26. H. Robert, Dicke Papers, Box 33; Department of Rare Books and Special Collections, Princeton University Library
27. R. Weiss, Private communication (Nov. 18, 2010)
28. R.H. Dicke, The sun's rotation and relativity. Nature **202**, 432 (1964)
29. R.H. Dicke, H.M. Goldenberg, Solar oblateness and general relativity. Phys. Rev. Lett. **18**, 313 (1967)
30. R.H. Dicke, H.M. Goldenberg, The oblateness of the sun. Astrophys. J. Supp. **27**, 131 (1974)
31. H. Robert, Dicke Papers, Box 4, Folder 2; Department of Rare Books and Special Collections, Princeton University Library
32. H.A. Hill, P.D. Clayton, D.L. Patz, A.W. Healy, Solar oblateness, excess brightness, and relativity. Phys. Rev. Lett. **33**, 1497 (1974)
33. N. Weiss, Solar seismology. Nature **259**, 78 (1976)
34. J. Christensen-Dalsgaard, Helioseismology. Rev. Mod. Phys. **74**, 1073 (2002)
35. J.R. Kuhn, K.G. Libbrecht, R.H. Dicke, Solar ellipticity fluctuations yield no evidence of g-modes. Nature **319**, 128 (1986)
36. I.I. Shapiro, A fourth test of general relativity. Phys. Rev. Lett. **13**, 789 (1964)
37. I.I. Shapiro, G.H. Pettengill, M.E. Ash, M.L. Stone, W.B. Smith, R.P. Ingalls, R.A. Brockelman, Fourth test of general relativity: preliminary results. Phys. Rev. Lett. **20**, 1265 (1968)
38. R.D. Reasenberg et al., Viking relativity experiment—verification of signal retardation by solar gravity. Ap. J. **234**, L219 (1979)
39. L.I. Schiff, Possible new experimental test of general relativity theory. Phys Rev. Lett. **4**, 215 (1960)
40. R.F. O'Connell, Schiff's proposed gyroscope experiment as a test of the scalar-tensor theory of general relativity. Phys. Rev. Lett. **20**, 69 (1968)
41. C.W.F. Everitt et al., Gravity probe B: final results of a space experiment to test general relativity. Phys. Rev. Lett. **106**, 221101 (2011)
42. R. Weiss, B. Block, A gravimeter to monitor the OSO mode of the earth. J. Geophys. Res. **70**, 5615 (1965)

43. R. Weiss, Electromagetically Coupled Broadband Gravitational Antenna. Quarterly Progress Report, Research Laboratory of Electronics, MIT, vol. 105, p. 54 (1972)
44. http://www.ligo.org
45. J.H. Taylor, J.M. Weisberg, A test of general relativity: gravitational radiation and the binary pulsar PSR 1913+16. Ap. J. **253**, 908 (1982)
46. P.J.E. Peebles, L.A. Page, Jr., R.B. Partridge, *Finding the Big Bang* (Cambridge University Press, Cambridge, 2009)
47. A.A. Penzias, R.W. Wilson, A measurement of excess antenna temperature at 4080 Mc/s. Ap. J. **142**, 419 (1965)
48. R.H. Dicke, P.J.E. Peebles, P.G. Roll, D.T. Wilkinson, Cosmic blackbody radiation. Ap. J. **142**, 414 (1965)
49. J.N. Bahcall, S. Neutrinos, Phys. Rev. Lett. **17**, 398 (1966)
50. C.M. Will, *Theory and Experiment in Gravitational Physics*, revised edn. (Cambridge University Press, Cambridge, 1993)
51. C.M. Will, The confrontation between general relativity and experiment. Living Rev. Relativ. **17**, 4 (2014). http://www.livingreviews.org/lrr-2014-4 (cited on 23/11/2015)

Part II
Mach's Principle and Bell's Inequality

"...In Part I it is conjectured that a "Mach's principle" might lead to a dependence of the local Newtonian gravitational constant, K, on universe structure, $K \sim \frac{M}{R}$. Einstein and others have suggested that general relativity predicts such a result. A closer analysis, however, including the carrying out of the geodesic equations to second order, seems to indicate that this is not true and that the apparent "Mach's principle" terms involving total universe structure are really only coordinate effects.

MACH'S PRINCIPLE

and

A VARYING GRAVITATIONAL CONSTANT

by

Carl H. Brans

A Dissertation

Submitted to the Physics Department of Princeton University
in Partial Fulfillment of Requirements for
the Degree Doctor of Philosophy

May, 1961

"...It is sometimes said that quantum theory saves free will: In the context of this paper, this might be reversed, so that free will saves quantum theory, at least in the sense of eliminating hidden variable alternatives. In other words, if there are any truly "free" events in the experiment, then there can be no classical determinism and hence no classical hidden variables. Conversely, given FCA (All aspects of the experiment; including detector settings, are determined by initial data at some sufficiently remote time.), there are no truly "free" or "random" events, although certain sets of variable values may be uncorrelated in any contemporary statistical sense. Thus, an FCA type of hidden variable theory can reproduce exactly the predictions of quantum theory, yet still preserve the apparent randomness of certain choices. ..."

International Journal of Theoretical Physics, Vol. 27, No. 2, 1988

Bell's Theorem Does Not Eliminate Fully Causal Hidden Variables

Carl H. Brans[1]

Received August 12, 1987

Chapter 10
Mach's Principle and the Origin of Inertia

Bahram Mashhoon

Abstract The current status of Mach's principle is discussed within the context of general relativity. The inertial properties of a particle are determined by its mass and spin, since these characterize the irreducible unitary representations of the inhomogeneous Lorentz group. The origin of the inertia of mass and intrinsic spin are discussed and the inertia of intrinsic spin is studied via the coupling of intrinsic spin with rotation. The implications of spin-rotation coupling and the possibility of history dependence and nonlocality in relativistic physics are briefly mentioned.

10.1 Introduction

Is motion absolute or relative? If the Newtonian absolute space and time are not real and therefore not responsible for the origin of inertia, then inertia must be due to immediate connections between masses. Thus one might expect that the inertial mass of a test particle increases in the vicinity of a large mass. Following this Machian line of thought, Einstein suggested that perhaps the physics of general relativity could be so interpreted as to allow for this possibility [1]. However, Brans showed that if one adopts the modern geometric interpretation of general relativity, then the inertial mass of a free test particle cannot change in a gravitational field. Indeed, this issue has since been completely settled as a result of Brans's thorough analysis [2, 3]. Moreover, Mach's principle played an important role in the scalar-tensor generalization of Einstein's tensor theory by Brans and Dicke [4].

Carl Brans has made basic contributions to gravitation theory and general relativity. *It is a great pleasure for me to dedicate this paper to Carl on the occasion of his eightieth birthday.*

B. Mashhoon (✉)
Department of Physics and Astronomy, University of Missouri,
Columbia, MO 65211, USA
e-mail: MashhoonB@missouri.edu

© Springer International Publishing Switzerland 2016
T. Asselmeyer-Maluga (ed.), *At the Frontier of Spacetime*,
Fundamental Theories of Physics 183, DOI 10.1007/978-3-319-31299-6_10

The connection between Mach's ideas [5] and Einstein's theory of gravitation [1] has been the subject of many interesting investigations [6]; furthermore, there is a diversity of opinion on this matter [6–9]. A more complete account of the views expressed in this brief treatment is contained in [10] and the references cited therein.

The special theory of relativity involves, among other things, the measurements of observers in Minkowski spacetime. The special class of *inertial* observers has played a pivotal role in the development of physics, since the fundamental laws of physics are expressed in terms of the measurements of these *hypothetical* observers. Indeed, inertial physics was originally established by Newton [11]. The measurements of inertial observers, each forever at rest in a given inertial frame in Minkowski spacetime, are related to each other by Lorentz invariance. Actual observers are all more or less accelerated. The measurements of accelerated observers must be interpreted in terms of the fundamental (but hypothetical) inertial observers.

The acceleration of an observer in Minkowski spacetime is independent of any system of coordinates and is in this sense *absolute*. Let $u^\mu = dx^\mu/d\tau$ be the 4-velocity of an observer following a timelike world line in Minkowski spacetime. Here, $x^\mu = (ct, \mathbf{x})$ denotes an event in spacetime, Greek indices run from 0 to 3, while Latin indices run from 1 to 3; moreover, the signature of the metric is +2 and τ is the proper time along the path of the observer. The observer's 4-acceleration $a^\mu = du^\mu/d\tau$ is orthogonal to u^μ, namely, $a_\mu u^\mu = 0$, since $u_\mu u^\mu = -1$. Thus a^μ is a spacelike 4-vector such that $a_\mu a^\mu = \mathcal{A}^2$, where the scalar $\mathcal{A}(\tau) \geq 0$ is the coordinate-independent magnitude of the observer's acceleration. An accelerated electric charge radiates electromagnetic waves. The existence of radiation is independent of inertial frames of reference and wave motion is in this sense absolute as well.

Relativity theory, namely, Lorentz invariance, is extended to accelerated observers in Minkowski spacetime in a pointwise manner. That is, at each instant along its world line, the accelerated observer is assumed to be momentarily equivalent to a hypothetical inertial observer following the straight tangent world line at that instant. The further extension of relativity theory to observers in a gravitational field is accomplished via Einstein's *local* principle of equivalence. Therefore, in general relativity spacetime is curved and the gravitational field is identified with spacetime curvature. Test particles and null rays follow timelike and null geodesics of the spacetime manifold, respectively. Moreover, spacetime is locally flat and Minkowskian. The global inertial frames of special relativity are thus replaced by local inertial frames; moreover, gravitation is rendered *relative* in this way due to its universality, a circumstance that does not extend to other fundamental interactions. The general equation of motion of a classical point particle in general relativity is given by

$$\frac{d^2 x^\mu}{d\tau^2} + \Gamma^\mu_{\alpha\beta} \frac{dx^\alpha}{d\tau} \frac{dx^\beta}{d\tau} = a^\mu, \tag{10.1}$$

where a_μ is the absolute acceleration of the particle due to nongravitational forces and $\Gamma^\mu_{\alpha\beta}$ are the Christoffel symbols. For instance, for a particle of mass m and charge q in an electromagnetic field $F_{\mu\nu}$, $a_\mu = (q/m) F_{\mu\nu} u^\nu$ by the Lorentz force law. At each event in spacetime, coordinates can be chosen such that the Christoffel symbols all vanish and Eq. (10.1) reduces to the corresponding relation in Minkowski spacetime.

Inertial forces appear in a system ("laboratory") that accelerates with respect to the ensemble of *local* inertial frames. These inertial forces are *not* due to the gravitational influence of distant masses, which would instead generate gravitational tidal effects in the laboratory [12–16].

10.2   Mach's Principle

Newton's *absolute* space and time refer to the ensemble of *inertial* frames, namely, Cartesian systems of reference that are homogeneous and isotropic in space and time and in which Newton's fundamental laws of motion are valid. Indeed, Newton's first law of motion, the principle of inertia, essentially contains the definition of an inertial frame. Newton argued that the existence of *inertial forces* provided observational proof of the reality of absolute space and time. Thus in classical mechanics, the motion of a Newtonian point particle is absolute, yet subject to the principle of relativity. However, the absolute motion of a particle, namely, its motion with respect to absolute space and time is not *directly* observable.

Mach considered *all* motion to be relative and therefore argued against the New-tonian conceptions of absolute space and time [5]. In his critique of Newtonian foundations of physics, Mach analyzed, among other things, the *operational* defi-nitions of time and space via masses and concluded that in Newtonian mechanics, masses are not organically connected to space and time [5]. In fact, in Chap. II of [5], on pp. 295–296 of Sect. VI, we find:

> "...Although I expect that astronomical observation will only as yet necessitate very small corrections, I consider it possible that the law of inertia in its simple Newtonian form has only, for us human beings, a meaning which depends on space and time. Allow me to make a more general remark. We measure time by the angle of rotation of the earth, but could measure it just as well by the angle of rotation of any other planet. But, on that account, we would not believe that the *temporal* course of all physical phenomena would have to be disturbed if the earth or the distant planet referred to should suddenly experience an abrupt variation of angular velocity. We consider the dependence as not immediate, and consequently the temporal orientation as *external*. Nobody would believe that the chance disturbance — say by an impact — of one body in a system of uninfluenced bodies which are left to themselves and move uniformly in a straight line, where all the bodies combine to fix the system of coördinates, will immediately cause a disturbance of the others as a consequence. The orientation is external here also. Although we must be very thankful for this, especially when it is purified from meaninglessness, still the natural investigator must feel the need of further insight — of knowledge of the *immediate* connections, say, of the masses of the universe...."

Thus Newton's absolute space and time are fundamentally different from their operational definitions by means of masses. Moreover, masses do not appear to "recognize" absolute space and time, since they have been "placed" in this arena without being immediately connected to it. In fact, the internal state of a Newtonian point particle, characterized by its mass m, has no direct connection with its external state in absolute space and time, characterized by its position and velocity $(\mathbf{x}, \mathbf{v})$ at a given time t. Thus only the *relative* motion of classical particles is directly observable. This epistemological shortcoming of Newtonian mechanics means that the external state of the particle m, namely, $(\mathbf{x}, \mathbf{v})$, can in principle be occupied by other particles *comoving* with it. It is indeed a prerequisite of the notion of *relativity* of motion of masses that an observer be capable of changing its perspective by comoving with each particle in turn. As particles can be *directly* connected with each other via interactions, we may say that particles have a *propensity* for relative motion.

With the advent of Maxwell's electrodynamics, Galilean relativity was gradually replaced by Lorentz invariance, in which the speed of light is the same in all inertial frames of reference. Indeed, it is impossible for an inertial observer to be comoving with light. Motion is either relative or absolute in classical physics. The motion of light in Minkowski's absolute spacetime is independent of inertial observers and is, in this sense, *nonrelative* or absolute. Thus in Lorentz invariance, the motion of inertial observers is relative, while the motion of electromagnetic radiation is absolute.

In classical physics, movement takes place via either particles or electromagnetic waves. As emphasized by Mach [5], the internal and external states of particles are not directly related, which leads to the notion of *relativity* of particle motion. However, the opposite is the case for electromagnetic waves. That is, the internal state of a wave, namely, its period, wavelength, intensity and polarization are all directly related to its external state characterized by its wave function, which is a solution of Maxwell's field equations. In this way, electromagnetic waves can "recognize" absolute spacetime and this leads to their propensity for absolute motion. Therefore, an inertial frame of reference can be characterized by standing electromagnetic radiation [17]. Indeed, ring lasers are now regularly employed in inertial guidance systems. Similarly, the wave nature of matter in quantum theory can be used to establish an inertial frame of reference. For instance, the rotation of the earth can be detected via superfluid helium [18]. Moreover, atom interferometers can function as sensitive inertial sensors, since they can measure acceleration and rotation to rather high precision [19–23].

10.3 Duality of Absolute and Relative Motion

Classical physics is the correspondence limit of quantum physics. It is therefore interesting to extend the quantum duality of classical particles and waves to their motions as well. That is, the motion of a quantum particle has complementary classical aspects in relative and absolute movements [13].

The epistemological shortcoming of Newtonian mechanics that was pointed out by Mach [5] essentially disappears in the quantum theory. That is, wave-particle duality makes it possible for a (quantum) particle to "recognize" absolute spacetime; moreover, it is impossible for a classical observer to be comoving with the particle in conformity with Heisenberg's uncertainty principle. Consider, for instance, the nonrelativistic motion of a particle of mass m in a potential V according to the Heisenberg picture. The Hamiltonian is

$$\hat{H} = \frac{1}{2m}\,\hat{p}^2 + V(\hat{\mathbf{x}})\,. \tag{10.2}$$

In this "particle" representation, the momentum operator of the particle is given by

$$\hat{\mathbf{p}} = m\,\frac{d\hat{\mathbf{x}}}{dt}\,, \tag{10.3}$$

so that the fundamental quantum condition, $[\hat{x}^j, \hat{p}^k] = i\,\hbar\,\delta_{jk}$, can be written as

$$\left[\hat{x}^j, \frac{d\hat{x}^k}{dt}\right] = i\,\frac{\hbar}{m}\,\delta_{jk}\,. \tag{10.4}$$

In sharp contrast to classical mechanics, the inertial mass of the particle is related, albeit in a statistical sense, to the observables corresponding to its position and velocity. This connection, through Planck's constant $\hbar$, disappears when $m \to \infty$. A macroscopically massive system behaves classically, since the perturbation experienced by the system due to any disturbance accompanying an act of observation would be expected to be negligibly small. Similarly, let $\hat{\mathbf{L}}$, $m\,\hat{L}^i = \epsilon_{ijk}\,\hat{x}^j\,\hat{p}^k$, be the specific orbital angular momentum operator of the particle; then,

$$[\hat{L}^j, \hat{L}^k] = i\,\frac{\hbar}{m}\,\epsilon_{jkn}\,\hat{L}^n\,. \tag{10.5}$$

The mechanical laws of classical physics pertaining to translational and rotational inertia hold in a certain average sense in quantum theory as well. Moreover, in terms of the Schrödinger picture, the state of the particle is characterized by a wave function $\Psi(t, \mathbf{x})$ that satisfies the Schrödinger equation. This equation is explicitly dependent upon the particle's inertial mass m, thereby connecting the internal and external states of the particle in the "wave" representation.

In relativistic quantum theory, rotational inertia involves the intrinsic angular momentum of the particle as well, thus leading to the inertia of intrinsic spin.

10.4   Inertia of Intrinsic Spin

Mass and spin describe the irreducible representations of the Poincaré group [24]. The state of a particle in spacetime is thus characterized by its mass and spin, which determine the inertial properties of the particle. In quantum theory, therefore, the inertial characteristics of a particle are determined by its inertial mass [25–27] as well as intrinsic spin [28–31].

To examine the inertia of intrinsic spin, we recall that the *total* angular momentum operator is the generator of rotations; therefore, we expect that intrinsic spin would couple to the rotation of a frame of reference in much the same way as orbital angular momentum. This means that in a macroscopic body rotating in the positive sense with uniform angular velocity $\boldsymbol{\Omega}$, the spins of the constituent particles do not naturally participate in the rotation and all instead tend to stay essentially fixed with respect to the local inertial frame. Thus relative to the rotating body, we have in the nonrelativistic approximation for each spin vector $\hat{\sigma}$,

$$\frac{d\hat{\sigma}^i}{dt} + \epsilon_{ijk}\, \Omega^j\, \hat{\sigma}^k = 0\,, \tag{10.6}$$

since the spin vector appears to precess with angular velocity $-\boldsymbol{\Omega}$ with respect to observers at rest with the rotating body. The Hamiltonian corresponding to this motion is

$$\hat{\mathcal{H}} = -\hat{\sigma} \cdot \boldsymbol{\Omega}\,, \tag{10.7}$$

because the Heisenberg equation of motion

$$i\hbar\, \frac{d\hat{\sigma}^k}{dt} = [\hat{\sigma}^k, \hat{\mathcal{H}}] \tag{10.8}$$

coincides with Eq. (10.6). For a discussion of the corresponding relativistic treatment and further developments of this subject, see [32–37]. Moreover, a review of this subject and a more complete list of references is contained in [38].

In general, the energy of an incident particle as measured by the rotating observer is given by

$$E' = \gamma\, (E - \hbar M\Omega)\,, \tag{10.9}$$

where E is the energy of the incident particle in the inertial frame and M is the total (orbital plus spin) "magnetic" quantum number along the axis of rotation. In fact, $M = 0, \pm 1, \pm 2, \ldots$, for a scalar or a vector particle, while $M \mp \frac{1}{2} = 0, \pm 1, \pm 2, \ldots$, for a Dirac particle. In the JWKB approximation, $E' = \gamma\, (E - \boldsymbol{\Omega} \cdot \mathbf{J})$, where $\mathbf{J} = \mathbf{r} \times \mathbf{P} + \boldsymbol{\sigma}$ is the total angular momentum of the particle and $\mathbf{P}$ is its momentum; hence, $E' = \gamma\, (E - \mathbf{v} \cdot \mathbf{P}) - \gamma\, \boldsymbol{\sigma} \cdot \boldsymbol{\Omega}$, where $\mathbf{v} = \boldsymbol{\Omega} \times \mathbf{r}$ is the velocity of the uniformly rotating observer with respect to the background inertial frame and γ is the Lorentz factor of the observer. The energy corresponding to spin-rotation coupling is naturally augmented by time dilation.

It is important to remark here that the spin-rotation coupling is completely independent of the inertial mass of the particle. Moreover, the associated spin-gravity coupling is an interaction of the intrinsic spin with the gravitomagnetic field of the rotating source that is also independent of the mass of the test particle. For instance, free neutral Dirac particles with their spins up and down (i.e., parallel and antiparallel to the vertical direction, respectively) in general fall differently in the gravitational field of the rotating earth [29].

The spin-rotation coupling has recently been measured for neutrons via neutron polarimetry [39]. Moreover, this general coupling has now been incorporated into the condensed-matter physics of spin mechanics and spin currents [40–48]. It is also expected to play a role in the emerging field of spintronics [49].

10.5 Helicity-Rotation Coupling

To illustrate the general nature of spin-rotation coupling, we now turn to the case of photons, see [50–56] and the references cited therein. Consider the measurement of the frequency of a plane monochromatic electromagnetic wave of frequency ω propagating along the axis of rotation of the observer. The result of the Fourier analysis of the measured field is

$$\omega' = \gamma \left(\omega \mp \Omega \right), \tag{10.10}$$

where the upper (lower) sign refers to positive (negative) helicity radiation. With $E = \hbar \omega$, our classical result (10.10) is consistent with the general formula (10.9) for spin 1 photons. The helicity-dependent contribution to the transverse Doppler effect in Eq. (10.10) has been verified via the GPS [57], where it is responsible for the phenomenon of *phase wrap-up* [57].

It is simple to interpret the coupling of helicity with rotation in Eq. (10.10), aside from the presence of the Lorentz factor that is due to time dilation. In a positive (negative) helicity wave, the electromagnetic field rotates with frequency ω $(-\omega)$ about the direction of propagation of the wave. The rotating observer therefore perceives positive (negative) helicity radiation with the electromagnetic field rotating with frequency $\omega - \Omega$ $(-\omega - \Omega)$ about the direction of wave propagation.

For the case of oblique incidence, the analog of Eq. (10.10) is

$$\omega' = \gamma \left(\omega - M \Omega \right), \tag{10.11}$$

where $M = 0, \pm 1, \pm 2, \ldots$ for the electromagnetic field. This exact classical result can be obtained by studying the electromagnetic field as measured by uniformly rotating observers. It is interesting to note that $\omega' = 0$ for $\omega = M \Omega$, a situation that is discussed in the next section, while ω' can be negative for $\omega < M \Omega$. The latter circumstance does not pose any basic difficulty, since it is simply a consequence of the absolute character of rotational motion.

10.6 Can Light Stand Completely Still?

The exact formula $\omega' = \gamma \left(\omega - \Omega \right)$ for incident positive-helicity radiation has a remarkable consequence that is not easily accessible to experimental physics: The incident wave stands completely still with respect to observers that rotate uniformly with frequency $\Omega = \omega$ about the direction of propagation of the wave. That is, helicity-rotation coupling has the consequence that a rotating observer can in principle be comoving with an electromagnetic wave; in fact, the wave appears to be oscillatory in space but stands completely still with respect to the rotating observer. The fundamental difficulty under consideration here is quite general, as it occurs for oblique incidence as well.

By a mere rotation, an observer can in principle stay completely at rest with respect to an electromagnetic wave. This circumstance is rather analogous to the difficulty with the pre-relativistic Doppler formula, where an inertial observer moving with speed c along a light beam would see a wave that is oscillatory in space but is otherwise independent of time and hence completely at rest. This issue, as is well known, played a part in Einstein's path to relativity, as mentioned in his autobiographical notes, see p. 53 of [58]. The difficulty in that case was eventually removed by Lorentz invariance; however, in the present case, the problem has to do with the way Lorentz invariance is extended to accelerated observers in Minkowski spacetime. In the special theory of relativity, Lorentz invariance is extended to accelerated systems via the *hypothesis of locality*, namely, the assumption that an accelerated observer is pointwise inertial [59, 60]. This circumstance extends to general relativity through Einstein's *local* principle of equivalence. The locality assumption originates from Newtonian mechanics, where the state of a particle is determined by $(\mathbf{x}, \mathbf{v})$ at time t. The accelerated observer shares this state with a comoving inertial observer; therefore, they are at that moment physically equivalent insofar as all physical phenomena could be reduced to pointlike coincidences of classical particles and rays of radiation. However, classical wave phenomena are in general intrinsically *nonlocal*.

According to the locality assumption, the world line of an accelerated observer in Minkowski spacetime can be replaced at each instant by its tangent and then Lorentz transformations can be pointwise employed to determine what the accelerated observer measures. To go beyond the locality postulate of special relativity theory, the past history of the observer must be taken into account. Thus the locality postulate must be supplemented by a certain average over the past world line of the observer. In this way, the observer retains the memory of its past acceleration. This averaging procedure involves a weight function that must be determined. In this connection, we introduce the fundamental assumption that *a basic radiation field can never stand completely still with respect to an observer*. On this basis a nonlocal theory of accelerated observers can be developed [61]. The nonlocal approach is in better correspondence with quantum theory than the standard treatment based on the hypothesis of locality [62].

History-dependent theories are nonlocal in the sense that the usual partial differential equations for the fields are replaced by integro-differential equations. Acceleration-induced nonlocality in Minkowski spacetime suggests that gravity could be nonlocal. This is due to the intimate connection between inertia and gravitation, implied by Mach (e.g., in his discussion of Newton's experiment with the rotating bucket of water on p. 279 of [5]) and developed in new directions by Einstein in his general theory of relativity [1]. That is, Einstein interpreted the principle of equivalence of inertial and gravitational masses to mean that an intimate connection exists between inertia and gravitation. One can follow Einstein's interpretation without postulating a *local* equivalence between inertia and gravitation as in general relativity; for instance, one can instead extend general relativity to make it history dependent. Recently, nonlocal theories of special and general relativity have been developed [63–67]. It turns out that nonlocal general relativity simulates dark matter. That is, according to this theory, what appears as dark matter in astrophysics [68–70] is essentially a manifestation of the nonlocality of the gravitational interaction [67].

10.7 Discussion

Classical relativistic mechanics and classical electrodynamics are mainly concerned with two types of motion, namely, local particle motion and nonlocal wave motion, respectively. These are brought together in geometric optics, where the *waves* are replaced by *rays* that can be treated in a similar way as classical point particles. With respect to Minkowski's absolute spacetime, particle motion is absolute; however, this absolute motion is not directly observable. On the other hand, the motion of classical particles naturally tends to be *relative*. Similarly, the motion of electromagnetic waves naturally tends to be *absolute*, though the corresponding wave equation is Lorentz invariant. In the quantum domain, this line of thought leads to the complementarity of absolute and relative motion; moreover, the notion of inertia must be extended to include intrinsic spin as well. The inertial coupling of intrinsic spin to rotation has recently been measured in neutron polarimetry [39]. The implications of the inertia of intrinsic spin are critically examined in the light of the hypothesis that an electromagnetic wave cannot stand completely still with respect to any accelerated observer.

References

1. A. Einstein, *The Meaning of Relativity* (Princeton University Press, Princeton, NJ, 1955)
2. C.H. Brans, Phys. Rev. **125**, 388 (1962)
3. C.H. Brans, Phys. Rev. Lett. **39**, 856 (1977)
4. C.H. Brans, R.H. Dicke, Phys. Rev. **124**, 925 (1961)
5. E. Mach, *The Science of Mechanics* (Open Court, La Salle, 1960)
6. J. Barbour, H. Pfister (eds.), *Mach's Principle: From Newton's Bucket to Quantum Gravity, Einstein Studies 6* (Birkhäuser, Boston, 1995)
7. B. Mashhoon, P.S. Wesson, Mach's principle and higher-dimensional dynamics. Ann. Phys. (Berlin) **524**, 63 (2012), arXiv:1106.6036 [gr-qc]
8. J. Barbour, Ann. Phys. (Berlin) **524**, A39 (2012), arXiv:1108.3057 [gr-qc]
9. B. Mashhoon, P.S. Wesson, Ann. Phys. (Berlin) **524**, A44 (2012), arXiv:1108.3059 [gr-qc]
10. H. Lichtenegger, B. Mashhoon, Mach's Principle, in *The Measurement of Gravitomagnetism: A Challenging Enterprise*, ed. by L. Iorio (NOVA Science, Hauppage, NY, 2007), Chap. 2. arXiv:physics/0407078 [physics.hist-ph]
11. I.B. Cohen, *The Birth of a New Physics* (Doubleday Anchor Books, Garden City, NY, 1960)
12. B. Mashhoon, F.W. Hehl, D.S. Theiss, Gen. Relativ. Gravit. **16**, 711 (1984)

13. B. Mashhoon, Complementarity of absolute and relative motion. Phys. Lett. A **126**, 393 (1988)
14. B. Mashhoon, Found. Phys. Lett. **6**, 545 (1993)
15. B. Mashhoon, in Directions in *General Relativity: Papers in Honor of Dieter Brill*, ed. by B.L. Hu, T.A. Jacobson (Cambridge University Press, Cambridge, UK, 1993), pp. 182–194
16. B. Mashhoon, Nuovo Cimento B **109**, 187 (1994)
17. H.A. Lorentz, Nature **112**, 103 (1923)
18. K. Schwab, N. Bruckner, R.E. Packard, Nature **386**, 585 (1997)
19. F. Riehle, Th. Kisters, A. Witte, J. Helmcke, Ch.J. Bordé, Phys. Rev. Lett. **67**, 177 (1991)
20. T.L. Gustavson, P. Bouyer, M.A. Kasevich, Phys. Rev. Lett. **78**, 2046 (1997)
21. T.L. Gustavson, A. Landragin, M.A. Kasevich, Class. Quantum Grav. **17**, 2385 (2000)
22. B. Dubetsky, M.A. Kasevich, Phys. Rev. A **74**, 023615 (2006)
23. S.M. Dickerson, J.M. Hogan, A. Sugarbaker, D.M.S. Johnson, M.A. Kasevich, Phys. Rev. Lett. **111**, 083001 (2013)
24. E.P. Wigner, Ann. Math. **40**, 149 (1939)
25. S.A. Werner, J.-L. Staudenmann, R. Colella, Phys. Rev. Lett. **42**, 1103 (1979)
26. G.F. Moorhead, G.I. Opat, Class. Quantum Grav. **13**, 3129 (1996)
27. H. Rauch, S.A. Werner, *Neutron Interferometry*, 2nd edn. (Oxford University Press, Oxford, UK, 2015)
28. B. Mashhoon, Phys. Rev. Lett. **61**, 2639 (1988)
29. B. Mashhoon, Phys. Lett. A **198**, 9 (1995)
30. B. Mashhoon, R. Neutze, M. Hannam, G.E. Stedman, Phys. Lett. A **249**, 161 (1998)
31. B. Mashhoon, H. Kaiser, Inertia of intrinsic spin. Phys. B **385–386**, 1381 (2006), arXiv:quantum-ph/0508182
32. F.W. Hehl, W.-T. Ni, Phys. Rev. D **42**, 2045 (1990)
33. L. Ryder, J. Phys. A **31**, 2465 (1998)
34. D. Singh, G. Papini, Nuovo Cimento B **115**, 223 (2000)
35. G. Papini, Phys. Rev. D **65**, 077901 (2002)
36. A. Randono, Phys. Rev. D **81**, 024027 (2010)
37. U.D. Jentschura, J.H. Noble, J. Phys. A: Math. Theor. **47**, 045402 (2014)
38. B. Mashhoon, Lect. Notes Phys. **702**, 112 (2006), arXiv:hep-th/0507157
39. B. Demirel, S. Sponar, Y. Hasegawa, Measurement of the spin-rotation coupling in neutron polarimetry. New J. Phys. **17**, 023065 (2015)
40. M. Matsuo, J. Ieda, E. Saitoh, S. Maekawa, Effects of mechanical rotation on spin currents. Phys. Rev. Lett. **106**, 076601 (2011)
41. M. Matsuo, J. Ieda, E. Saitoh, S. Maekawa, Spin-dependent inertial force and spin current in accelerating systems. Phys. Rev. B **84**, 104410 (2011)
42. M. Matsuo, J. Ieda, K. Harii, E. Saitoh, S. Maekawa, Mechanical generation of spin current by spin-rotation coupling. Phys. Rev. B **87**, 180402(R) (2013)
43. M. Matsuo, J. Ieda, S. Maekawa, Renormalization of spin-rotation coupling. Phys. Rev. B **87**, 115301 (2013)
44. J. Ieda, M. Matsuo, S. Maekawa, Theory of mechanical spin current generation via spin-rotation coupling. Solid State Commun. **198**, 52 (2014)
45. D. Chowdhury, B. Basu, Effect of spin rotation coupling on spin transport. Ann. Phys. **339**, 358 (2013)
46. J.-Q. Shen, S.-L. He, Geometric phases of electrons due to spin-rotation coupling in rotating C60 molecules. Phys. Rev. B **68**, 195421 (2003)
47. J.R.F. Lima, F. Moraes, The combined effect of inertial and electromagnetic fields in a fullerene molecule. Eur. Phys. J. B **88**, 63 (2015)
48. M. Hamada, T. Yokoyama, S. Murakami, Spin current generation and magnetic response in carbon nanotubes by the twisting phonon mode. Phys. Rev. B **92**, 060409(R) (2015)
49. G. Papini, Spin currents in non-inertial frames. Phys. Lett. A **377**, 960 (2013)
50. P.J. Allen, Am. J. Phys. **34**, 1185 (1966)
51. B. Mashhoon, Phys. Lett. A **139**, 103 (1989)

52. B. Mashhoon, Optics of rotating systems. Phys. Rev. A **79**, 062111 (2009), arXiv:0903.1315 [gr-qc]
53. B. Mashhoon, Nonlocal Special Relativity: Amplitude Shift in Spin-Rotation Coupling, in *Proceedings of Mario Novello's 70th Anniversary Symposium*, ed. by N. Pinto Neto, S.E. Perez Bergliaffa (Editora Livraria da Fisica, Sao Paulo, 2012), pp. 177–189, arXiv:1204.6069 [gr-qc]
54. K.Y. Bliokh, Y. Gorodetski, V. Kleiner, E. Hasman, Phys. Rev. Lett. **101**, 030404 (2008)
55. K.Y. Bliokh, J. Opt. A: Pure Appl. Opt. **11**, 094009 (2009)
56. K.Y. Bliokh, A. Aiello, J. Opt. **15**, 014001 (2013)
57. N. Ashby, Relativity in the global positioning system. Living Rev. Relativ. **6**, 1 (2003), http://www.livingreviews.org/lrr-2003-1
58. P.A. Schilpp, *Albert Einstein: Philosopher-Scientist* (Library of Living Philosophers, Evanston, Illinois, 1949)
59. B. Mashhoon, The hypothesis of locality in relativistic physics. Phys. Lett. A **145**, 147 (1990)
60. B. Mashhoon, Limitations of spacetime measurements. Phys. Lett. A **143**, 176 (1990)
61. B. Mashhoon, Nonlocal theory of accelerated observers. Phys. Rev. A **47**, 4498 (1993)
62. B. Mashhoon, Phys. Rev. A **72**, 052105 (2005), arXiv:hep-th/0503205
63. B. Mashhoon, Nonlocal special relativity. Ann. Phys. (Berlin) **17**, 705 (2008), arXiv:0805.2926 [gr-qc]
64. F.W. Hehl, B. Mashhoon, Nonlocal gravity simulates dark matter. Phys. Lett. B **673**, 279 (2009), arXiv:0812.1059 [gr-qc]
65. F.W. Hehl, B. Mashhoon, Formal framework for a nonlocal generalization of Einstein's theory of gravitation. Phys. Rev. D **79**, 064028 (2009), arXiv:0902.0560 [gr-qc]
66. B. Mashhoon, Nonlocal gravity: the general linear approximation. Phys. Rev. D **90**, 124031 (2014), arXiv:1409.4472 [gr-qc]
67. S. Rahvar, B. Mashhoon, Observational tests of nonlocal gravity: galaxy rotation curves and clusters of galaxies. Phys. Rev. D **89**, 104011 (2014), arXiv:1401.4819 [gr-qc]
68. V.C. Rubin, W.K. Ford, Astrophys. J. **159**, 379 (1970)
69. M.S. Roberts, R.N. Whitehurst, Astrophys. J. **201**, 327 (1975)
70. Y. Sofue, V. Rubin, Rotation curves of spiral galaxies. Annu. Rev. Astron. Astrophys. **39**, 137 (2001)

Chapter 11
The Significance of Measurement Independence for Bell Inequalities and Locality

Michael J. W. Hall

Abstract A local and deterministic model of quantum correlations is always possible, as shown explicitly by Brans in 1988: one simply needs the physical systems being measured to have a suitable statistical correlation with the physical systems performing the measurement, via some common cause. Hence, to derive no-go results such as Bell inequalities, an assumption of measurement independence is crucial. It is a surprisingly strong assumption—less than $1/15$ bits of prior correlation suffice for a local model of the singlet state of two qubits—with ramifications for the security of quantum communication protocols. Indeed, without this assumption, any statistical correlations whatsoever—even those which appear to allow explicit superluminal signalling—have a corresponding local deterministic model. It is argued that 'quantum nonlocality' is bad terminology, and that measurement independence does not equate to 'experimental free will'. Brans' 1988 model is extended to show that no more than $2 \log d$ bits of prior correlation are required for a local deterministic model of the correlations between any two d-dimensional quantum systems.

11.1 Introduction

Various no-go results exist for models of quantum phenomena, based on various more or less plausible assumptions for the structure of such models. Such results support a longstanding view that quantum mechanics is more or less implausible—indeed, Niels Bohr was famously quoted as saying [1]:

> Those who are not shocked when they first come across quantum theory cannot possibly have understood it.

M.J.W. Hall (✉)
Centre for Quantum Computation and Communication Technology
(Australian Research Council), Centre for Quantum Dynamics, Griffith University,
Brisbane, QLD 4111, Australia
e-mail: michael.hall@griffith.edu.au

© Springer International Publishing Switzerland 2016
T. Asselmeyer-Maluga (ed.), *At the Frontier of Spacetime*,
Fundamental Theories of Physics 183, DOI 10.1007/978-3-319-31299-6_11

This has led not only to much philosophical discussion on which assumption(s) should be relaxed, but also to surprising applications of what might be termed quantum implausibility, such as quantum cryptography and quantum computation.

The most remarkable of these no-go results are Bell inequalities, which imply that at least one of the plausible properties of determinism, locality and measurement independence must be given up to successfully describe quantum correlations between distant measurement regions [2–4]. Here measurement independence denotes the statistical independence of (i) any physical parameter influencing the selection of measurement procedures from (ii) any physical parameter influencing measurement outcomes, and is typically justified by an appeal to experimental free will [5].

The question of which property should be relaxed is not just a matter of idle speculation: the security of quantum cryptographic protocols, for example, relies on there being no deterministic description underlying correlations between distant measurement outcomes—an eavesdropper possessing such a description would be able to determine the cryptographic key [6]. Hence, any unconditionally secure protocol based on violation of Bell inequalities must, to ensure there is no deterministic description available, assume that the properties of locality and measurement independence hold. A similar requirement applies to device-independent protocols for randomness generation [6].

While most discussion in the literature focuses on choosing between locality or determinism to model quantum correlations, it was pointed out by Brans in 1988 that there is in fact an explicit local *and* deterministic model, obtained by relaxing measurement independence [7]. Brans further observed that there is an inherent conflict in assuming that both determinism and measurement independence hold: if the physical world is deterministic, then correlations between physical parameters are generic.

There has been a recent upsurge of interest in local deterministic models, including their construction [8–10]; the derivation of generalised Bell inequalities incorporating a given level of measurement dependence [9–17]; impacts on device-independent quantum communication protocols [13–15]; and new experimental tests [18, 19].

In this contribution I briefly review the assumptions leading to Bell inequalities (Sect. 11.2); pause to urge replacement of the misleading terms 'quantum nonlocality' and 'Bell nonlocality' in the literature by the more neutral term 'Bell nonseparability' (Sect. 11.3); compare the degrees of measurement dependence of various local deterministic models for the singlet state, and extend the Brans model to show that local deterministic models for two d-dimensional quantum systems require no more than $2 \log_2 d$ bits of measurement-dependent correlation (Sect. 11.4); and discuss the relevance of local deterministic models to questions of causality and free will—including a demonstration of the *prima facie* paradoxical existence of a local deterministic model for superluminal correlations (Sect. 11.5).

11.2 The Well Trod Path to Bell Inequalities

11.2.1 Bayes Theorem

Consider an experiment in which

- Preparation procedure P is carried out.
- Measurement procedure M is performed.
- Outcome m is recorded.

In a joint measurement scenario one is interested in the case where the measurement procedure M decomposes into two subprocedures x and y, with respective outcomes a and b, i.e.,

$$M \equiv (x, y), \qquad m \equiv (a, b).$$

Statistical correlations between these outcomes are represented by some joint probability density $p(m|M, P) = p(a, b|x, y, P)$, which can be in principle measured via many repetitions of the experiment. Part of the physicist's job is to find an underlying model for these correlations, for a given set of experiments $\{(x, y, P)\}$.

In particular, an underlying model introduces additional physical variables of some sort. Denoting these underlying variables by λ, Bayes theorem immediately tells us that

$$p(m|M, P) = p(a, b|x, y, P) = \int d\lambda \, p(a, b|\lambda, x, y, P) \, p(\lambda|x, y, P), \quad (11.1)$$

with integration replaced by summation over any discrete ranges of λ. A given model must therefore specify the type of information encoded in λ, and the underlying probability densities $p(a, b|\lambda, x, y, P)$ and $p(\lambda|x, y, P)$.

For example, in standard quantum mechanics λ may be taken to range over a set of density operators, with

$$p_Q(a, b|\lambda, x, y, P) = \text{tr}[\lambda E_{ab}^{xy}], \qquad p_Q(\lambda|x, y, P) = \delta(\lambda - \lambda_P), \quad (11.2)$$

for some density operator λ_P associated with preparation procedure P and some positive operator valued measure (POVM) $\{E_m^M \equiv E_{ab}^{xy}\}$ associated with the joint measurement procedure $M = (x, y)$.

11.2.2 Bell Separability

A given underlying model may or may not satisfy certain physically plausible properties, such as determinism, causal correlations, etc. In the scenario typically consid-

ered for Bell inequalities [2], one requires the quantities in Eq. (11.1) to satisfy the
following three properties in particular:

Statistical completeness: *All statistical correlations arise from ignorance of the
underlying variable*, i.e.,

$$p(a, b|\lambda, x, y, P) = p(a|\lambda, x, y, P)\, p(b|\lambda, x, y, P). \tag{11.3}$$

Thus, all correlations between measurement outcomes vanish when the additional
information encoded in λ is specified [20]. This property is also known as outcome
independence, and is guaranteed to hold for *deterministic* models, i.e., for

$$p(a, b|\lambda, x, y, P,) \in \{0, 1\}.$$

Indeed, the existence of a deterministic model for a given set of correlations is
equivalent to the existence of a statistically complete model [11, 21]. The original
motivation for statistical completeness was in fact outcome determinism, via an
appeal to the existence of an underlying reality in which all measurement outcomes
are predetermined [2].

Statistical locality: *Distant measurement subprocedures do not influence each
other's underlying outcome probability distributions*, i.e.,

$$p(a|\lambda, x, y, P) = p(a|\lambda, x, P), \qquad p(b|\lambda, x, y, P) = p(b|\lambda, y, P). \tag{11.4}$$

Thus, an observer cannot distinguish, via any local measurement x, whether a distant
observer has carried out measurement y or y', even given knowledge of the underlying
variable λ. This property, also known as parameter independence, is justified by
the principle of relativity when the measurement subprocedures are carried out in
spacelike separated regions [2].

Measurement independence: *The measurement procedure $M = (x, y)$ is not correlated with the underlying variable*, i.e.,

$$p(\lambda|x, y, P) = p(\lambda|P). \tag{11.5}$$

Thus, knowledge of the underlying variable gives no information about the measurement procedure, and vice versa. This property is often justified by an appeal to
'experimental free will' [5], as will be discussed in some detail further below.

   The combination of all three properties is equivalent, via Eqs. (11.1) and (11.3)–
(11.5), to:

Bell separability: The joint probabilities of distantly-performed measurement procedures have an underlying model of the form

$$p(a, b|x, y, P) = \int d\lambda \, p(\lambda|P) \, p(a|\lambda, x, P) \, p(b|\lambda, y, P), \qquad (11.6)$$

i.e., a model satisfying statistical completeness, statistical locality and measurement independence.

Bell separable models were first introduced by Bell [2] (see Ref. [22] for a recent review), and capture the notion that statistical correlations between distant regions separate into independent contributions, distinguished by their dependencies on the measurement subprocedures as per Eq. (11.6).

11.2.3 Bell Inequalities

Statistical correlations which have a Bell separable model satisfy various inequalities, known as Bell inequalities. For example, if each measurement outcome is labelled by ±1, Bell separability implies that the Clauser-Horne-Shimony-Holt (CHSH) inequality [3]

$$E(x, y, P) + E(x, y', P) + E(x', y, P) - E(x', y', P) \leq 2 \qquad (11.7)$$

holds for any four pairs of distantly performed measurement procedures (x, y), (x, y'), (x', y) and (x', y'), where $E(x, y, P) := \sum_{a,b=\pm 1} a b \, p(a, b|x, y, P)$.

It is now well known that not only do the predictions of standard quantum mechanics violate such Bell inequalities: so does nature [23, 24]. Hence, our world is Bell-nonseparable, and one or more of the three properties in Eqs. (11.3)–(11.5) must be relaxed in any underlying model thereof.

It is worth noting that the existence of a Bell separable model as per Eq. (11.6), for some given set of joint measurement procedures $\{(x, y)\}$ and preparation procedure P, is also equivalent to the existence of a formal joint probability distribution for any finite subset $(x_1, \ldots, x_m, y_1, \ldots y_n)$—whether or not this subset has an experimental joint implementation. In particular, one may define [21]

$$p_F(a_1, \ldots, a_m, b_1, \ldots b_n|x_1, \ldots, x_m, y_1, \ldots y_n, P) := \int d\lambda p(\lambda|P) \prod_{j=1}^{m} p(a_j|\lambda, x_j, P)$$

$$\times \prod_{k=1}^{n} p(b_k|\lambda, y_k, P). \qquad (11.8)$$

Bell inequalities correspond to boundary inequalities for the space (polytope) of such formal joint probability distributions [6, 25–27].

11.3 Why 'Quantum Nonlocality' Is Bad Terminology

The violation of Bell inequalities by quantum systems is often referred to as 'quantum nonlocality' or 'Bell nonlocality'. To do so is quite misleading, however, as it implicitly—and incorrectly—suggests that some sort of mysterious action-at-a-distance is necessarily involved in quantum correlations.

In particular, only *one* of the three assumptions in Sect. 11.2.2 makes reference to notions of locality: *statistical locality* requires that a distant measurement procedure cannot be identified from local statistics. The other two assumptions, statistical completeness and measurement dependence, do not require any notion of acausal information transfer between distant regions (as shown explicitly in Sects. 11.4 and 11.5). Hence, by dropping either one of these two assumptions, Bell inequality violations may be modelled in a perfectly local manner.

Further, the property of statistical locality is automatically satisfied in standard quantum mechanics, via the usual tensor product representation

$$E_{ab}^{xy} = E_a^x \otimes E_b^y \qquad (11.9)$$

of the POVM in Eq. (11.2) for measurements in separated regions [28]. Hence, quantum mechanics is in fact *local*, with respect to the only sense in which this concept makes an appearance in the derivation of Bell inequalities!

This has led Mermin to conclude that use of the term 'quantum nonlocality' is no more than "fashion at a distance" [29], and Kent to lament it as a "confusingly oxymoronic phrase" that conflicts with the notion of locality in quantum field theory [30]. An extended critique has been given recently by Żukowksi and Brukner [31] (see also Ref. [32]).

While 'Bell nonlocality' is preferable to 'quantum nonlocality', insofar as the adjective vaguely implies some sort of special qualification, essentially the same criticisms apply. Moreover, since quantum communication protocols that rely on the violation of Bell inequalities *require* the assumption of statistical *locality* (and measurement independence), to ensure indeterminism (see Sect. 11.1), it is similarly 'confusingly oxymoronic' to assert that such protocols rely on Bell *nonlocality*, as is commonly done [6].

I therefore strongly urge adoption of the more neutral term 'Bell nonseparability'.

11.4 Local Deterministic Models of Quantum Correlations

11.4.1 Relaxing Measurement Independence

As noted previously, violation of a Bell inequality, and hence of Bell separability, implies that at least one of the properties in Eqs. (11.3)–(11.5) must be relaxed in any underlying model. The degree to which these properties need to be individually or jointly relaxed, relative to various measures, has been recently reviewed [11].

For example, the standard quantum mechanics model satisfies the property of statistical locality in Eq. (11.4), as noted in the previous section. Comparison of Eqs. (11.2) and (11.5) shows that it also satisfies the property of measurement independence. Hence, since it predicts violations of Bell inequalities, it follows that the standard quantum mechanics model must relax the property of statistical completeness. This is indeed so: the joint probability $p_Q(a, b|\lambda, x, y, P)$ in Eq. (11.2) is only guaranteed to factorise, as per Eq. (11.3), for tensor product states $\lambda = \lambda_1 \otimes \lambda_2$.

A major contribution by Brans was to provide, in contrast, the first explicit *local deterministic* model of quantum correlations, by instead relaxing the assumption of measurement independence [7]. The existence of such a fully causal model for Bell-nonseparable correlations further emphasises the point made in Sect. 11.3, that the properties of statistical completeness and measurement independence do not rely on any concept of locality *per se*.

Brans' model for the singlet state of two spin-1/2 particles or qubits, together with two subsequent models, are briefly described in Sect. 11.4.2, before discussing the generalisation of Brans' model to arbitrary quantum correlations in Sect. 11.4.3. However, it is of interest to first quantify the degree of measurement dependence of any given model, so as to be able to make quantitative comparisons between different models.

Several measures of measurement independence have been discussed in the literature [9–16]. Attention here will be confined to the 'measurement dependence capacity', which directly quantifies the correlation between the joint measurement procedure and underlying variable in terms of the maximum mutual information between them [10]:

$$C_{MD} := \sup_{p(x,y)} H(\Lambda : X, Y) = \sup_{p(x,y)} \int d\lambda dx dy\, p(\lambda, x, y|P) \log_2 \frac{p(\lambda, x, y|P)}{p(\lambda|P)\, p(x, y)}.$$

(11.10)

Here the supremum is over all possible probability densities $p(x, y)$ for the joint measurement procedure; $H(\Lambda : X, Y)$ denotes the mutual information; and $p(\lambda, x, y, |P) := p(\lambda|x, y, P)\, p(x, y)$. Note that the mutual information quantifies the average information gained about the measurement procedure from knowledge of the underlying variable, and vice versa, in terms of the number of bits required to represent the information [33].

The above measure vanishes if and only the measurement independence condition in Eq. (11.5) is satisfied. A useful upper bound follows via [11]

$$C_{MD} = \sup_{p(x,y)} \left[H[\Lambda] - \int dx dy\, p(x, y) H_{x,y}(\Lambda) \right] \leq H_{max}(\Lambda) - \inf_{x,y} H_{x,y}(\Lambda),$$

(11.11)

where $H(\Lambda)$ denotes the entropy of the underlying variable λ, with maximum possible value $H_{max}(\Lambda)$, and $H_{x,y}(\Lambda)$ denotes the entropy of $p(\lambda|x, y, P)$.

11.4.2 Singlet State Models

Brans model: Letting P_S denote a preparation procedure for the singlet state $|\Psi_S\rangle$, the Brans model in its simplest form corresponds to choosing $\lambda \equiv (\lambda_1, \lambda_2)$, with $\lambda_1, \lambda_2 = \pm 1$, and identifying the labels x and y with measurement of spin in the corresponding unit directions x and y. The corresponding probabilities in Eq. (11.1) are then specified, via Eq. (11) of Ref. [7], by

$$p_B(a, b|\lambda, x, y, P_S) := \delta_{a,\lambda_1} \delta_{b,\lambda_2}, \qquad p_B(\lambda|x, y, P_S) := \frac{1 - \lambda_1 \lambda_2 \, x \cdot y}{4}.$$
$$(11.12)$$

This trivially reproduces the correct singlet state probabilities $p(a, b|x, y, P_S) = (1 - abx \cdot y)/4$, via Eq. (11.1). The model is deterministic, and clearly satisfies the properties of statistical completeness and statistical locality in Eqs. (11.3) and (11.4)—but not the measurement independence property in Eq. (11.5). Brans further showed that the correlation existing between the underlying variable λ and the measurement procedures x and y can be simulated causally and deterministically [7], as will be discussed in Sect. 11.5.

To evaluate C_{MD} for this model, note first that λ takes only 4 distinct values, implying that $H_{max}(\Lambda) = \log_2 4 = 2$ bits. A straightforward calculation further gives

$$H_{x,y}(\Lambda) = \log_2 2 + h(x \cdot y) \geq 1 \text{ bit,}$$

where $h(a)$ denotes the entropy of the probability distribution $\{(1 \pm a)/2\}$. Hence, using Eq. (11.11), the degree of measurement dependence in Eq. (11.10) is bounded by 1 bit of correlation. It is straightforward to check that this bound is achieved by the choice $p(x, y) = [\delta(x + y) + \delta(x - y)]/(8\pi)$ in Eq. (11.10), where x and y range over all directions on the unit sphere, yielding

$$C_{MD}^B = 1 \text{ bit.} \qquad (11.13)$$

Thus, no more than one bit of correlation is required in the Brans model for the singlet state. This result is generalised in Sect. 11.4.3.

Degorre et al. model: The local deterministic model of the singlet state due to Degorre et al. takes the underlying variable λ to be a point on the unit sphere, with [8]

$$p_D(a, b|\lambda, x, y, P_S) := \delta_{a,A(\lambda,x)} \delta_{b,B(\lambda,y)}, \qquad p_D(\lambda|x, y, P_S) := \frac{|\lambda \cdot x|}{2\pi}, \quad (11.14)$$

where $A(\lambda, x) := \text{sign} \, \lambda \cdot x$ and $B(\lambda, y) := -\text{sign} \, \lambda \cdot y$ determine the local outcomes. Thus, these outcomes correspond to the projections of 'classical' spin vectors, λ and $-\lambda$, onto the measurement directions x and y respectively. As in the Brans model, the properties of statistical completeness and statistical locality are clearly satisfied, while the property of measurement independence is clearly not. Note from Eq. (11.14) that λ is only correlated with *one* of the local measurement directions, x. This model was also independently put forward by Barrett and Gisin [10].

To calculate C_{MD} for the Degorre et al. model, note that the entropy of the underlying variable is maximised by a uniform distribution over the unit sphere, with $H_{\max}(\Lambda) = \log_2 4\pi$. Further, letting θ denote the angle between λ and x, one has

$$H_{x,y}(\Lambda) = -\int_0^\pi d\theta \int_0^{2\pi} d\phi \, \sin\theta \, \frac{|\cos\theta|}{2\pi} \log_2 \frac{|\cos\theta|}{2\pi} = \log_2 2\pi e^{1/2}.$$

Hence, the upper bound in Eq. (11.11) is $\log_2 2/\sqrt{e}$. This bound is achieved, for example, by $p(x, y) = (4\pi)^{-1} p(y)$, yielding [10]

$$C_{MD}^{D} = \log_2 \frac{2}{\sqrt{e}} \approx 0.279 \text{ bits.} \tag{11.15}$$

Thus, noting Eq. (11.13), the Degorre et al. model requires less correlation between the measurement directions and the underlying variable, in comparison to the Brans model. Moreover, this correlation is only required between λ and *one* of the local measurement directions.

Hall model: Finally, it is of interest to consider a local deterministic model with an even smaller degree of measurement dependence [9]. The underlying variable is again a point on the unit sphere, corresponding to a 'classical' spin vector, and again the local outcomes are determined by the projection of this spin vector onto the local measurement directions, via

$$p_H(a, b | \lambda, x, y, P_S) := \delta_{a, A(\lambda, x)} \, \delta_{b, B(\lambda, y)}, \tag{11.16}$$

with $A(\lambda, x) = \text{sign } \lambda \cdot x$ and $B(\lambda, y) = -\text{sign } \lambda \cdot y$ as for the Degorre et al. model. However, in contrast to the latter model,

$$p_H(\lambda | x, y, P_S) := \frac{1}{4\pi} \frac{1 + (x \cdot y) \, \text{sign } [(\lambda \cdot x)(\lambda \cdot y)]}{1 + (1 - 2\phi_{xy}/\pi) \, \text{sign } [(\lambda \cdot x)(\lambda \cdot y)]}, \tag{11.17}$$

where $\phi_{xy} \in [0, \pi]$ denotes the angle between directions x and y.

The interesting aspect of this model is its low degree of measurement dependence, with [11]

$$C_{MD}^{H} \approx 0.0663 \text{ bits.} \tag{11.18}$$

Thus, in comparison to Eqs. (11.13) and (11.15), a remarkably low degree of correlation, less than $1/15$ of a bit, is required to model the singlet state.

11.4.3 Generalising the Brans Model to Arbitrary Quantum States

While the Brans model of the singlet state is not optimal with respect to the degree of measurement dependence required, it does have the significant advantage of being easily generalisable to a local deterministic model for *all* quantum correlations, with a corresponding simple upper bound for the degree of measurement dependence required.

In particular, consider a preparation procedure P corresponding to some quantum density operator ρ describing two quantum systems, where these systems have a d_1-dimensional and a d_2-dimensional Hilbert space respectively. Further, consider an arbitrary joint measurement of two Hermitian observables, x and y, on these systems, with corresponding POVM $\{E_{ab}^{xy}\}$. The joint outcome (a, b) may be labelled by elements of the set $O := \{1, 2, \ldots, d_1\} \times \{1, 2, \ldots, d_2\}$, without any loss of generality. To construct a corresponding local deterministic model, we choose the underlying variable to be $\lambda = (\lambda_1, \lambda_2) \in O$, and generalise Eq. (11.12) to

$$p_B(a, b|\lambda, x, y, P) := \delta_{a,\lambda_1} \delta_{b,\lambda_2}, \qquad p_B(\lambda|x, y, P) := \text{tr}[\rho E_{\lambda_1\lambda_2}^{xy}] \qquad (11.19)$$

Substitution into Eq. (11.1) immediately recovers the quantum probability density $p(a, b|x, y, P) = \text{tr}[\rho E_{ab}^{xy}]$, as required.

To obtain an upper bound for the degree of measurement dependence required in this model, note that $0 \leq H(\Lambda) \leq \log_2 d_1 d_2$, with the upper bound corresponding to a uniform distribution over λ. Hence, from Eq. (11.11), one immediately has the upper bound

$$C_{MD}^B \leq \log_2 d_1 + \log_2 d_2 \qquad (11.20)$$

for the degree of measurement independence. In particular, no more than $2\log_2 d$ bits of correlation are required to model the statistics of all Hermitian observables on two d-dimensional quantum systems. Note that the bound holds irrespective of whether the corresponding POVMs factorise as per Eq. (11.9). It would be of interest to determine whether the bound also holds for non-Hermitian observables, i.e., for arbitrary joint POVMs.

11.5 Questions of Causality and Free Will

11.5.1 Causality in Measurement Dependent Models

Brans did more than give the first explicit local and deterministic model for singlet state correlations. He also showed that the corresponding probability distribution $p_B(\lambda|x, y, P)$ was compatible with a fully causal explanation, and that it did not contradict the notion of 'experimental free will' in any operational sense [7]. The causal aspect will be discussed here, and the free will aspect in Sect. 11.5.2. The surprising existence of a local deterministic model for *superluminal* correlations is given in Sect. 11.5.3.

The violation of measurement independence, i.e., a correlation such that

$$p(\lambda|x, y, P) \neq p(\lambda|P),$$

may at first sight suggest that the joint measurement procedure (x, y) has a causal effect on the statistics of λ. However, this is not so: using Bayes rule the above equation can equivalently be written in either of the forms

$$p(x, y|\lambda, P) \neq p(x, y|P), \qquad p(\lambda, x, y|P) \neq p(\lambda|P) \, p(x, y|P),$$

where the latter form is seen to be perfectly symmetrical with respect to the measurement procedure and λ. Correlation does not specify causation.

In fact, for any measurement-dependent correlation, $p_0(\lambda|x, y, P) \neq p_0(\lambda|P)$ say, and any distribution of joint measurement procedures, $p_0(x, y|P)$ say, the corresponding probability density $p_0(x, y|\lambda, P) \neq p_0(x, y|P)$ is uniquely determined by the laws of probability. It is straightforward to construct a causal model for this probability density, of the form

$$p_0(x, y|\lambda, P) = \int d\mu \, p(x|\mu) \, p(y|\mu) \, p(\mu|\lambda, P), \qquad (11.21)$$

where μ is a further underlying variable. It is clear from this equation that the correlation can be causally implemented via generation of the distribution of μ by λ and P, with subsequent local generation of the distribution of the measurement subprocedures x and y by μ, with no retrocausal or superluminal propagation required.

As an explicit example, choose $\mu \equiv (\mu_1, \mu_2)$, where (μ_1, μ_2) labels the set of possible joint measurement procedures $\{(x, y)\}$, with $p(x|\mu) := \delta(x - \mu_1)$, $p(y|\mu) := \delta(y - \mu_2)$, and

$$p(\mu|\lambda, P) := \frac{p_0(\lambda|\mu_1, \mu_2, P) \, p_0(\mu_1, \mu_2|P)}{\int\int d\mu_1 d\mu_2 \, p_0(\lambda|\mu_1, \mu_2, P) \, p_0(\mu_1, \mu_2|P)}. \qquad (11.22)$$

It is straighforward to check that these choices reproduce Eq. (11.21), as desired. Thus no violation of causality is required by measurement dependent models, such as those in Sect. 11.4.

11.5.2 Free Will and Conspiracy

Brans noted that the assumption of measurement independence is fundamentally inconsistent with a fully deterministic world [7]. In such a world even the preparation procedure P, along with the measurement $M = (x, y)$ and the outcomes $m = (a, b)$, will be predetermined by suitable underlying variables, and hence are generically all correlated. Thus, in a superdeterministic world, assuming that x and y are *not* correlated with the underlying variables, i.e., measurement independence, amounts to conspiracy! In such a world, measurement dependence—and hence the possibility of Bell inequality violation—is only to be expected [34].

However, as previously remarked, measurement independence is typically justified by an appeal to 'experimental free will': surely the selection of measurement procedures is independent of any underlying physical variables that determine the outcomes? For if they were not, surely this would compromise our perception of having the free will to make such a selection?

There are several responses that can be made in this regard, in addition to the obvious point that the subjective experience of free will does not imply its objectivity. The first is practical: in actual tests of Bell nonseparability and Bell inequalities, physical systems rather than physicists are used to 'randomly' select measurement procedures [23, 24]. This fact has practical relevance for the security of commercial quantum cryptographic devices that contain such systems: how can we trust random number generators built by a third party? [13–15]. We certainly do not expect these devices to have 'free will', and their degree of measurement dependence is easily manipulated by the device manufacturer.

The second is operational: there is no experimental distinction that can be made between models satisfying measurement independence and models which do not [7]—at least, under the proviso that the distribution of measurement procedures is independent of the preparation procedure, $p(x, y|P) = p(x, y)$. In such a case, all that is operationally accessible to the 'free will' of the experimenter(s) in this regard is the choice of $p(x, y)$. However, *any* such choice is compatible with measurement-dependent models: it merely implies that the operationally-inaccessible joint distribution $p(\lambda, x, y|P)$ is given by $p(\lambda|x, y, P) p(x, y)$ [11].

The third is rhetorical: suppose that experimenters were informed that there was a physical quantity they could not change: no matter what choices of preparation and measurement procedures they made, using their 'free will', the quantity mysteriously came out to be the same—even for joint measurements in spacelike separated regions. Would this necessarily represent a lack of 'free will'? No, not if the quantity was the total energy! Conservation laws are not considered to be conspiratorial. This suggests the intriguing possibility of a local deterministic model for quantum systems in which $p(x, y)$ emerges as a conserved quantity [35].

Thus, there is no *a priori* reason why the behaviour of experimenters or random generators should *not* be statistically correlated with a given system to some degree, reflecting a common causal dependence on some underlying variable, even in the absence of superdeterminism and/or in the presence of 'free will'. However, it must be admitted that a measurement-dependent model in which $p(x, y)$ emerged as a conserved quantity would be far more compelling than those presented in Sect. 11.4.

11.5.3 Local Deterministic Models of Superluminal Correlations

It is remarkable to note that even correlations which appear to allow superluminal signalling can be modelled in a local and deterministic manner, by relaxing the assumption of measurement independence.

For example, consider some preparation procedure P and two joint measurement procedures $M = (x, y)$ and $M' = (x', y)$, with corresponding experimental joint probability distributions satisfying

$$p(b|x, y, P) \neq p(b|x', y, P). \tag{11.23}$$

Thus, knowledge of the local outcome distribution of procedure y provides information about whether the procedure x or x' was performed. This is an example of a 'signalling' correlation. Such a correlation is not surprising in the case that the measurement subprocedure y is performed in the future lightcone of subprocedure x or x'—this would simply represent the possibility of signalling from the past to the future. However, such a correlation would be very surprising in the case of spacelike-separated subprocedures x and y, as it would appear to amount to the possibility of superluminal signalling.

Surprisingly, perhaps, a local deterministic model for such signalling correlations is easily obtained, via a straightforward adaptation of the extended Brans model discussed in Sect. 11.4.3. In particular, for a given set of experimental joint probability distributions $\{p_E(a, b|x, y, P)\}$, choose the underlying variable to range over the set of possible joint measurement outcomes, with $\lambda = (\lambda_1, \lambda_2) \in \{(a, b)\}$, and define

$$p(a, b|\lambda, x, y, P) := \delta_{a,\lambda_1} \delta_{b,\lambda_2}, \qquad p(\lambda|x, y, P) := p_E(\lambda_1, \lambda_2|x, y, P). \tag{11.24}$$

This model is explicitly deterministic, and clearly satisfies both statistical completeness and statistical locality, *whether or not* the experimental correlations are signalling! Further, a causal description of the measurement-dependent correlation $p(\lambda|x, y, P)$ can always be given, as per the discussion in Sect. 11.5.1.

The resolution of this paradoxical result is that the very notion of 'signalling' logically requires some degree of measurement independence: if one has, for example, no control at all over the choice of measurement x or x' in Eq. (11.23), then one has no ability at all to signal—e.g., 'buy' or 'sell'—via such a choice [14].

It follows that one should not refer to 'no-signalling' or 'signal locality' without a simultaneous commitment to measurement independence (I must recant having done so previously [11]). Moreover, a simple tweak of the above model implies that it is possible to replace any underlying model that violates the property of statistical locality by one that instead violates measurement independence. Indeed, in this regard Barrett and Gisin have previously shown that any underlying deterministic model that requires at most m bits of superluminal communication can be replaced by one that requires $C_{MD} \leq m$ bits of measurement-dependent correlation [10].

11.6   Conclusions

One of the most remarkable discoveries in physics is the violation of Bell separability by quantum phenomena: any underlying model of such phenomena must relax at least one of the properties of statistical completeness, statistical locality or measurement independence. There is a strong intuition among physicists that perfect correlations between distant measurement outcomes, such as singlet state correlations, should be deterministically and locally mediated, independently of the joint measurement procedure. However, this intuition fails in the light of Bell inequality violation.

Given that standard quantum mechanics satisfies statistical locality and measurement independence, Occam's razor suggests that it is the intuition behind determinism (and thus statistical completeness) that must be given up. On the other hand, it may be argued that relaxing measurement dependence is relatively far more efficient: only 1/15 of a bit of measurement dependence is required to model the singlet state, in comparison to 1 bit of communication in nonlocal models, and 1 bit of shared randomness in nondeterministic models [11]. In the end, however, whether or not one's personal preference is guided by simplicity or efficiency, the consideration of all three properties cannot be avoided—and is of practical relevance in assessing the reliability of device-independent quantum communication protocols.

It is a pleasure to be able to acknowledge the seminal contribution of Carl Brans to this ongoing debate, as part of this Festschrift to mark his 80th birthday. His explicit local deterministic model for quantum correlations has led to a better understanding of the significance of measurement (in)dependence, and has stimulated many new results and ideas. He will no doubt be pleased that one of the latter [18] may lead to an experimental connection with his many cosmological interests: the recent proposal to test measurement independence in a Bell inequality experiment by using the light from distant quasars that have never been in causal contact.

Acknowledgments I thank Cyril Branciard, Antonio Di Lorenzo, Jason Gallicchio, Nicolas Gisin, Valerio Scarani, Joan Vaccaro, Mark Wilde and Howard Wiseman for various stimulating discussions on this topic over the past few years. This work was supported by the ARC Centre of Excellence CE110001027.

References

1. W. Heisenberg, *Physics and Beyond* (Harper and Row, New York, 1971), p. 206
2. J.S. Bell, On the Einstein Podolsky Rosen paradox. Physics **1**, 195–200 (1964)
3. M.F. Clauser, M.A. Horne, A. Shimony, R.A. Holt, Proposed experiment to test local hidden-variable theories. Phys. Rev. Lett. **23**, 880–884 (1969)
4. J.S. Bell, The theory of local beables. Preprint CERN-TH 2053/75, reprinted in. Dialectica **39**, 86–96 (1985)
5. J.S. Bell, A. Shimony, M.A. Horne, J.F. Clauser, An exchange on local beables. Dialectica **39**, 85–110 (1985)
6. N. Brunner, D. Cavalcanti, S. Pironio, V. Scarani, S. Wehner, Bell nonlocality. Rev. Mod. Phys. **86**, 419–478 (2014)

7. C. Brans, Bell's theorem does not eliminate fully causal hidden variables. Int. J. Theoret. Phys. **27**, 219–226 (1988)
8. J. Degorre, S. Laplante, J. Roland, Simulating quantum correlations as a distributed sampling problem. Phys. Rev. A **72**, 062314 (2005)
9. M.J.W. Hall, Local deterministic model of singlet state correlations based on relaxing measurement independence. Phys. Rev. Lett. **105**, 250404 (2010)
10. J. Barrett, N. Gisin, How much measurement independence is needed in order to demonstrate nonlocality? Phys. Rev. Lett. **106**, 100406 (2011)
11. M.J.W. Hall, Relaxed Bell inequalities and Kochen-Specker theorems. Phys. Rev. A **84**, 022102 (2011)
12. M. Banik, M.D.R. Gazi, S. Das, A. Rai, S. Kunkri, Optimal free will on one side in reproducing the singlet correlation. J. Phys. A **45**, 205301 (2012)
13. D.E. Koh, M.J.W. Hall, J.E. Pope, C. Marletto, A. Kay, V. Scarani, A.E. Ekert, Effects of reduced measurement independence on Bell-based randomness expansion. Phys. Rev. Lett. **109**, 160404 (2012)
14. L.P. Thinh, L. Sheridan, V. Scarani, Bell tests with min-entropy sources. Phys. Rev. A **87**, 062121 (2013)
15. J.E. Pope, A. Kay, Limited measurement dependence in multiple runs of a Bell test. Phys. Rev. A **88**, 032110 (2013)
16. G. Pütz, D. Rosset, T.J. Barnea, Y.-C. Liang, N. Gisin, Arbitrarily small amount of measurement independence is sufficient to manifest quantum nonlocality. Phys. Rev. Lett. **113**, 190404 (2014)
17. R. Chaves, R. Kueng, J.B. Brask, D. Gross, Unifying framework for relaxations of the causal assumptions in Bell's theorem. Phys. Rev. Lett. **114**, 140403 (2015)
18. J. Gallicchio, A.S. Friedman, D.I. Kaiser, Testing Bell's inequality with cosmic photons: closing the setting-independence loophole. Phys. Rev. Lett. **112**, 110405 (2014)
19. D. Aktas, S. Tanzilli, A. Martin, G. Pütz, R. Thew, N. Gisin, Demonstration of quantum nonlocality in the presence of measurement dependence. Phys. Rev. Lett. **114**, 220404 (2015)
20. J.P. Jarrett, On the physical significance of the locality conditions in the Bell arguments. Noûs **18**, 569–589 (1984)
21. A. Fine, Joint distributions, quantum correlations, and commuting observables. J. Math. Phys. **23**, 1306–1310 (1982)
22. H.M. Wiseman, E.C.G. Cavalcanti, Causarum Investigatio and the two Bell's theorems of John Bell (2015). Eprint: arXiv:1503.06413 [quant-ph]
23. A. Aspect, G. Dalibard, G. Roger, Experimental test of Bell inequalities using time-varying analyzers. Phys. Rev. Lett. **49**, 1804–1807 (1982)
24. B. Hensen, H. Bernien, A.E. Dréau, N. Reiserer et al., Loophole-free Bell inequality violation using electron spins separated by 1.3 kilometres. Nature **526**, 682–686 (2015)
25. A. Fine, Hidden variables, joint probability, and the Bell inequalities. Phys. Rev. Lett. **48**, 291–295 (1982)
26. A. Garg, N.D. Mermin, Farkas's lemma and the nature of reality: statistical implications of quantum correlations. Found. Phys. **14**, 1–39 (1984)
27. I. Pitowsky, George Boole's 'conditions of possible experience' and the quantum puzzle. Brit. J. Phil. Sci. **45**, 95–125 (1994)
28. M.J.W. Hall, Imprecise measurements and non-locality in quantum mechanics. Phys. Lett. A **125**, 89–91 (1987)
29. N.D. Mermin, What do these correlations know about reality? Nonlocality and the absurd. Found. Phys. **29**, 571–587 (1999)
30. A. Kent, Locality and reality revisited, in *Quantum Locality and Modality*, ed. by T. Placek, J. Butterfield (Kluwer, Dordrecht, 2002), pp. 163–171
31. M. Żukowski, C. Brukner, Quantum nonlocality–it ain't necessarily so. J. Phys. A **47**, 424009 (2014)
32. M.J.W. Hall, Comment on 'Non-realism: deep thought or soft option?', by N. Gisin. Eprint: arXiv:0909.0015 [quant-ph] (2009)
33. R.E. Blahut, *Principles and Practice of Information Theory* (Addison Wesley, New York, 1987)

34. G. 't Hooft, Models on the boundary between classical and quantum mechanics. Phil. Trans. R. Soc. A **373**, 20140236 (2015)
35. A. Di Lorenzo, Private communication (2011)

Part III
Exotic Smoothness and Space-Time Models

"... This result [the existence of an exotic $\mathbb{R}^4$] could have great significance in all fields of physics, not just relativity, Some model of spacetime underlies every field of physics. It has now been proven that we cannot infer that space is necessarily smoothly standard from investigating what happens at spatial infinity, even for topologically trivial $\mathbb{R}^4$. It seems very clear that this is potentially very important to all of physics since it implies that there is another possible obstruction, in addition to material sources and topological ones, to continuing external vacuum solutions for any field equations from infinity to the origin. Of course, in the absence of any explicit coordinate patch presentation, no example can be displayed. However, this leads naturally to a conjecture, informally stated as:

Class. Quantum Grav. 11 (1994) 1785–1792. Printed in the UK

Localized exotic smoothness

Carl H Brans†
Institute for Advanced Study, Princeton, NJ 08540, USA
and
Physics Department, Loyola University, New Orleans, LA 70118, USA

Received 6 December 1993, in final form 5 April 1994

General Relativity and Gravitation, Vol. 34, No. 10, October 2002 (© 2002)

LETTER

Cosmological Anomalies and Exotic Smoothness Structures

Torsten Aßelmeyer-Maluga[1] and Carl H. Brans[2]

Received March 6, 2002

Conjecture 1. This localized exoticness can act as a source for some externally regular field, just as matter or a wormhole can."

"... In summary, what we want to emphasize is that without changing the Einstein equations or introducing exotic, yet undiscovered forms of matter, or even without changing topology, there is a vast resource of possible explanations for recently observed surprising astrophysical data at the cosmological scale provided by differential topology. ..."

Chapter 12
Exotic Smoothness, Physics and Related Topics

Jan Słdkowski

Abstract In 1854 Riemann, the father of differential geometry, suggested that the geometry of space may be more than just a mathematical tool defining a stage for physical phenomena, and may in fact have profound physical meaning in its own right. Since then various assumptions about the spacetime structure have been put forward. But to what extent the choice of mathematical model for spacetime has important physical significance? With the advent of general relativity physicists began to think of the spacetime in terms a differential manifolds. In this short essay we will discuss to what extent the structure of spacetime can be determined (modelled) and the possible role of differential calculus in the due process. The counterintuitive discovery of exotic four dimensional Euclidean spaces following from the work of Freedman and Donaldson surprised mathematicians. Later, it has been shown that exotic smooth structures are especially abundant in dimension four—the dimension of the physical spacetime. These facts spurred research into possible the physical role of exotic smoothness, an interesting but not an easy task, as we will show.

12.1 Introduction

The outcomes of physical measurements are expressed in rational numbers: all meaningful measurements are performed with certain accuracy and it is even hard to imagine how can they produce an irrational number. Nevertheless, we believe that all possible values of physical variables constitute the set of real numbers $\mathbb{R}$. Most of physical theories, including quantum gravity, use the concept of spacetime as scene of physical processes, at least approximately. We also suppose that the spacetime, whatever it actually is, can be faithfully modeled as a manifold. At this stage the algebra of real continuous functions $C(M)$ on the spacetime manifold M comes to play [16]. This algebra play central rôle in physics, although this fact is not always

J. Słdkowski (✉)
Institute of Physics, University of Silesia, Uniwersytecka 4,
PL 40-007 Katowice, Poland
e-mail: jan.sladkowski@us.edu.pl

© Springer International Publishing Switzerland 2016
T. Asselmeyer-Maluga (ed.), *At the Frontier of Spacetime*,
Fundamental Theories of Physics 183, DOI 10.1007/978-3-319-31299-6_12

stressed or even perceived. It is tightly intertwined with one of the most important and fundamental open problems in theoretical physics: to explain the origin and structure of spacetime and to analyse how faithful our theoretical models of the spacetime can be. We will suppose that it is possible to determine the algebra of, say continuous, $K-$valued functions $C(M, K)$ defined on the spacetime (assumed to be a topological space) with sufficient for our aim accuracy for various K. Actually, if we confine our aspirations to the analysis of only local properties of M the algebra in question can be substantially smaller. This does not mean that we have to be able to find all elements of $C(M, K)$ "experimentally": some abstraction or inductive construction would be sufficient. With obvious abuse of language, we will call elements of $C(M, K)$ observables. Our aim is to find out to what extent the structure of the mathematical model of spacetime is determined by $C(M, K)$ for M being a topological space. We will also analyse what happens if we admit of M being a differential manifold or to have no topology. Finally, we will show how $C(M, K)$ can be used to construct a field theory of fundamental interactions in the A. Connes' noncommutative geometry formalism [10].

It is a great honor and pleasure to be able to contribute to this volume celebrating Carl Brans eightieth birthday and his scientific achievements that influenced the development of the theory of gravitation in such a significant way.

12.2   The Topology of Spacetime

A lot of information on a given topological space M is encoded in the associated algebras $C(M, \mathbb{R})$ of continuous real functions defined on M. For any set M the family $\mathcal{C}$ of real functions $M \rightarrow \mathbb{R}$ determines a minimal topology $\tau_\mathcal{C}$ on M such that all function in $\mathcal{C}$ are continuous [16, 28]. It is less known that the reals can be replaced by some other algebraic structures [28, 29]. Therefore, we will also consider $C(M, K)$, the algebras of K-valued functions, K being a topological ring, field, algebra and so on. Suppose that our experimental technique is powerful enough to reconstruct $C(M, K)$ acting on our model of the spacetime M. What information about M provides us with knowledge of $C(M, K)$ information concerning M can be extracted from these data? If M is a set and $\mathcal{C}$ a family of real functions $M \rightarrow \mathbb{R}$ then $\mathcal{C}$ determines a (minimal) topology $\tau_\mathcal{C}$ on M such that all function in $\mathcal{C}$ are continuous [16]. In general, there would exist real continuous functions on M that do not belong to $\mathcal{C}$ and other families of real functions on M can define the same topology on M. The topological space represented by $(M, \tau_\mathcal{C})$ would be a Hausdorff space if and only if for every pair of different points $p_1, p_2 \in M$ there is a function $f \in \mathcal{C}$ such that $f(p_1) \neq f(p_2)$. If the "space" or "time" are continuous by their nature we can hardly imagine any experiment that would be able to discover or distinguish two points "unseparable points".[1] From the physical point of view, to

---

[1] Actually, one can impose various inequivalent forms of such separation "axioms".

be able to distinguish x from y in our model of spacetime we have to find such an observable $f \in C(M)$ that for $x, y \in M$ $f(x) \neq f(y)$. Therefore, sooner or later it seems reasonable to accept that

$$f(x) = f(y) \quad \forall f \in C(M) \quad \Rightarrow \quad x = y. \tag{12.1}$$

From the mathematical point of view, we have to identify all points that are not distinguished by $C(M)$ in the above sense. It is then easy to show that such spaces are Hausdorff spaces. This means that we should look for the topological representation of the spacetime in the class of Hausdorff spaces. Note, that it is standard to postulate even more—one assumes paracompactness of space time manifold—this gives us the powerful tools of differential geometry. To proceed, let us define [16, 19, 28, 29]:

Definition 12.1 Let E be a topological space. A topological Hausdorff space X is called E-compact (E-regular) if it is homeomorphic to a closed (arbitrary) subspace of E^Y, for some Y.

Here E^Y is the space mappings $Y \to E$ (Tychonoff power). One can prove that for a topological space X (not necessarily a Hausdorff one!) we can construct an E-regular space $\tau_E X$ and its E-compact extension $\upsilon_E X$ so that we have [16, 28]

$$C(X, E) \cong C(\tau_E X, E) \cong C(\upsilon_E X, E) \cong C(\upsilon_E \tau_E X, E), \tag{12.2}$$

where $\cong$ denotes isomorphism. The spaces $\tau_E X$ and $\upsilon_E \tau_E X$ have the assumed property (1)! It should be obvious that, in general, our theoretical model of the spacetime would not be unique. This important result also says that we can always model our spacetime as a subset of some Tychonoff power of $\mathbb{R}$ provided $C(M)$ is known. But it also says that we can model it on a subset of a Tychonoff power of a different topological space e.g. the rational numbers $\mathbb{Q}$. Actually, it is our choice. The topological number fields $\mathbb{R}$ and $\mathbb{Q}$ have the very important property of determining uniquely (that is up to a homeomorphism) $\mathbb{R}$- and $\mathbb{Q}$-compact sets provided the appropriate algebras of continuous functions are known:

$$C(X, E) \cong C(Y, E) \iff X \text{ is homeomorphic to } Y, \quad E = \mathbb{R} \text{ or } \mathbb{Q}. \tag{12.3}$$

Other topological rings might also have this property but it is far from being a rule. But this does not mean that the spacetime modelled on $C(M, E)$ is homeomorphic to the one modelled on $C(M, E')$. Hewitt have shown that $\mathbb{R}$-compact spaces are determined up to a homeomorphism by $C(X, E)$, where $E = \mathbb{R}$, $\mathbb{C}$ or $\mathbb{H}$ are the topological fields of real, complex numbers and quaternions, respectively [19]. This means that if we are interested in modelling spacetime on an $\mathbb{R}$-compact space then we can use $C(M, \mathbb{R})$, $C(M, \mathbb{C})$ or $C(M, \mathbb{H})$ to determine it. Unfortunately, such conclusion is false for rational numbers. It a serious obstacle as we probable never get more than rational numbers out of any feasible experiment.

To sum up, $\mathbb{R}$-compact are determined up-to homomorphism from $C(M)$. Consider any space M. There exists the smallest $\mathbb{R}$-compact space υM in which M is dense and υM would be the actual spacetime that we would probably reconstruct. υM need not to be compact—every subset of euclidean space is $\mathbb{R}$-compact, e.g. $\mathbb{R}^4$. $\mathbb{R}$-compact for a sufficiently wide class of functions for our aims: it is not easy to construct a space that is not $\mathbb{R}$-compact. Discrete spaces are $\mathbb{R}$-compact, practically all metrizable spaces are $\mathbb{R}$-compact. The reader is referred to [16, 28, 29] for details.

We will also need to deal with the technical problem of deciding whether we are dealing with the algebra $C(X, E)$ or only with the algebra of all bounded E-valued functions on X, $C^*(X, E)$ (if this concept of boundness make any sense). For a compact space X we have $C(X, E) = C^*(X, E)$, but in general, they are distinct. Spaces on which all continuous real functions are bounded are called pseudocompact. An $\mathbb{R}$-compact pseudocompact space is compact. We might get hints that some observables may in fact be unbounded but we are unlikely to be able to observe infinities. Moreover, with a high probability physical resources are not unlimited but an unbounded observable would be necessary for spacetime to be noncompact. Therefore, if we suppose that we can only recover $C^*(M, \mathbb{R}) \equiv C^*(M)$, then we can as well suppose that M is compact (for an $\mathbb{R}$-compact M). We often compactify configuration spaces by adding extra points or imposing appropriate boundary conditions (e.g. by demanding that all relevant fields vanish at infinity is practically equivalent to the one point compactification of the spacetime[2] A topological space X has more then one such an extension (compactification). Although mathematically one compactification can be differentiated from another with help of regular subrings this is unlikely to be done on the physical ground. Therefore, we will be forced to make various assumptions to choose one among the possible compactifications.

We have argued for looking for the spacetime model in the class of $\mathbb{R}$-compact M spaces. But what if we consider a more general field? It is often conjectured that at the sub-Plankian scales spacetime is non-Archimedean. Such spaces are much less tame do deal with. The Archimedean axiom says that, for any $x \in K$, there is a natural number n such that $|nx| > 1$. An Archimedean field is one in which this axiom holds. Examples are the real numbers and the complex numbers. Technically, there are no other examples: the only complete Archimedean fields are $\mathbb{R}$ and $\mathbb{C}$.[3] A non-Archimedean field is a complete normed field. For a given a non-Archimedean K, the cartesian product K^n is disconnected and it is not easy to follow the insight gained from $\mathbb{R}$ or $\mathbb{C}$. Note that filling in gaps between points (completion) of $\mathbb{Q}$ results in $\mathbb{R}$ but the initial set $\mathbb{Q}$ becomes practically negligible: we can more or less ignore the rational numbers when we do analysis. Therefore, the gap-filling process is performed in a more abstract way by introducing a Grothendieck topology on K^n.

The problems of dimension, density and "tightness" of the spacetime can also be addressed in terms of rings of real continuous functions with various topologies

---

[2]That is we impose that the field in question vanish at the extra point glued to the space.

[3]The Archimedean axiom is also satisfied by C, with powers of the usual norm, and restrictions of these norms to subfields.

although experimental verification of these properties except dimension seems to be unlikely. The reader is referred to [1, 2] for details. The cardinality of the spacetime seem to be to abstract to have any practical significance, there are suggestions that such conclusions may be wrong [31]. Such problems might lead us outside the standard axioms of set theory.

One may also wonder if the knowledge of some symmetries might be of any help. Unfortunately, a topological space X is not determined by its symmetries considered as continuous maps $X \to X$ [15, 22, 27, 30]. Of course it sometimes can provide us with useful information. For example, if we know that some group G acts transitively on X then the cardinality of X cannot be greater than that of G [6]. For example, if we are pretty sure that the Poincare or Galilean symmetry groups act transitively on the spacetime we have got an upper bound on the cardinality of the spacetime. The situation is better if some extra structures are imposed, e.g. demanding existence of lorentzian structure is quite restrictive.

12.3 What if There Is No Topology on the Spacetime?

Up to now we have considered the arbitrariness of our mathematical model X of the spacetime as determined by $C(X, \mathbb{R})$. This means that we assume that M is a topological space. We can also ask to what extent algebras that we identify as an algebra of physical observables on the spacetime actually define a topological space. A commutative algebra must fulfill various sets of conditions to represent a $C(X, \mathbb{R})$ of some topological space X. If we suppose that our model of the spacetime is not a topological space we can deal with $\mathbb{R}^X$, the algebra of all real functions on X. But to have any "selective" power we have to demand the existence of some additional structure on X, for example to distinguish a collection of subsets of X or fix an algebraic structure on the class of functions we consider [22]. If (X, τ) is a pair consisting of a set X and a family τ of its subsets then we can define sort of "continuity" and "homeomorphisms" by replacing topology by the family τ. In this such cases the following theorem holds [16, 28].

Theorem 12.1   *Let X and Y be two sets and τ and σ families of their subsets containing the empty set, closed with respect to finite intersections and summing up to X and Y, respectively. Then X and Y are "homeomorphic" if and only if there is an isomorphism of the semigroups D^X and D^Y such that "$C(X, D)$" is mapped onto "$C(Y, D)$".*

Such generalized space are more difficult to deal with than ordinary topological spaces therefore we think that spacetime should be modelled in the class of topological spaces.

12.4   Differential Structure?

Existence of a differential structure on the spacetime manifold is a nice property.
It certainly is not indispensable. Not every topological space or even topological
manifold can support differential structures and demanding the existence of such
structures severely restricts our choice of models. A differential structure of class
k on a manifold M can be defined by specifying the (sub-)algebra of differentiable
functions $C^k(M, \mathbb{R})$ of the algebra $C(M, \mathbb{R})$. The algebra $C^\infty(M)$ of smooth real
functions on M determines M up to a diffeomophism (the points of M are in one-
to-one correspondence with maximal ideals in $C^\infty(M)$). The algebra of continuous
function on M is much larger than $C^k(M, \mathbb{R})$. If the laws of physics are "smooth"
then the spacetime should be modelled as a smooth manifold. Therefore, in this case
then $C^\infty(M, \mathbb{R})$ is be sufficient to construct M. As any real manifold M can be
embedded in $\mathbb{R}^n$ for some n M is $\mathbb{R}$-compact. The most popular models of spacetime are pseudo-Riemannian manifolds. Such spaces are metrizable. This means that
these manifolds are as topological spaces determined by $C(X, E)$, where $E = \mathbb{R}$, $\mathbb{C}$
or $\mathbb{H}$ but additional knowledge of the subalgebra of differentiable functions is needed
to determine the differential structure. Differential structures might not be unique If
this is the case the "additional" differential structures are usually referred to as *fake*
or *exotic* ones. Surprisingly, they are specially abundant in the four dimensions (one
only needs to remove one point from a given manifold to get a manifold with exotic
structures [12–14, 17, 18]. More astonishing is the fact that the four dimensional
euclidean space $\mathbb{R}^4$ supports uncountably many exotic structures. We have to interpret these mathematical results in physical terms. H. Brans was first who realized this
fact. He has conjectured that exoticness can be a source for some gravitational field,
just as standard matter can [7–9]. One can put forward many arguments that exotic
smoothness might have physical sense [7–9, 24, 25], the actual lack of any tractable
description hinders physical predictions. Nevertheless, recent results of Asselmeyer-
Maluga, Brans and Król shed a new interesting light on this important issue [3–5].
There are suggestions that existence of exotic smoothness has "something" to do
with quantization of various models. Inflation can also be spurred by exoticness.
In this context one can also ask if there is an gravitational analogue of the Bohm-
Aharonov effect. That is if some points are removed or excluded from the spacetime
exotic differential structures do emerge. The "standard" metric tensor and related
structures defined by the distribution of matter might not be smooth with respect to
some exotic differential structures. Such effect could a priori be detected say in cos-
molological/astrophysical observations. This would mean that there is "additional"
curvature required by consistency of differential structures.

The existence of exotic differential structures is certainly a challenge to physicists.
These problems are involved and it is very difficult to distinguish between cause and
effect.

It should be noted here non-Archimedean (p-adic, nonstandard) analysis is also
considered to be the correct mathematical formalism to cope with dynamics at sub-
Planckian levels [23]. Presently, no concrete data can be used for or against such
ideas.

12.5 What if the Spacetime Is Pointless?

We have seen that topology and differential structures can be reconstructed from algebras of functions. These algebras are necessary commutative. Connes [11] made a step forward and showed how to differential geometry without the topological background. The basic structures are a C^*-algebra $\mathcal{A}$ represented in some Hilbert space H and an operator $\mathcal{D}$ acting in H. The differential da of an $a \in \mathcal{A}$ is defined by the commutator $[\mathcal{D}, a]$ and the integral is replaced by the Diximier trace, Tr_ω, with an appropriate inverse n-th power of $|\mathcal{D}|$ performing the role of the volume element $d^n x$:

$$\int a = \frac{Tr_\omega a |\mathcal{D}|^{-n}}{V}, \tag{12.4}$$

where V is some constant majorizing the eigenvaules λ_j of $\mathcal{D}$ ($\lambda_j < \frac{V}{j}$, $j \to \infty$). The Diximier trace of an operator O is, roughly speaking, the logarithmic divergence of the trace:

$$Tr_\omega O = \lim_{n \to \infty} \frac{\lambda_1 + \cdots + \lambda_n}{\log n}, \tag{12.5}$$

where λ_i is the i-th eigenvalue of O. See Ref. [10, 11] for details. The differential geometry is build up as follows. The notions of covariant derivative (∇), connection (A) and curvature (F) forms are defined so that standard properties are conserved:

$$\nabla = d + A, \quad F = \nabla^2 = dA + A^2, \tag{12.6}$$

where $A \in \Omega_\mathcal{D}^1$ is the algebra of one forms defined with respect to d. Fiber bundles became projective modules on $\mathcal{A}$ in this language. For example, an n-dimensional Yang-Mills action can be given by the formula:

$$\mathcal{L}(A, \psi, \mathcal{D}) = Tr_\omega \left(F^2 |\mathcal{D}|^{-n} \right) + \langle \psi | \mathcal{D} + A | \psi \rangle, \tag{12.7}$$

where $\langle | \rangle$ denotes the inner product in the corresponding Hilbert space. For $\mathcal{A} = C^\infty(M, \mathbb{R})$ and $\mathcal{D}$ being the ordinary Dirac operator we recover the ordinary Riemannian geometry of the (spin-) manifold M. In, general, noncommutative C^*-algebras do not "produce" topological spaces. In some "almost trivial" cases M be multiplied by some discrete space. In this approach gravitation is hidden in the metric tensor that "enters" the Dirac operator. This means that we may not know the structure of the spacetime with satisfactory precision but nevertheless fundamental interactions determine it in a quite unique way. Of course, it is possible that the C^* algebra $\mathcal{A}$ that describes correctly fundamental interactions do not correspond to any topological space and spacetime can only approximately be described as a topological space or that fundamental interactions does not determine it uniquely. It should be stressed here that matter fields (fermions) and their interactions are essential in the process determining the spacetime structure and the notion of spacetime is not a fundamental one. The noncommutative geometry formalism actually says that fermions create the spacetime at least on the theoretical level. The pure gauge sector is insufficient because two E-compact spaces X and Y are homeomorphic if and only if the categories of all modules over $C(X, E)$ and $C(Y, E)$ are equivalent. Some models of set theory can be constructed which admit "sets" that have no points. They might not be empty but still are pointless [21]. Toposes are regarded as the simplest generalization of classical set theory. They can be used as the stage for gravitational interactions [20]. Such objects are much more difficult to identify and differentiating among them might be possible only at the aesthetic or philosophical levels.

12.6 Gravitational Interactions

Up to now, we have deliberately avoided any physical interpretation of its points. Although spacetime plays the role of stage in almost all physical models it is inextricably linked with gravity. Since the birth of general relativity, spacetime is understood as the set of all physical events and its geometry is governed by distribution of matter and its dynamics. The metric field $\mathbf{g}$ unlike other physical fields represents nothing else but a class of properties a more or less substantial spacetime M. Mathematically, the spacetime geometry is an additional structure over the topological space M, e.g. for defining the metric field $\mathbf{g}$ and deriving from it the curvature (connection) and so on. Einstein's general relativity requires that M is a smooth manifold and the metric $\mathbf{g}$ is from the Hilbert-Einstein action:

$$S = \int d^4x \sqrt{-g} \left[\frac{R}{2\kappa^2} + L_m(\mathbf{g}, \psi) \right], \tag{12.8}$$

where R is the curvature scalar and $L_m(\mathbf{g}, \psi)$ is the matter lagrangian with matter fields collectively denoted by ψ. This can be generalized to

$$S = \int d^4x \sqrt{-g} \left[\frac{f(\mathfrak{R})}{2\kappa^2} + L_m(\mathbf{g}, \psi) \right], \tag{12.9}$$

where $f(\mathfrak{R})$ is an arbitrary function of the curvature tensor $\mathfrak{R}$. Therefore, we have a plethora of acceptable theories of gravitational interaction and among them a wide class of virtually indistinguishable theories with local observations. This class would probably be extended by some generalized noncommutative geometry models. We can hardly believe that we would ever be able recover the correct theory of gravity from observational data, cg the discussion in [26]

12.7 Conclusions

We have described the problem of mathematical modelling of the spacetime structure. A priori we should be able to build a faithful and unique model of the spacetime in the class of $\mathbb{R}$-compact spaces. Nevertheless, some of the features would have to be conjectured. We have to find a phenomenon that cannot be described in terms of the algebra $C(M, \mathbb{R})$ to reject the assumption of $\mathbb{R}$-compactness. If we restrict ourselves only to topological methods, we will not be able to construct the topological model M of the spacetime uniquely—extra assumptions of, say, "minimality" should be made (Occam's razor). Some of popular assumptions about might never be provable. For example, an unbounded observable is necessary to prove noncompactness of spacetime manifold. In the general case, we will be able to construct only the Stone-Čech compactification βM of the space M and such spaces are in some sense maximal and could be enormous.[4] The existence of a differential structure on M allows for the identification of M with the set of maximal ideals of $C^\infty(M, \mathbb{R})$, although we anticipate that the determination of the differential structure may be problematic, especially if there is a lot of them. Note that the construction of the standard model of electroweak interactions imply that fundamental interactions determine the model of spacetime in the class $\mathbb{R}$-compact space in a unique way because they are specified by $C(M, \mathbb{H})$. This is not true for other symmetry groups, e.g. GUTs, are lacking in such a determinative power. Matter fields are fundamental for defining and determining the spacetime properties and the associated geometries. If we are not able to determine $C(M, \mathbb{R})$ or $C(M, \mathbb{Q})$ then our knowledge of the spacetime structure is significantly less accurate. In general, we have a bigger class of spaces "at our disposal" and we are more free in making assumptions about the topology and even about the cardinality of the model of spacetime.

References

1. A.V. Arhangelskii, *Topologiceskye Prostranstva Funkcii (in Russian)* (Moscow Univ. Press, Moscow, 1989)
2. A.V. Arhangelskii (ed.), *General Topology III* (Springer, Berlin, 1995)
3. T. Asselmeyer-Maluga, J. Król, Adv. High Energy Phys. Article Number: 867460 (2014)
4. T. Asselmeyer-Maluga, J. Król, Mod. Phys. Lett. **28** Article Number: 1350158 (2013)
5. T. Asselmeyer-Maluga, J. Król, Int. J. Geom. Met. Mod. Phys. **10** Article Number: 1350055 (2013)
6. P. Bankstone, J. Pure Appl. Algebra **97**, 221 (1994)
7. C.H. Brans, Class. Quant. Gravity **11**, 1785 (1994)
8. C.H. Brans, J. Math. Phys. **35**, 5494 (1995)
9. C.H. Brans, T. Asselmeyer-Maluga, *Exotic Smoothness and Physics: Differential Topology and Spacetime Models* (World Scientific, Singapore, 2007)
10. A. Connes, J. Lott, Nucl. Phys. B Proc. Suppl. **18B** 29 (1990)

---

[4]For example, $\beta\mathbb{N}$ has the cardinality $2^{2^{\aleph_0}}$ and still is totally disconnected; any Hausdorff space with a basis of cardinality $\leq \aleph_1$ is a continuous image of $\beta\mathbb{N} \setminus \mathbb{N}$.

11. A. Connes, *Noncommutative Geometry* (Academic Press, London, 1994)
12. S. DeMichelis, M. Freedman, J. Differ. Geom. **35**, 219 (1992)
13. S.K. Donaldson, J. Differ. Geom. **18**, 279 (1983)
14. M. Freedman, J. Differ. Geom. **17**, 357 (1982)
15. I.M. Gelfand, A.N. Kolgomorov, Dokl. Acad. Nauk SSSR **22**, 11 (1939)
16. L. Gillman, M. Jerison, *Rings of Continuous Functions* (Springer, Berlin, 1986)
17. R.E. Gompf, J. Differ. Geom. **18**, 317 (1983)
18. R.E. Gompf, J. Differ. Geom. **37**, 199 (1993)
19. E. Hewitt, Trans. Am. Math. Soc. **64**, 45 (1948)
20. J. Król (2008). arXiv:0804.4217v2
21. R.S. Lubarsky (2015). arXiv:1510.00988
22. S. Mrówka, Acta Math. Acad. Sci. Hung. **21**, 239 (1970)
23. E.E. Rosinger, A. Khrennikov, AIP Conf. Proc. 03/2011 **1327**(1), 520–526. doi:10.1063/1. 3567483
24. J. Sładkowski, Acta Phys. Pol. B **27**, 1649 (1996)
25. J. Sładkowski, Int. J. Mod. Phys. D **9**, 311 (2001)
26. L. Sokołowski, Acta Phys. Pol. B (2015). In press
27. E.S. Thomas, Trans. Am. Math. Soc. **126**, 244 (1967)
28. E.M. Vechtomov, Itogi Nauki i Tekhniki, ser Algebra, Topologia, Geometria (in Russian), vol. 28, p. 3 (1990)
29. E.M. Vechtomov, Itogi Nauki i Tekhniki, ser. Algebra, Topologia, Geometria (in Russian), vol. 29, p. 119 (1990)
30. J.V. Whittaker, Ann. Math. **62**, 74 (1963)
31. O. Yaremchuk, (1999). arXiv: quant-ph/9902060

Chapter 13
Model and Set-Theoretic Aspects of Exotic Smoothness Structures on $\mathbb{R}^4$

Jerzy Król

Abstract Model-theoretic aspects of exotic smoothness were studied long ago uncovering unexpected relations to noncommutative spaces and quantum theory. Some of these relations were worked out in detail in later work. An important point in the argumentation was the forcing construction of Cohen but without a direct application to exotic smoothness. In this article we assign the set-theoretic forcing on trees to Casson handles and characterize small exotic smooth R^4 from this point of view. Moreover, we show how models in some Grothendieck toposes can help describing such differential structures in dimension 4. These results can be used to obtain the deformation of the algebra of usual complex functions to the noncommutative algebra of operators on a Hilbert space. We also discuss the results in the context of the Epstein-Glaser renormalization in QFT.

13.1 Infinite Geometric Constructions and Set-Theoretic Forcing

Currently it is a bit of a folklore to say that dimension 4 is exceptional both in physics and mathematics. On the one hand this is the dimension where Einstein theories of relativity were formulated, where the physics of particles and quantum fields found their marvelous realization on (curved) Minkowski spacetimes, and where the cosmological evolution of our world is to be described. On the other hand, many curious mathematical facts, like the existence of exotic R^4, or in fact, of a continuum many of them, take place exactly in this dimension. It was a big effort of many mathematicians in 1980s like Donaldson, Freedman, Gompf, Taubes and many others whose work on topology and geometry of manifolds in dimension 4 opened our eyes on the unique 4-dimensional topological and 'smooth' world and help in its understanding. However, taking seriously advanced and technical mathematical findings as applicable to physics, required much scientific imagination and courage in those days. It was Carl Brans who took the step in a series of papers [1–4].

J. Król (✉)
Institute of Physics, University of Silesia, ul. Uniwersytecka 4, 40-007 Katowice, Poland
e-mail: jerzy.krol@us.edu.pl

© Springer International Publishing Switzerland 2016
T. Asselmeyer-Maluga (ed.), *At the Frontier of Spacetime*,
Fundamental Theories of Physics 183, DOI 10.1007/978-3-319-31299-6_13

Soon after, there appeared the work of Torsten Asselmeyer-Maluga (e.g. [5]) and Jan Sładkowski (e.g. [6, 7]) who approached the role of exotic $\mathbb{R}^4$'s in physics from various perspectives. Carl's Brans ideas and the papers above were an inspiration to me and I have been lucky as a researcher to work together with Torsten and Jan within the recent years. It is a big honor and pleasure to me to contribute to the volume celebrating the work of Carl Brans.

Exotic smoothness structures on $\mathbb{R}^4$ are just Riemannian, curved smooth 4-manifolds (exotic R^4) which topologically are (homeomorphic to) $\mathbb{R}^4$. In this chapter, I will show that the perspective of set theory and Grothendieck toposes, hence foundations of mathematics, is the right one when considering physical applications of exotic, open 4-smoothness. Even though this is neither obvious nor widely accepted approach, the use of model and set-theoretic methods in physics has a firm and vivid tradition arisen from the foundations of mathematics (e.g. [8–11]). That was developed substantially further in recent years (e.g. [12–17]).

In physics, set theory is usually considered informally as unchanged eternal background which goes together with the classical 2-valued logic. However, when one allows for variations in such background more formal, axiomatic formulation is needed. That is why set theory is understood as the first order axiomatic Zermelo-Fraenkel (ZF) theory of sets with possible addition of the axiom of choice (AC)—ZFC. Similarly arithmetic is usually described as an axiomatic first order theory—Peano arithmetic (PA) (see the discussion regarding the order of formal theories versus set theory in [18]). The variations in the theories can be grasped by considering various models of these theories. Classically such models (Tarski) are built in the category Set of sets and functions between them. All models of first order theories undergo usual limitations and benefits which follow the Gödel or Löwenheim-Skolem-like theorems (and much more, see e.g. [19]). We also will be using more general models of (intuitionistic) set theory in other categories like toposes where the logic becomes intuitionistic [20].

The forcing method is known from the independence results in set theory since 1960s [21] and allows for changing the models. In general, forcing in mathematics is a very rich, technical and advanced subject (see e.g. [19, 22]). For the purpose of this work it is a method for studying the real numbers line. Thus Cohen forcing in a narrow sense used in the chapter can be seen as a mechanism of adding real numbers to the model and thus changing the model of ZF(C) and the real line. This is also a tool for exploring the exotic smooth R^4's (see e.g. [17, 23, 24]).

We start with infinities appearing in some geometric constructions in dimensions 3 and 4 like Casson handles and Alexander horned sphere (wild embeddings). These infinities are the inevitable and intrinsic features of the constructions. On the other hand, infinity by itself is a natural and central topic in set theory. The key for understanding this relation is precisely the Cohen forcing. On the algebraic level a forcing is generated by some complete atomless Boolean algebra—in this case the forcing is nontrivial and can eventually add some reals to the ground model M of ZFC. In the case of Cohen forcing the algebra is the *unique* atomless Boolean algebra with a dense countable subset. In fact it holds true:

Lemma 13.1 (Corollary 25.4, p. 189 [25]) *Let A be a complete atomless Boolean algebra that contains a countable dense subset. Then A is isomorphic to the algebra RO(CS) of regular open subsets of the Cantor set CS.*

Any (signed) tree canonically generates a partial order (partially ordered set). A partial order $(\mathbb{P}, \leq)$ is called *separative* if for all $p, q \in \mathbb{P}$ such that $p \not\leq q$ there exists $r \leq p$ with $r \perp q$. Here $r \perp q$ means incompatibility relation i.e. there does not exist k that neither $q \leq k$ nor $r \leq k$ is true. Then, the important lemma holds true:

Lemma 13.2 (Lemma 13.33, [25]) *Every separative partial order $\mathbb{P}$ can be completed to a complete Boolean algebra B such that $\mathbb{P}$ is dense in $B \setminus \{0\}$ and the partial order in $\mathbb{P}$ agrees with $\leq_B$. B is unique up to isomorphism.*

Next we ask the question: which rooted trees do represent a separative partial order? One easily finds that the full binary tree (the one which has precisely 2 branches at every node) does. Moreover:

Lemma 13.3 *The full binary tree represents the countable dense subset (partial order) of some complete atomless Boolean algebra.*

This is because the full binary tree represents the Cantor set in $(0, 1)$ interval: one assigns to every branch 0 or 2 numbers which appear in the three-mal decompositions $0.x_1 x_2 x_3 \ldots$ of numbers in $(0, 1)$. Then missing numbers correspond precisely to $x_i = 1, i = 1, 2, 3, \ldots$. The nodes of the tree represent the members of the countable partial order which is dense in the partial order of the tree hence in the corresponding Boolean algebra. The algebra is RO(CS) which is atomless and generates the nontrivial Cohen forcing.   □

Now the point is that the Cantor set generated by the binary tree is frequently realized geometrically by Casson handles construction in dimension 4 and by wild embeddings of spheres in dimension 3 (see e.g. [26]). Casson handles (CH) (see e.g. [27–29]) appear in the handle-body decompositions of small exotic smooth open 4-manifolds [29] and are also represented by the infinite signed rooted trees [29, 30]. If the tree was finite and the CH smooth, the Casson handle would be the ordinary smooth 2-handle.[1]

Let me quote an important and elementary observation by Kato ([31], p. 114) which ensures that given a signed tree we have a Casson handle spanned on that tree:

> There are sufficiently many Casson handles. In fact to each infinite signed tree, one can associate a Casson handle.

Let M be a model of ZFC and $M[G]$ its generic extension by Cohen forcing [19, 22]. Then we can prove the following:

Theorem 13.1 *A general Casson handle appearing in the handlebody of a small exotic R^4 determines a nontrivial Cohen forcing adding a Cohen real in some generic model $M[G]$ of ZFC.*

---

[1]Every CH is topologically (as a pair) homeomorphic to the standard 2-handle which was shown by Freedman [28].

Proof First, any Casson handle can be embedded in the simplest CH which is the linear tree with only one, positive or negative, self-intersection at each level. This follows from the fact that every CH with a bigger signed tree than the tree of another CH is embeddable in this 'smaller-tree-CH'. One should respect the rule that the smaller tree is homeomorphically embedded into the bigger one. Adding self-intersections on any level and killing the generators by gluing kinky handles determines the embedding. Moreover, the resulting embeddings of CH's preserves the attaching areas of CH (or at least attaching circles and their framings). The last means that whenever the simplest CH were exotic (the attaching circle determines the non-smooth slice) the embedded CH with a bigger tree would be exotic too [30].

Second, instead of attaching an arbitrary CH let us attach the simplest one (see Figs. 13.1 and 13.2) with the linear signed tree in which we know the bigger one is embeddable. In general we do not know whether the CH with such a tree is exotic although we know it is exotic for the 'only $+$' or 'only $-$' trees.

Next, let us consider the Casson handle determined by the full binary tree (BT) with one infinite branch identical with the linear one above. Such 'binary-tree-CH' embeds in the linear CH and let us forget the signs in the binary CH. Then from Lemmas 13.1 and 13.3 the algebra $RO(CS)$ is the unique Cohen forcing algebra generated by BT. $\square$

Note that every CH determines the same (up to isomorphism) Cohen algebra thus the nontrivial Cohen forcing in a generic model $M[G]$. In dimension 3 given wildly embedded 3-sphere, say horned Alexander sphere, a 'grope' is assigned naturally to it which is spanned on the infinite binary tree again ([26], pp. 18–19). Thus Cohen forcing can be built also in this case. We do not discuss the meaning of it here but only note that wild embeddings in dimension 4 is the other side of exotic open 4-smoothness and this can be understood physically as a quantum state [32, 33].

Cohen forcing changes the real line substantially, namely the reals in the model M constitute merely measure zero subset of the extended real line in $M[G]$, hence of $\mathbb{R}$. As shown above it is also assigned to replacing the standard smooth 2-handles by an exotic Casson handle, hence to changing the smoothness structures on $\mathbb{R}^4$. If the forcing acted over $\mathbb{R}$-line in $\mathbb{R}^4$ and resulted in exotic R^4 the following important question would arise: Can an extension of the real line by forcing be a valid tool when exploring exotic smoothness in dimension 4? In some sense this kind of forcing should add reals to the full $\mathbb{R}$ resulting in the same $\mathbb{R}$ since R^4 is again the Riemannian smooth real manifold. We will analyze this problem in the next and subsequent sections.

13.2 From the Standard to Categorical $\mathbb{R}^4$

One needs 'adding' more real numbers to the already full $\mathbb{R}$. What is the meaning of such procedure? We will show that the modification of logic and set theory is needed.

From the external absolute point of view a set-theoretic forcing adds reals (if any at all) to subsets R_M of $\mathbb{R}$ where R_M is a set of real numbers in some model M of ZFC. Internally there is no difference between (1st order) properties of real lines R_M and $\mathbb{R}$. Suppose that we already have a well defined model of the standard real line $\mathbb{R}$.[2] Starting with $\mathbb{R}$ can one add consistently more reals to the line? More precisely: can one construct a bigger real line which would have *the same* properties as $\mathbb{R}$ but be different as a set (thus containing more reals)? Our general motivation for considering such questions, as observed in the 1st section, is that we expect such procedure to possibly modify the smoothness of manifolds.

Reducing the properties of the real line to its 1st order properties, and the logic to first order logic, Robinson showed [34] that there are non-standard models of arithmetic $^*N$ and analysis $^*R$. They are end-extensions of the standard $\mathbb{N}$ and $\mathbb{R}$ respectively and contain infinite natural and real numbers. Moreover, $^*R$ contains infinitesimal invertible real elements. Now, every true 1st order formula ϕ about natural numbers is fulfilled in $^*N$ iff it is fulfilled in $\mathbb{N}$, i.e. $^*N \models \phi \equiv \mathbb{N} \models \phi$. We say that $^*N$ and $\mathbb{N}$ are elementary equivalent and write:

$$^*N \simeq_1 \mathbb{N} \ (^*R \simeq_1 \mathbb{R}), \tag{13.1}$$

meaning, one can not distinguish the two models just by their 1st order properties. We would like to strengthen the indistinguishability as above and consider something like $^*N \simeq_{2,3,...} \mathbb{N} \ (^*R \simeq_{2,3,...} \mathbb{R})$.[3] It is seemingly a trivial task, since 2nd order theory of natural or real numbers are categorical and the real line $\mathbb{R}$ is the only (up to isomorphism) model allowed, hence indeed $^*N \simeq_{2,3,...} \mathbb{N}$.

That is why we are rather looking for an environment (the twist) where non-standard models for arithmetic and analysis may exist, are nontrivial, i.e. different, and are valid for higher order theories, i.e. some second order properties of the models become identical after the twisting. Without any twist these particular properties would not coincide. As noted above we can not achieve the nontrivial realization of the full classical indistinguishability $^*N \simeq_2 \mathbb{N} \ (^*R \simeq_2 \mathbb{R})$ since 2nd order arithmetic has isomorphic models.

To imagine how the twist could work one can introduce three parameters (w, α, ϵ) controlling the twist—w corresponds to the weakening of the arithmetic and/or the logic, and the other two to the fractions (belonging to $(-1, 1)$) of the numbers of all true formulas of the first and second orders correspondingly. $\alpha = 0$ and $w = 0$ mean that all true 1st order formulas of both models, ($^*R$ and $\mathbb{R}$), are determined with respect to the first order (i.e. $\alpha = 0$) classical (i.e. $w = 0$) predicate logic. Similarly,

---

[2] A formal theory giving rise to the unique up to isomorphism model of real numbers should use the 2nd order logic. Such theories are called categorical (in $\aleph_1$). The theory of Archimedean complete ordered field is categorical. It is a second order theory.

[3] It would be sufficient to consider $^*N \simeq_2 \mathbb{N}$ since there are theorems reducing the higher order to 2nd order logic (e.g. [35, 36]).

$\epsilon = 0$ and $w = 0$ mean that all second order formulas of the models are determined w.r.t. the classical second order logic. Thus one writes

$$^*N \simeq^w_{1-\alpha,1+\epsilon} \mathbb{N} \; (^*R \simeq^w_{1-\alpha,1+\epsilon} \mathbb{R}) \qquad (13.2)$$

when the logic is weakened and the sets of the first order formulas and second order formulas have been modified and especially some 2nd order formulas become identical in both models after the twist. The $+, -$ signs indicate the twist or the rotation in the parameter space. The value of the parameters depends on the degree of how much of weak and nonclassical logic is used. We do not need to determine the relation between the parameters more precisely here. Instead, let us consider the important example. We will weaken the logic and arithmetic considerably and take the models in a constructive set-up, i.e. in toposes.

This weak Peano arithmetic was recognized in detail by Moerdijk and Reyes [37] when they considered the non-standard models of numbers in smooth toposes and built the smooth topos model for synthetic differential geometry. We present the discussion of the elements of their construction important for us in the Appendix 13.4.2.

The important point is that the objects of natural numbers (NNO) in smooth toposes like Zariski ($\mathcal{Z}$) and Basel topos ($\mathcal{B}$) determined by the natural embedding of manifolds from Set to the toposes, i.e. the map $s : \mathbb{M} \to \mathcal{Z}$, sends $\mathbb{N}$ to the standard natural numbers $s(\mathbb{N})$ in $\mathcal{Z}, \mathcal{B},$[4] fails to generate a proper object of real numbers $s(\mathbb{R}) = R_{\mathcal{Z}}$ (or $R_{\mathcal{B}}$). For example: $R_{\mathcal{Z}}$ is nonarchimedean with respect to $s(\mathbb{N})$ so thus (13.21) does not hold. Besides $[0, 1] \subset R_{\mathcal{Z}}$ is noncompact with respect to $s(\mathbb{N})$. As the consequence this last property devastates the homology theory of manifolds in $\mathcal{Z}$ ([37], pp. 280–284.).

To cure this one should turn to the modified object of natural numbers $N_{\mathcal{Z}}$ (smooth natural numbers) which is not the canonical standard NNO $s(\mathbb{N})$ in $\mathcal{Z}$. As shown by Moerdijk and Reyes the axioms of the weak logic (13.17)–(13.19) are fulfilled in $\mathcal{Z}$ however the type N is interpreted now as $N_{\mathcal{Z}}$ i.e. it is the smooth NNO. $R_{\mathcal{Z}}$ is now Archimedean w.r.t. $N_{\mathcal{Z}}$, $[0, 1]$ is compact (smooth compact, or s-compact), the homologies of manifolds are tractable and in particular the internal topologies of manifolds in $\mathcal{Z}$ are well-defined. Internal in $\mathcal{Z}$ constructions and theories are formulated such as the true natural numbers are $N_{\mathcal{Z}}$ rather than the standard $s(\mathbb{N})$. The shift $s(\mathbb{N}) \to N_{\mathcal{B}}$ changes some second order properties of real and natural numbers such that now in $\mathcal{Z}$ internal constructions are more like the external ones.

The construction of $N_{\mathcal{Z}}$ follows the filterproduct construction. Namely, the object $R_{\mathcal{Z}} \simeq s(\mathbb{R})$:

$$R_{\mathcal{Z}} = s(\mathbb{R}) = \mathbb{L}(-, lC^{\infty}(\mathbb{R}))$$

is the representable object of $\mathcal{Z}$ [37]. It is non-archimedean with respect to $s(\mathbb{N})$ as said above. Instead one defines the object of smooth natural numbers $N_{\mathcal{Z}}$ thus allowing for the modification of *finiteness*. Let $(\sin(\pi x))$ be the ideal in $C^{\infty}(\mathbb{R})$. The

---

[4]As the object in a topos this standard NNO is the constant sheaf of natural numbers.

representable object in $\mathcal{Z}$ of smooth integer numbers $Z_{\mathcal{Z}}$ is now defined as ([37], p. 252)[5]:

$$Z_{\mathcal{Z}} = l(C^\infty(\mathbb{R})/(\sin \pi x)), \quad N_{\mathcal{Z}} = l(C^\infty(\mathbb{R})/(\sin \pi x, x \geq 0)). \quad (13.3)$$

Taking the ideal F of functions which are non-zero only on finite initial segments of $\mathbb{N}$, then the quotient $l(C^\infty(\mathbb{N})/F)$ represents a non-standard infinite natural number in $\mathcal{Z}$.

To have the standard $s(\mathbb{N}) \simeq \mathbb{N}$ one can define it as the subtype of N_Z:

$$\mathbb{N} = \{n \in N_Z : \forall_{S \in P(N_Z)} (0 \in S \wedge \forall_{n \in N_Z} (m \in S \to m+1 \in S) \to n \in S)\} \quad (13.4)$$

which means $\mathbb{N}$ fulfills the strong induction scheme we know from Peano arithmetic [37]. However, when logic is weakened (in the metatheory) the 'true' natural numbers are defined with respect to the coherent induction scheme (13.17) in which case one does not distinguish $\mathbb{N}$ and N_Z. We do not dwell upon such metatheoretic considerations here (see however [17]).

Even if the subtype $\mathbb{N} \subset N_Z$ can be defined as in (13.4) still it is undecidable[6]:

$$\mathcal{Z} \models (\mathbb{N} \neq N_{\mathcal{Z}}) \to (\mathbb{N} \text{ is not decidable in } N_{\mathcal{Z}}).$$

The important question is the extent up to which one can consistently replace $\mathbb{N}$ by N_Z. What is crucial here is that the 2nd order property of $R_{\mathcal{Z}}$ of being Archimedean is again retrieved with respect to N_Z. Similarly, the interval $[0, 1]$ is compact again with respect to N_Z. The twist (13.2) is realized by the shift:

$$s(\mathbb{N}) \to N_{\mathcal{Z}} \quad (13.5)$$

which allows for the retrieving of some internal higher order properties of theories in $\mathcal{Z}$ which were lost when the canonical standard NNO was in use.

We will demonstrate how this intuitionistic model for weak arithmetic and especially the shift (13.5) is related to both smoothness structures in dimension 4 and the procedure of adding reals by forcing.

13.2.1 Smooth Natural Numbers in $\mathcal{B}$

Weak logic as described in the previous section (and in the Appendix) guarantees that there is a NNO different than $s(\mathbb{N})$, i.e. $N_{\mathcal{Z}}$ which replaces consistently the standard NNO in the intuitionistic set-up. The crucial point is that $N_{\mathcal{Z}}$ contains also

---

[5]$l( )$ is the member of $\mathbb{L}$—the category of loci which is opposite to the category of (finitely generated) smooth rings ([37], p. 58).

[6]A subset $A \subset B$ is decidable when $a \in A$ is decidable property, i.e. when $\forall_{a \in B}(a \in A \vee a \notin A)$.

non-standard natural numbers what indicates that $N_\mathcal{Z}$ is an intuitionistic analogue of $*N$ known from the non-standard analysis (NA). Internal in the toposes, higher order intuitionistic theories are formulated internally in $\mathcal{Z}$ w.r.t. $N_\mathcal{Z}$ and $R_\mathcal{Z}$ leaving aside their standard counterparts. But such radical departure from standardness modifies finiteness such that infinite big non-standard natural numbers are considered as s-finite.

In general there are two kinds of infinitesimal elements in $R_\mathcal{Z}$: invertible ($\mathbb{I} \subset R_\mathcal{Z}$) and nilpotent ones. Nilpotent elements are required by the synthetic differential geometry approach and they represent forms like dx ($d^2 = 0$), while invertible elements are predicted by the non-standard analysis of Abraham Robinson which can be generated by taking inverses of infinite non-standard natural numbers. The smooth topos unifies both kinds of infinitesimals in the one real line R where they exist as real numbers. Moreover, most internal higher order theories perceive the smooth numbers as true real and natural numbers. The important class of such theories are differentiable manifolds whose category $\mathbb{M}$ is mapped into the smooth toposes via s transform, and they require s-numbers to define their topology, compactness, connectedness or homologies.

However, do there really exist 'non-standard' and invertible infinitesimal elements of R_Z, i.e. $\mathbb{I}$ in $\mathcal{Z}$? In fact it holds [37]:

$$Z \models \neg\neg[\exists_x x \in R_Z \cap \mathbb{I}] \tag{13.6}$$

which is a rather weak version of the existence of invertible infinitesimals (recall that the logic in $\mathcal{Z}$ is intuitionistic and double negation does not cancel in general). To strengthen this result the Authors of [37] proposed to modify the topos $\mathcal{Z}$ towards $\mathcal{B}$ such that now one proves:

$$\mathcal{B} \models \exists_x x \in R_Z \cap \mathbb{I}. \tag{13.7}$$

To obtain this result one has to modify the Grothendieck topology in $\mathcal{Z}$ and then to be sure invertible infinitesimals do exist, one adds them by the *forcing on stages* (see the Appendix 1 in [37]). Thus, indeed in the internal environment of $\mathcal{B}$ the non-standard real numbers are added by forcing. This is the extension of the real line by adding new reals which we discussed in Sects. 13.1 and 13.2. Such procedure is not in general possible in higher orders and in the classical $\{0, 1\}$ logic, but it is possible in the weaker logic of the topos $\mathcal{B}$ realizing the twist (13.2) by the shift (13.5).

13.2.2 The Smooth Topos $\mathcal{B}$ Localized on $\mathbb{R}^n$

Here we want to show that smoothness structures on $\mathbb{R}^4$ can have their origins at the level of models of the real line. Moreover, continuum many different exotic smoothness structures R^4's can be understood at that level. Given the real line (higher order, classical) $\mathbb{R}$ it is Archimedean with respect to $\mathbb{N}$. To have such a unique model $\mathbb{R}$ we can think of it as the model for the second order theory of real numbers or the theory of an Archimedean complete ordered field, both having unique (up to isomorphisms) models. On the contrary, reducing the properties of $\mathbb{N}$ or $\mathbb{R}$ to the first order we get a plurality of non-standard models $*N$ and $*R$ in every infinite cardinality. Can one have different non-standard models $*R$ all having the cardinality of continuum? The answer is the following:

Lemma 13.4 *Under the Continuum Hypothesis (or under $2^{<c} = c$) there are 2^c different non-isomorphic models $*R$ all having the cardinality c.*

The part of the proof important to us is the observation that every non-principal ultrafilter $\mathcal{U}$ on the set $\mathbb{N}$ generates a non-standard $*R_{\mathcal{U}}$ of the cardinality continuum as an ultrapower construction, and two such ultrapowers are isomorphic if and only if the ultrafilters generating them are isomorphic w.r.t. a permutation of $\mathbb{N}$. Finally there are 2^c non-isomorphic ultrafilters on $\mathbb{N}$. ☐

Thus starting with the higher order $\mathbb{R}$ one has up to 2^c possibilities to choose its 1st order continuous reducts $\mathbb{R} \to *R$. This extends to the relation basic to us (especially for $n = 4$) with 1 to 2^c possibilities:

$$\mathbb{R}^n \xrightarrow{2nd \to 1st} *R^n. \tag{13.8}$$

Let us complete this correspondence with another one as follows:

$$\mathbb{R}^n \xrightarrow{2nd \to 1st} *R_1^n \xrightarrow{sh} *R_2^n \xleftarrow{2nd \to 1st} \mathbb{R}^n. \tag{13.9}$$

We would like to have (13.9) realized as smooth correspondence also in the middle arrow, and valid in the higher orders. This is the point where the topos $\mathcal{B}$ and the twist (13.5) come into play. We are further extending the correspondence (13.9) into the following $\mathcal{B}$-modified one:

$$\mathbb{R}^n \xrightarrow{2nd \to 1st} *R_1^n \xrightarrow{e_1} R_{\mathcal{B}}^n \xrightarrow{[d]} R_{\mathcal{B}}^n \xleftarrow{e_2} *R_2^n \xleftarrow{2nd \to 1st} \mathbb{R}^n. \tag{13.10}$$

We are going to determine the internal in $\mathcal{B}$ [d]-continuous and even differentiable map. Let Fin be the ideal in $P(\mathbb{N})$ of finite subsets of $\mathbb{N}$. The algebra $P(\mathbb{N})/\text{Fin} = P(\omega)/\text{Fin}$ is an atomless Boolean algebra. Moreover, all nonprincipal ultrafilters on $\mathbb{N}$ are the members of the Stone remainder $\beta[\omega] \setminus \omega$ of the algebra $P(\omega)/\text{Fin}$. Recall that the Frechet cofinite filter $\mathcal{F}$ on $\mathbb{N}$ is defined as:

$$\mathcal{F} = \{F \in P(\mathbb{N}) : \mathbb{N} \setminus F \in \text{Fin}.\} \tag{13.11}$$

The following obvious but important lemma holds true:

Lemma 13.5 *Every nonprincipal ultrafilter $\mathcal{U}$ on $\mathbb{N}$ contains the Frechet cofinite filter $\mathcal{F}$.*

Let us consider now the specific relation of non-standard models $*N, *R$ in classical logic (Set) and in toposes (higher order intuitionistic logic).

Lemma 13.6 *In $\mathcal{B}$ and $\mathcal{Z}$ the non-standard models are built as filterproduct constructions based on the Frechet filter $\mathcal{F}$ rather than on ultrafilters.*

This follows from the direct construction of smooth natural numbers in $\mathcal{B}$ (see [37], p. 252). Moreover, to respect constructivism in toposes one cannot base on the AC (especially using ultrafilters strongly depends on AC). In [38] Moerdijk showed explicitly that the constructive non-standard PA in the topos $Sh(\mathbb{F})$ of sheaves on the category of filters is based on the smooth natural numbers constructed with respect to the Frechet filter $\mathcal{F}$.

Corollary 13.1 *All non-standard models $^*N$ ($^*R$) are mapped by e_1, e_2 in (13.10), into the single intuitionistic non-standard model $N_{\mathcal{B}}$ ($R_{\mathcal{B}}$) in $\mathcal{B}$.*

This is the consequence of: (1) All ultrafilters are the extensions of the unique Frechet filter (Lemma 13.5). (2) Different nonstandard models of R (with the cardinality continuum) are constructed on the base of non-isomorphic nonprincipal ultrafilters on $\mathbb{N}$. (3) Lemma 13.6.   □

Let us consider relations on $\tilde{\mathbb{N}}$ modulo the ideal of finite subsets Fin, e.g. the equality becomes $A =^* B$ meaning $A \triangle B = A \setminus B \cup B \setminus A \in$ Fin. We call a $1:1$ function $f : D_f \rightarrow \mathrm{Im}_f$, $D_f, \mathrm{Im}_f \subset \mathbb{N}$ an almost permutation of $\mathbb{N}$ whenever domain of f, D_f, and its image Im_f are almost $\mathbb{N}$, i.e. $D_f =^* \mathbb{N} =^* \mathrm{Im}_f$.

Each such almost permutation f of $\mathbb{N}$ gives rise to the automorphism d_f of the Boolean algebra $P(\omega)/\mathrm{Fin}$. Namely

$$d_f([A]) = [f(A \cap D_f)] \text{ for } [A] \in P(\omega)/\mathrm{Fin}. \tag{13.12}$$

Even though there can be up to 2^c nontrivial automorphisms of the algebra $P(\omega)/\mathrm{Fin}$ [39], it is still valid that:

Lemma 13.7 *There are c automorphisms of $P(\omega)/\mathrm{Fin}$ which give rise to almost permutations of $\mathbb{N}$.*

This is crucial for us to consider such trivial automorphisms since they forbid $\mathbb{N}$, hence $\mathbb{R}$, to be constant and give definite transformations of $\mathbb{N}$. Moreover, as shown by Shelah [40], the statement that there are only c automorphisms of $P(\omega)/\mathrm{Fin}$ (only trivial) is consistent with ZF. So in the above sense we restrict our considerations to the trivial automorphisms case. Let us note that:

Lemma 13.8 *Every trivial automorphism of $P(\omega)/\mathrm{Fin}$ represented by a permutation $\sigma : \omega \rightarrow \omega$ corresponds to a mapping (shift) between non-isomorphic non-standard models of $\mathbb{R}$ of the cardinality c.*

This is a direct consequence of the relation of the nonprincipal ultrafilters and non-standard models of $\mathbb{R}$, and the fact that the Stone space of $P(\omega)/\mathrm{Fin}$, i.e. $\beta[\omega]$, contains all nonprincipal ultrafilters on ω, i.e. $\beta[\omega] \setminus \omega$. Every permutation of $\mathbb{N}$ extends to a homeomorphism $\beta(\sigma) : \beta[\omega] \rightarrow \beta[\omega]$ and to an automorphism of $\beta[\omega] \setminus \omega$ (e.g. 3.41, p. 88 in [39]). This last defines the shift between the non-standard models.   □

In fact we need the following converse relation:

Corollary 13.2 *For every automorphism of $P(\omega)/\mathrm{Fin}$ there exists the shift-map between non-isomorphic non-standard $\mathfrak{c}$-models of $\mathbb{R}$ such that the automorphism realizes this shift between the models.*

Now given the shift-map $sh : {}^*R_1 \to {}^*R_2$ as in (13.9) we can think of it as determined by some automorphism of $P(\omega)/\mathrm{Fin}$. Note that this correspondence is obviously non-unique. Taking an internal in $\mathcal{B}$ extension $[d]$ of the shift as in (13.10) gives rise to the following:

Theorem 13.2 *Every external shift $sh : {}^*R_1 \to {}^*R_2$ determines the internal s-differentiable maps $[d]_{1,2}, [d]_{2,1} : R_{\mathcal{B}}^n \to R_{\mathcal{B}}^n, n = 1, 2, 3, \ldots$.*

Note that $[d]_{1,2}$ and $[d]_{2,1}$ are generated in Set by the 'inverse' almost permutations of $\mathbb{N}$.

Proof First, any non-standard model ${}^*R_i$ is obtained via the ultraproduct construction w.r.t. an ultrafilter $\mathcal{U}_i$. $\mathcal{U}$ is the extension of the Frechet filter $\mathcal{F}$. In $\mathcal{B}$ the 'non-standard' real line is $R_{\mathcal{B}}$ obtained via the *filter construction* w.r.t. $\mathcal{F}$. Hence we have $[d]_{1,2} : R_{\mathcal{B}} \to R_{\mathcal{B}}$. Second, every internal $[d]$ is continuous in $\mathcal{B}$ (see Theorem 3.6, p. 270 in [37]). Next, since $\mathcal{B}$ is the model of synthetic differential geometry (there exist nilpotent infinitesimals $D \subset R_{\mathcal{B}}$) it holds true the Kock-Lawvere axiom in $\mathcal{B}$, which gives ([37], p. 302):

$$\forall_{f \in R^R} \forall_{x \in R} \exists!_{f'(x) \in R} \forall_{h \in D} f(x+h) = f(x) + hf'(x)$$

where R stands for $R_{\mathcal{B}}$. Note that $f'(x)$ is just the symbol for the unique $y = f'(x)$ such that $y \in R$. Repeating the procedure we determine subsequently $f''(x), f'''(x), \ldots$. Thus $f \in R^R$ is a standardly infinitely many times differentiable internal function. Finally, we apply again the Kock-Lawvere axiom to the 'inverse' map $[d]_{2,1}$ which leads to a similar differentiability. □

Definition 13.1 The pair $([d]_{1,2}^n, [d]_{2,1}^n)$, or $[d]_{1,2}^n$ to shorten, is called an *internal diffeomorphism* or *s-diffeomorphism* of $R_{\mathcal{B}}^n, n = 1, 2, 3$.

Note that any internal diffeomorphism as above is generated by the shift between the non-standard $\mathfrak{c}$-models of $\mathbb{R}$. One could wonder whether the s-diffeomorphisms can be non-identity maps since they all are generated w.r.t. the Frechet filter. However, due to Lemma 13.7 there is precisely $\mathfrak{c}$ shifts which guarantee that $\mathbb{N}$ hence $R_{\mathcal{B}}$ are not constant.

Now we are ready to define the central object in this section (cf. [41]):

Definition 13.2 Let M^n be a smooth n-dimensional manifold and $\{U_\alpha : U_\alpha \in \mathcal{O}\}$ its regular open cover. We call $^{(\mathcal{B})}M^n$ *a n-dimensional manifold M^n locally modified by the topos $\mathcal{B}$*, or the *smooth $\mathcal{B}$ structure on M^n*, whenever it holds:

- For every regular open cover $\{U_\alpha\}$ of M^n there exists some $U_\alpha \in \{U_\alpha\}$ such that U_α is internal object of the internal in $\mathcal{B}$ topology of $s(M^n)$.

- If two such open U_α, U_β are internal in $\mathcal{B}$ their nonempty internal meet defines the local change of coordinates in $\mathcal{B}$ which contain the s-diffeomorphisms: $\eta_{\alpha\beta} = [d]_{1,2} : U_\alpha \cap U_\beta \to U_\alpha \cap U_\beta$.

Next we would like to ensure that s-diffeomorphisms do not arise from a Set-based diffeomorphism. To this end let the class of trivial automorphisms of $P(\omega)/\mathrm{Fin}$ be suitably limited: one allows only those trivial automorphisms whose almost permutations of $\mathbb{N}$ contain at least one non-identical almost cycle—*cyclic almost permutations* and we will call them cyclic permutations if it does not cause any confusion.[7] There still exist continuum many such almost permutations and none of them is extendable in Set to any orientation preserving diffeomorphism of $\mathbb{R}$.

Summarizing:

1. s-diffeomorphisms are not images of diffeomorphisms from Set, hence the local modification by $\mathcal{B}$ of the smoothness structure of M^n is nontrivial and categorical.
2. s-diffeomorphism is generated in Set by a cyclic almost permutation of $\mathbb{N} \subset \mathbb{R}$ so it is not extendable to any orientation-preserving diffeomorphism of $\mathbb{R}$.
3. In $\mathcal{B}$ each permutation of $s(\mathbb{N}) \subset R_{\mathcal{B}}$ gives rise to the s-diffeomorphism $([d]_{ij}, [d]_{ji})$.

13.3 From the Categorical to Exotic $\mathbb{R}^4$

Given the local $\mathcal{B}$-modification of the smooth structure on M^n we are interested in its impact on the actual classical smoothness of manifolds. One obvious classical limit (this which does not depend on $\mathcal{B}$) of the $\mathcal{B}$-modified structure on M^n is just the smooth structure of M^n we started with. In this case all open $U_\alpha \in \mathcal{O}$ become (again) Set based external objects. There is however another, more refined possibility.

Definition 13.3 1. We say that a classical limit of the $\mathcal{B}$-deformed smooth structure on M^n *factors through the non-standard models* $^*R_1$ *and* $^*R_2$ whenever they are c-models of $\mathbb{R}$ and the $\mathcal{B}$-deformation was performed according to (13.9) and (13.10) where now $\mathcal{O} \ni U_\alpha \simeq \mathbb{R}^n$ on the l.h.s. and $\mathcal{O} \ni U_\beta \simeq \mathbb{R}^n$ on the r.h.s. of these relations.
2. A *nontrivial classical limit* of the $\mathcal{B}$-deformed smooth structure of M^n is a smooth structure on M^n which factors through some non-standard models of $\mathbb{R}$ while reaching the Set and higher order levels.

The point is that even though local $\mathcal{B}$-modifications of M^n take all almost permutations of $\mathbb{N}$ into internal s-diffeomorphisms, hence a single $\mathcal{B}$-deformed structure emerges, on the Set level it is not so. Namely we can prove the important result:

---

[7] An almost cyclic permutation is an almost permutation p, i.e. $p : A \xrightarrow{1:1} B$, $A =^* \mathbb{N} =^* B$, which reverses the order of elements of some $C \subset A$ when compared to the order of $p(C)$.

Theorem 13.3 *For different non-isomorphic c-models* $^*R_1$ *and* $^*R_2$ *with the cyclic automorphism shifting them, the classical nontrivial limit (if it exists!) of the local B-deformed structure of* $\mathbb{R}^4$ *is some exotic smooth* $R^4_{1,2}$.

To fix the result we need the simple but crucial observation:

Lemma 13.9 *Given a smooth structure on* $\mathbb{R}^4$ *if there does not exist any open cover of* $\mathbb{R}^4$ *containing a single coordinate patch* $\mathbb{R}^4$ *this structure has to be exotic.*

If a smooth $\mathbb{R}^4$ has a single coordinate patch $U \simeq \mathbb{R}^4$ it is diffeomorphic to the standard $\mathbb{R}^4$. If none of its open covers contains a single element, such R^4 can not be diffeomorphic to the standard $\mathbb{R}^4$. □

Proof (Theorem 13.3) We will show that any coordinate patch of the B-modified $\mathbb{R}^4$ can not contain the single chart. On the contrary let there exist a single coordinate patch $\mathbb{R}^4$ for the classical limit of the B-modified $\mathbb{R}^4$ as above. But in this case any open cover can be deformed by diffeomorphisms to a cover whose transition functions are identities. However, the factorization of some U_α through $^*R_1$ and U_β through $^*R_2$ and the cyclic condition on the permutations of $\mathbb{N}$ excludes the identities. □

Note that this proof works in the case of B-modified $\mathbb{R}^n$ since in this case for the standard $\mathbb{R}^n$ one can have a single coordinate patch. It is known that exotic R^n's exist only in dimension $n = 4$ so that means that classical smooth limits of the categorical B-modifications do not exist for $n \neq 4$. What is so special in dimension 4 that enables the existence of the limit as above? Some explanation comes from the special relation of Casson handles and geometric constructions in dimension 4 with the smooth NNO in B.

13.3.1 Casson Handles in B

When proving that emerged smooth R^4 is exotic we left aside the case when there are external diffeomorphisms which can be mapped onto the internal ones. The reason is that they could be 'gauged out' to the identity on the intersections by some external diffeomorphisms. However, making the additional assumption, which is also partly and implicitly present in so far analysis, we can include some external diffeomorphisms as generating exotic smooth R^4's. Namely, assume explicitly that *natural numbers N are generated as different objects by non-isomorphic c-models of* $\mathbb{R}$. This means that given the almost permutations of $\mathbb{N}$ generated by different c-models of $\mathbb{R}$ they can not be 'gauged out' to the identity whenever the models of $\mathbb{R}$ are non-isomorphic. This is rather strong low level assumption which reverses our 'natural thinking' about the relation of real and natural numbers. However, in B we had a similar situation: given the canonical object of real numbers R_B which is the image $s(\mathbb{R})$ from Set, we had to modify the NNO $s(\mathbb{N})$ to the smooth N_B. The real numbers determined the NNO. Now we want to follow this line of reasoning and show that Casson handles are related with the smooth NNO in B.

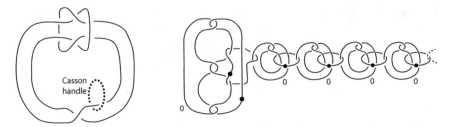

Fig. 13.1 The simplest small exotic R^4 with the simplest possible Casson handle attached to the Akbulut cork

Fig. 13.2 The simplest possible Casson handle which gives rise to an exotic R^4

Let us consider one example. The simplest known exotic $\mathbb{R}^4$ can be represented in the Kirby calculus language as a handle-body with a single Casson handle (Fig. 13.1, [29], p. 363). The simplest possible Casson handle with a single positive intersection at each level (Fig. 13.2, see [29], p. 363). Let us assign a partial non-cyclic permutation p of $\mathbb{N}$ to this CH, namely define it by: the number of level, i.e. n, plus the number of intersections at each level, i.e. 1. partial permutation:

$$p : n \to n + 1, n \in \mathbb{N}. \tag{13.13}$$

Such a permutation defines the automorphism of $P(\omega)/\text{Fin}$ according to (13.12) and thus corresponds to the shifts between the c-models of $\mathbb{R}$. Based on the assumption we indeed arrive at the exotic $\mathbb{R}^4$.

This simple example justifies the assumption as a basic rule in the context of exotic smooth structures on $\mathbb{R}^4$. It also shows that Casson handles are nontrivially related with the object of NN in $\mathcal{B}$. One can make this relation even more direct by interpreting the trees spanning CH's as built w.r.t. the smooth rather than standard NN. In this case we say that a Casson handle is spanned by a tree in $\mathcal{B}$. Let us turn again to the simplest CH represented by its Kirby diagram in the Fig. 13.2 (see [29], p. 363). The tree is just infinite $+$-signed linear order of levels. The crucial information is its infiniteness resulting from the geometric construction. More precisely:

Lemma 13.10 *If the smooth Casson handle construction terminated after finitely many steps it is the standard smooth 2-handle.*

This means that any smooth $\mathbb{R}^4$ with a handle-body containing *all* smooth finite CH's becomes the standard smooth $\mathbb{R}^4$. Let us now associate the smooth NN to the levels of the simplest CH:

$$\text{\# of level} \to n \in \mathbb{N} \subset N_\mathcal{B}$$

just by taking the infinite set of levels as complementary to the finite set $\{0\}$ thus becoming a member of the Frechet filter $\mathcal{F}$. But this means that the infinite tree of this CH is just s-finite in $\mathcal{B}$. When one performs similar enumerating of infinite number of levels in an arbitrary CH the result is the following:

Lemma 13.11 *Infinite Casson handles are spanned in $\mathcal{B}$ by s-finite trees.*

This together with Lemma 13.10 indicates that indeed the internal arithmetic of $\mathcal{B}$ has something to do with exotic smoothness, since one can state:

Corollary 13.3 *Exotic smoothness structures on $\mathbb{R}^4$ (smooth 4-manifolds), while transformed into $\mathcal{B}$ by s, belong to the class of s-standard smooth $\mathbb{R}^4$. They all are internally s-diffeomorphic.*

This result in fact agrees with our previous observation that externally distinct, even discontinuous maps lead to internal s-diffeomorphisms. What was crucial in establishing it was the shift (replacement) from the standard $\mathbb{N}$ to the smooth $N_\mathcal{B}$. The same shift is crucial in the above seeing CH's as s-finite objects. Observe that turning to the locally modified by $\mathcal{B}$ structures of manifolds, allows for the shifting between various exotic $\mathbb{R}^4$'s, not necessarily between exotic and the standard ones. Namely it holds:

Theorem 13.4 *Let R^4 be a small exotic $\mathbb{R}^4$ whose handle-body contains k many CH's for some $k \in \mathbb{N}$. Let a local $\mathcal{B}$-modification of R^4 be performed such that $l < k, l \in \mathbb{N}$ l-many CH's belong to the local open neighborhood which is internal in $\mathcal{B}$. Then, there exists a classical limit of this modification which is an exotic R_{k-l}^4 (with only $k - l$ nontrivial CH's).*

Proof Observe that internally l CH's becomes s-finite CH's (those corresponding to the s-finite spanning trees). It is enough to define the classical limit as R_{k-l}^4 by requiring that s-finite CH's are sent to the actually finite ones. ☐

Now, we see that the local modification of the manifold smooth structures by $\mathcal{B}$ and taking classical limits, works as an analogue of the large diffeomorphism where the actual smooth exotic R^4's represent a kind of *generalized* isotopy classes of embeddings (or small, coordinate-like diffeomorphisms). Working entirely in Set one can not realize exotic R^4's as merely isotopy classes of embeddings since there is no diffeomorphism at all connecting different exotic R^4's. Moreover, in this generalized set-up, one can study a class of topological and smooth manifolds allowing for the local categorical modifications (and the resulting new concept of equivalence). The local character of the modification leads to *generalized manifolds* which are partially both in Set and $\mathcal{B}$.

13.4 Some Consequences to Physics

Starting with $\mathbb{R}$ and $^*R$ and creating the pairs of such reals for both models we arrive at the isomorphic fields of complex numbers, even though $\mathbb{R}$ and $^*R$ are non-isomorphic. This is connected with the fact that in $\mathbb{C}$ one can not define the NNO $\mathbb{N}$ (starting from the axioms of the complete ordered algebraically closed field of characteristic zero). But this means that we can use $^*R$ instead of $\mathbb{R}$ in the case of the complete ordered algebraically closed field of characteristic zero, i.e. $\mathbb{C}$.

Given the divergent expression $1 + 2 + 3 + 4 + \cdots = \sum_{i=1}^{\infty} i$ it is bigger than any $n \in \mathbb{N}$ so this sum, if existed as the natural number (and in 1st order language), corresponds to a non-standard number of some model $^*N$ hence $^*R$. Moreover, such non-standard element exists in any non-standard $\mathfrak{c}$-model of $\mathbb{R}$, since every non-standard model of $\mathbb{N}$ is the (conservative) end-extension of $\mathbb{N}$.

Note that we get the same $\mathbb{C}$ (up to isomorphism) starting from any $^*R$ by building the space of pairs with the algebraic operations of $\mathbb{C}$. This is the consequence of categoricity of $\mathbb{C}$. Thus, possibly the non-standard big values, like the infinite sum above, should correspond (via the isomorphisms of models) to some finite value in $\mathbb{C}$.

Indeed, suppose such value does not exist, then each pair of the form $(\sum_{i=1}^{\infty} i, b)$, $b \in {}^*R$ can not correspond to any standard complex number. But it does since every $^*C \overset{\text{iso}}{\simeq} \mathbb{C}$ ($\mathbb{C}$ is $\mathfrak{c}$-categorical). Moreover it has to correspond via the isomorphism to some standard pair $z \in \mathbb{C}, z = (x, y); x, y \in \mathbb{R}$. The point is the following: $\mathbb{C}$ allows $2^{\mathfrak{c}}$ nontrivial automorphisms and they give rise to the isomorphisms $^*C \overset{\text{iso}}{\simeq} \mathbb{C}$ for every $^*C$ generated via the ultrafilter constructions. On the other hand there are only 2 automorphisms of $\mathbb{C}$ that send $\mathbb{R}$ to $\mathbb{R}$—the identity and the complex conjugation. This, together with the fact that fixed points of all automorphisms of $\mathbb{C}$ are all rational numbers, i.e. $\forall_{\phi \in \text{Aut}(\mathbb{C})} \forall_{r \in \mathbb{Q}} \phi(r) = r$, give that the image of $\sum_{i=1}^{\infty} i$ under any isomorphism $^*C \overset{\text{iso}}{\simeq} \mathbb{C}$ has to be irrational pair $(x, y) \in \mathbb{C} : x, y \in \mathbb{I}$. This is in fact result of a very discontinuous and wild behavior of the (wild) automorphisms of $\mathbb{C}$ realizing the above isomorphism. On the other hand if one would like to have a finite value assigned to this iso which would not be dependent on the choice of the non-standard model $^*R$ it had to be rational number as it is a fixed point of every automorphism. In what follows we would like to consider this model-theoretic mechanism for assigning finite values to divergent expressions in context of exotic smoothness structures on $\mathbb{R}^4$. Then we try to understand this phenomenon in context of renormalization and regularization ever-present in perturbative quantum field theories.[8]

Lemma 13.12 *For any exotic smooth R^4 (which is topologically $\mathbb{R}^4$) any diffeomorphic image of it can not send smooth coordinate line $\mathbb{R}$ to the smooth $\mathbb{R}$.*

If there were such diffeomorphism the exotic R^4 would factorize as $\mathbb{R} \times \mathbb{R}^3$ which is necessarily standard. $\square$

---

[8]This part of the work was performed in the cooperation with Krzysztof Bielas.

One can equivalently state the lemma as: *If the topological* $\mathbb{R}$ *is smooth line in a smooth* R^4 *this has to be standard* $\mathbb{R}^4$. Thus, when a smooth diffeomorphism of R^4 preserves $\mathbb{R}$ as the factor this can happen only for the standard $\mathbb{R}^4$. In the case of automorphisms of $\mathbb{C}$ when $\mathbb{R}$ is send to $\mathbb{R}$ then the automorphism can not be wild. Otherwise, any wild automorphism scatters in a very discontinuous way the real line in the complex plane (leaving the rational numbers fixed). For any exotic diffeomorphism of R^4 it can not smoothly send the line $\mathbb{R}$ to itself, though continuously it does. As we explained in the previous sections and in this one, both situations are connected with non-standard $\mathfrak{c}$-models of $\mathbb{R}$.

Let us consider the non-standard $^*C$ (though isomorphic to $\mathbb{C}$) as generated by pairs of the non-standard reals, i.e. $^*C \simeq \{(a, b) \in {}^*R \times {}^*R\}$. Then, make the product: $^*C^2 \simeq {}^*R^4$. When turning to the higher orders one gets the unique (up to isomorphisms) standard real field and the equality reads: $\mathbb{C}^2 \simeq \mathbb{R}^4$. Instead, one can use an automorphism of $\mathbb{C}$ to obtain (non-canonical) isomorphism $^*C \overset{\text{iso}}{\simeq} \mathbb{C}$ and thus $\mathbb{C}^2 \simeq_{\text{iso}} {}^*R^4$. Given different $^*R^4$'s one gets different automorphisms of $\mathbb{C}$ and thus different realizations of the isomorphism above. It follows that one can use different wild automorphisms of $\mathbb{C}$ to distinguish (index) different non-standard models of $\mathbb{R}$. Given $\mathbb{R}^4$ locally modified by $\mathcal{B}$ and taking its classical limit which factors through $^*R_1, {}^*R_2$, this results in the exotic $R^4_{1,2}$ and thus the correspondence follows:

Corollary 13.4 *Pairs* (α_1, α_2) *of automorphisms of* $\mathbb{C}$, *where at least one automorphism is wild, distinguishes different exotic* $R^4_{1,2}$'s.

This relation can be expressed in terms of Eq. (13.10) which for $n = 4$ and by turning to the $\mathbb{C}$ leads to the fully external description:

$$\mathbb{C}^2 \simeq \mathbb{R}^4 \overset{2nd \to 1st}{\longrightarrow} {}^*R_1{}^4 \to {}^*C_1 \overset{(\text{iso},0)}{\to} \mathbb{C} \times \mathbb{C} \overset{(0,\text{iso})}{\leftarrow} {}^*C_2 \leftarrow {}^*R_2{}^4 \overset{2nd \to 1st}{\longleftarrow} \mathbb{R}^4 \simeq \mathbb{C}^2.$$

$$(13.14)$$

The middle $\mathbb{C} \times \mathbb{C}$ product emerges from the component-wise automorphisms of $\mathbb{C}$ giving rise to the isomorphisms $\alpha_i : {}^*C_i \simeq \mathbb{C}, i = 1, 2$ and this is the pair (α_1, α_2) which represents exotic $R^4_{1,2}$. The relation is, however, highly non-constructive, similarly to the wild automorphisms of $\mathbb{C}$ and the ultrafilters constructions.

As we observed the wild automorphisms of $\mathbb{C}$ should somehow allow for the assignment of finite values to some divergent expressions. We can make this point more tractable by turning to the relation with exotic $R^4_{1,2}$ and making use of the very special properties of $\mathcal{B}$. So we turn again to (13.10) from (13.14). The point is that $\mathcal{B}$ locally modifies $\mathbb{R}^4$ and the theory of distributions in $\mathcal{B}$ looks very special, namely all distributions in $\mathcal{B}$ are regular (constructive and w.r.t. the smooth real line and natural numbers) and each external distribution is canonically mapped into the internal one. In fact it holds ([37], Theorem 3.6, p. 324 and Remark on p. 322):

Theorem 13.5 [37] *In* $\mathcal{B}$ *for every distribution* μ *on* R^n *there exists a predistribution (function)* $\mu_0 : R^n \to R$ *such that for all* $f \in F_n$:

$$\mu(f) = \int f(x)\mu_0(x)\mathrm{d}x.$$

Here F_n denotes the internal space of test functions in dimension n. Also as stated by Theorem 3.15.3, p. 336 in [37], there exists a bijection between the external distributions in Set and the internal in $\mathcal{B}$ given by the global section functor Γ : $\mathcal{B} \to$ Set. In particular, the product and the square roots of distributions are thus well-defined in $\mathcal{B}$ as operations on the representing internal functions.

13.4.1 Renormalization in the Coordinate Space

Now we can discuss the problem of renormalization in perturbative quantum field theory based on this special representation of distributions in $\mathcal{B}$. Note also that in $\mathcal{B}$ the standard NNO, i.e. $\mathbb{N}$, is replaced with the smooth NNO, $N_\mathcal{B}$, such that 'finite' in $\mathcal{B}$ is 'infinite' externally in Set. Thus indeed $\mathcal{B}$ is a natural category for addressing renormalization questions. Given the interaction Lagrangian $\mathcal{L} = \frac{\lambda}{k!}\phi_I^k$ of the ϕ^k neutral scalar massive quantum field theory its S-matrix is determined in Dyson series representation, as [42]:

$$S = \sum_{n=0}^{\infty} \frac{i^n}{n!} \int_{M^n} T(\mathcal{L}_I(x_1)\mathcal{L}_I(x_2)\ldots\mathcal{L}_I(x_n))dx_1 dx_2 \ldots dx_n \qquad (13.15)$$

where $M = \mathbb{R}^{1,3}$, $M^n = M \times \cdots \times M$ n-times and T stays for the time-ordered products of the operator-valued distributions $\mathcal{L}_I$, hence S is the operator-valued distribution either. The time ordering is defined for two operator valued functions A, B on M as (we follow the presentation in [42]):

$$T(A(x_1)B(x_2)) = \Theta(x_1^0 - x_2^0)A(x_1)B(x_2) + \Theta(x_2^0 - x_1^0)B(x_2)A(x_1) \quad (13.16)$$

where $\Theta(x)$ is the Heaviside function on $\mathbb{R}$, i.e. $\Theta(x) = 0, x < 0$ and $\Theta(x) = 1, x \geq 0$. Here $x_i^0, i = 1, 2$ are time coordinates of $x_i \in M, i = 1, 2$. However, as noted in [42] for general operator-valued irregular distributions one cannot create the products of them by discontinuous functions like Θ. If one, however, works outside the thick diagonal, $D_n = \{x \in M^n : \exists_{i \neq j} x_i = x_j\}$, then the Θ is continuous hence the product (13.16) is well-defined. This is the core of the various problems in perturbative QFT, let us quote the opinion of Authors of [42]:

> In fact the mathematical origin for the appearance of short-distance singularities in perturbation theory is the ill-defined notion of time-ordering reviewed above. Epstein and Glaser proposed a way to construct well-defined time ordered products T_n, one for each power n of the coupling constant, that satisfy a set of suitable conditions explained below, the most prominent being that of locality or micro-causality. The power series S constructed by (13.15) using the Epstein-Glaser time-ordered product T is a priori finite in every order, and renormalization corresponds then to stepwise extension of distributions from $M_n \backslash D_n$ to M_n. In general, distributions can not be extended uniquely onto diagonals. The resulting degrees of freedom are in one-to-one correspondence with the degrees of freedom (finite

renormalizations) in momentum space renormalization programs like BPHZ and dimensional regularization.

Instead of reviewing the Epstein-Glaser construction let us observe that for regular distributions the problem in (13.16) does not arise since they can be represented by the operator-valued functions, their product is well-defined and they can be multiplied by Θ. True problem arises for irregular distributions like Dirac δ. Moreover, if all distributions were regular the Epstein-Glaser construction would give as the extension over $M_n \setminus D_n$ just the regular distributions we started with.

Now recall that in $\mathcal{B}$: 1. *every* distribution is regular (Theorem 13.5), 2. *every* distribution in Set can be naturally mapped to a distribution in $\mathcal{B}$ (Theorem 3.15.3, p. 336, [37]), and 3. the function $^{\mathcal{B}}\Theta : R \rightarrow R$ is continuous. This observations motivate the following procedure:

Given M^n (n-product of the Minkowski spacetime, $n = 1, 2, 3 \ldots$) let the diagonal $D_n \subset U_\beta \in \mathcal{O}$ for some regular open cover $\{U_\alpha\}_{\alpha \in I}$ where $U_\beta \in \{U_\alpha\}_{\alpha \in I}$. Then, one locally modifies M^n by $\mathcal{B}$ such that $U_\beta \in \mathcal{B}$ according to Definition 13.2.

Under the procedure above one indeed has well-defined extensions of distributions over the diagonals D_n in a sense of internal logic of $\mathcal{B}$ localized on M^n. Observe also that the local modification of M^n implies some local modification of spacetime M itself (if not, all factors in M^n are not modified, hence M^n neither).

Corollary 13.5 *Varying the underlying geometry of a spacetime manifold by the local modification of its smooth structure by $\mathcal{B}$, i.e. $^{(\mathcal{B})}M$, gives rise to the renormalization of some perturbative QFT when formulated on such modified manifolds.*

Let us introduce the following additional suppositions:

All the local deformations of M are generated by the underlying local deformations by $\mathcal{B}$ of $\mathbb{R}^4$, and let the classical limit of them factorize through some $^*R_i, i = 1, 2$, thus leading to exotic $R_{1,2}^4$. Then it follows:

Corollary 13.6 *The renormalization problem of some perturbative QFT can be translated into the geometry of some (Euclidean) exotic R^4 background which complements the Minkowski flat spacetime.*

One can restate the corollary as: *Ultraviolet (UV) divergences in some perturbative QFT determine exotic smoothness of the Euclidean $\mathbb{R}^4$ background.* We expect that ultraviolet divergences counterterms of some perturbative QFT's on Minkowski spacetime are expressible in terms of the Riemannian (sectional) curvature of $R_{1,2}^4$. This Euclidean curved 4-background complements the Minkowski's one. Recall that exotic R^4's are just Riemannian smooth 4-manifolds which can not be flat. Thus the Corollary 13.6 indicates that a curvature in spacetime, hence nonzero density of gravitational energy emerges, when renormalization problem is solved geometrically. This connection with gravity is a rather universal, non-perturbative phenomenon of different perturbative QFT's and it is an important feature of the approach.

13.4.2 QM on Smooth **R⁴** and Model Theory

The specific model-theoretic approach to exotic smoothness of open 4-manifolds like $\mathbb{R}^4$ presented here has also the advantage that one can still think in terms of local differentiation and (global) functions and arrive at the model-theoretic set-up. This is complementary to the approach via Riemannian structures and curvature, which anyway indicates that exotic R^4's are 'normal' smooth 4-manifolds and functions are local objects on them. We follow the work [24] and the case of exotic R^4's is again crucial here. We work in the complementary picture and the analysis is based on model-theoretic tools but it is worth mentioning that strong connection of small exotic smooth R^4's with QM formalism, noncommutative spaces, QFT and quantum gravity, was indeed shown and developed by purely geometric and topological methods (see e.g. [32, 43, 44]).

Lemma 13.13 *Let R^4 be some exotic smoothness structure on $\mathbb{R}^4$. There has to exist a continuous (non-standard smooth) real-valued function on $\mathbb{R}^4$ which would be smooth on R^4, or a continuous real-valued function smooth on $\mathbb{R}^4$ but merely continuous on R^4.*

If such function did not exist precisely the same functions would be smooth in both structures and the smoothness structures would be equivalent and manifolds diffeomorphic (being homeomorphic).   □

So, let $f : \mathbb{R}^4 \to \mathbb{R}$, $f \in C^0(\mathbb{R}^4)$ and $f \in C^\infty(R^4)$ so f is exotic smooth. f can not be everywhere standardly differentiable on $\mathbb{R}^4$ but, when changing the smoothness structure into R^4, it can. Moreover, the differentiation is locally the same as a standard one, since R^4 is a Riemannian smooth 4-manifold. What would happen if one tried to differentiate globally any nondifferentiable continuous function? One should follow the pattern of generalized differentiation of functions or distributions. Outside the domains where the function is not continuous, the differentiation agrees with normal local differentiation. We are looking for the model-theoretic compensation (representation) for such global 'non-standard' distributional differentiation. The result is precisely the $\mathbb{R}^4$ locally modified by $\mathcal{B}$.

Namely, it is always possible to choose open neighborhoods containing the non-smooth domains of the function f such that in these domains the functions would be represented by *regular* distributions. However, iterating differentiation of them leads to irregular distributions as well, like Dirac δ-distribution. Then, we can turn to a $\mathbb{R}^4$ locally modified by $\mathcal{B}$ such that the neighborhoods are internal in $\mathcal{B}$ and every external distribution, also irregular, is represented internally by regular one (Theorem 3.15.3, p. 336, [37]), i.e. by some internal *smooth* function. This is the model-theoretic smoothing of continuous functions on $\mathbb{R}^4$. Taking the classical nontrivial limit of this local modification by $\mathcal{B}$ the result is some exotic R^4 as in Theorem 13.3. On the contrary, every local modification by $\mathcal{B}$ sends some irregular distributions to the internal smooth functions. Thus the following definition is natural and direct in this context: we call the modification by $\mathcal{B}$ the *model-theoretic representation of an exotic smooth structure on $\mathbb{R}^4$* [24] provided it sends some irregular distributions to the internal smooth functions. Let exotic smooth $R_{1,2}^4$ be the classical limit of our $^{(\mathcal{B})}R^4$ which factorizes through $^*R_i$, $i = 1, 2$.

Lemma 13.14 *Let the model-theoretic representation of the exotic smooth $R_{1,2}^4$ be $^{(\mathcal{B})}R^4$. In the classical trivial, i.e. standard $\mathbb{R}^4$ limit, the space of exotic smooth functions on $R_{1,2}^4$ contains some irregular external distributions on the standard $\mathbb{R}^4$.*

First, in the classical limit we do not have the dependence on $\mathcal{B}$ any longer. Next, suppose that classical limit as in the formulation of the lemma does not contain the distribution. Then the global differentiation of every smooth function on $R_{1,2}^4$ agrees with the global standard differentiation on $\mathbb{R}^4$. So, the smoothness structure of $R_{1,2}^4$ has to be the standard one.   □

Next consider the Fourier transform of smooth functions FT : $C^\infty(\mathbb{R}^4) \to C^\infty(\mathbb{R}^4)$. Let us represent the discontinuous functions in some open neighborhood by the corresponding irregular distributions as before. FT extends over the space of L^2-functions and distributions on $\mathbb{R}^4$ thus over $C^\infty(R_{1,2}^4)$ in the $\mathbb{R}^4$ representation. The image of such Fourier operator is again $C^\infty(R_{1,2}^4)$. This is the core of the interpretation of QM formalism on exotic R^4.

Lemma 13.15 *The FT of δ and δ-distribution itself, they both belong to the model-theoretic representation of an exotic smooth R^4 in the standard $\mathbb{R}^4$ limit.*

We would like to interpret this result directly on exotic R^4. Note that the FT of δ is ~ 1 and it is geometrically a straight line, say coordinate axes, in the standard structure. However, this line can not be any smooth coordinate line in any exotic R^4, since this would give the factorization and the collapse of the structure to the standard one. However, the tangent space of every exotic R^4 is trivial, i.e. $TR^4 \simeq T\mathbb{R}^4 \simeq T_0\mathbb{R}^4$ (R^4 is contractible) and we consider this $1(x)$ as the coordinate line in the tangent space TR^4 [24].[9] This coordinate line is spanned by $\sim\partial_x$ in the generator tangent space. Thus FT mixes the standard tangent space with coordinate space R^4 and thus ∂_x is sent to the multiplication operation in the model-theoretic representation of exotic R^4. Given a large exotic R^4 (which can not be embedded into the standard $\mathbb{R}^4$) its contraction to a ball in $\mathbb{R}^4$ gives rise to:

Theorem 13.6 (Corollary 4, [24]) *One can interpret the noncommutative relations of the position and momentum operators in the, contracted to a 4-ball, classical limit of the model-theoretic representation of a large exotic smooth R^4.*

Based on this interpretation the mechanism of decoherence in spacetime was proposed where QM effects disappear by taking uncontracted limit of such contracted R^4 [24].

Acknowledgments The author appreciates much the important and fruitful discussions with Torsten Asselmeyer-Maluga and Krzysztof Bielas within the years about the wide range of topics appearing in the Chapter.

---

[9]One could also think about such structures as having the generalized tangent spaces like e.g. $TR^4 \oplus {}^*TR^4$. Indeed, one can relate [45] some small exotic R^4's with deformations of Hitchin structures (gerbes) defined on $TS^3 \oplus {}^*TS^3$, $S^3 \subset \mathbb{R}^4$.

Appendix

Weak Arithmetic in Smooth Toposes

In order to work constructively in arbitrary topos the correct logic is intuitionistic—one avoids the axiom of choice (AC) and the law of excluded middle (e.g. [20]). Next, instead of the axiom of choice, and even finite AC, one has the axiom of bounded search [37] as in (13.19), recursion rule is replaced by the finitely presented type recursion (13.18) and full induction is replaced by the following (13.17) coherent induction scheme[10]:

$$\text{Ind}: \ \phi(0) \wedge \forall_{x \in N}(\phi(x) \to \phi(x+1)) \to \forall_{x \in N}\phi(x), \ \text{for } \phi \text{ coherent} \quad (13.17)$$

$$\text{Rec}: \ \forall_{f \in S^{S \times T}} \forall_{a \in S^T} \exists!_{g \in S^{N \times T}} \forall_{x \in T}(g(0, x)) = \quad (13.18)$$

$$a(x) \wedge \forall_{n \in N} g(n+1, x) = f(g(n, x))$$

$$\text{weak AC}: \ \forall_{A \in P(N \times N)}(\forall_{n \in N} \exists_{m \in N} A(n, m) \to \quad (13.19)$$

$$\forall_{n_0 \in N} \exists_{m_0 \in N} \forall_n n \le n_0 \exists_m m \le m_0 \wedge A(n, m)).$$

The type S in (13.18) has to be finitely presented and the formula ϕ in (13.17) coherent (e.g. [37], pp. 297–298) which results in further weakening of the logic.[11] Given such substantial weakening of the logic and arithmetic one gains the degree of indistinguishability of the standard and certain non-standard models of natural numbers. These weak properties are augmented by the usual subset of PA axioms (still in the intuitionistic logic):

$$N \text{ is a subtype of } R; \quad (13.20)$$

$$R \text{ is Archimedean}: \forall_{x \in R} \exists_{n \in N} x < n; \quad (13.21)$$

$$0 \in N \text{ and } \forall_{x \in R}(x \in N \to x+1 \in N) \text{ and} \quad (13.22)$$

$$\forall_{x \in R}(x \in N \wedge x+1 = 0 \to \bot).$$

As shown by Moerdijk and Reyes [37] these properties characterizing weak intuitionistic arithmetic along with coherent formulas and type restrictions as in (13.17) and (13.18) above, are fulfilled in some *smooth* toposes like smooth Zariski topos $\mathcal{Z}$ or Basel topos $\mathcal{B}$.

References

1. C.H. Brans, in *Roles of space-time models*, ed. by A. Marlow. Quantum Theory and Gravitation (Academic Press, New York, 1980)
2. C.H. Brans, Exotic smoothness and physics. J. Math. Phys. **35**, 5494 (1994)

---

[10]These axioms are written within the varying types formalism of Feferman [37, 46].

[11]$P(N \times N)$ is the power set of $N \times N$ and $A(n, m)$ means $(n, m) \in A \in P(N \times N)$.

3. C.H. Brans, Localized exotic smoothness. Class. Quant. Gravity **11**, 1785 (1994)
4. C.H. Brans, Absolute spacetime: the twentieth century ether. Gen. Relativ. Gravit. **31**, 597 (1999)
5. T. Asselmeyer, Generation of source terms in general relativity by differential structures. Gen. Rel. Grav. **34**, 597 (1996)
6. J. Sładkowski, Exotic smoothness, noncommutative geometry and particle physics. Int. J. Theor. Phys. **35**, 2075 (1996)
7. J. Sładkowski, Gravity on exotic R^4 with few symmetries. Int. J. Mod. Phys. D **10**, 311 (2001)
8. P. Benioff, Models of Zermelo Frankel set theory as carriers for the mathematics of physics I, II. J. Math. Phys. **17**(5), 618 (1976)
9. A. Kock, Synthetic differential geometry. Lond. Math. Sci. Lect. Notes **51** (1981)
10. F.W. Lawvere, "Continuously varying sets: Algebraic geometry = Geometric logic", in Logic Coll. **73**, 135 (North Holland, Bristol, 1975)
11. G. Takeuti, *Two applications of logic to mathematics* (Math. Soc. Japan **13**, Kano Memorial Lecture 3, 1978)
12. P. Benioff, Language is physical. Quant. Inf. Proc. **1**, 495 (2002)
13. P. Benioff, Space and time dependent scaling of numbers in mathematical structures: Effects on physical and geometric quantities, will appear in Quantum Information Processing, Howard Brandt memorial issue
14. A. Doering, C.J. Isham, A topos foundation for theories of physics: I. Formal languages for physics. J. Math. Phys. **49**, 053515 (2008)
15. C. Heunen, N.P. Landsman, B. Spitters, A topos for algebraic quantum theory. Comm. Math. Phys. **291**(1), 63 (2009)
16. C.J. Isham, in *Topos Methods in the Foundations of Physics* ed. by H. Halvorson. Deep Beauty (Cambridge University Press, 2010)
17. J. Król, Background independence in quantum gravity and forcing constructions. Found. Phys. **34**,(3), 361 (2004)
18. J. Vaananen, Second-order logic and foundations of mathematics. Bull. Symbolic Logic **7**, 504 (2001)
19. T. Jech, *Set Theory* (Springer, New York, 2003)
20. S. MacLane, I. Moerdijk, *Sheaves in Geometry and Logic*, A First Introduction to Topos Theory (Springer, New York, 1992)
21. P.J. Cohen, Proc. Nat. Acad. Sci. USA **50**, 1143 (1963)
22. T. Bartoszyński, H. Judah, *Set Theory: On the Structure of the Real Line* (A.K. Peters, Massachusets, 1995)
23. J. Król, Set theoretical forcing in quantum mechanics and AdS/CFT correspondence. Int. J. Theor. Phys. **42**(5), 921 (2003). arXiv:quant-ph/0303089
24. J. Król, Exotic smoothness and noncommutative spaces. The model-theoretical approach. Found. Phys. **34**(5), 843 (2004)
25. W. Just, M. Weese, *Discovering Modern Set Theory*, Set-Theoretic Tools for Every Mathematician II (AMS, Providence, 1997)
26. M.H. Freedman and team Freedman. Bing Topology and Casson Handles 2013. Santa Barbara/Bonn Lectures (2013)
27. T. Asselmeyer-Maluga, C.H. Brans, *Exotic Smoothness and Physics*, Differential Topology and Spacetime Models (World Scientific, Singapore, 2007)
28. M.H. Freedman, The topology of four-dimensional manifolds. J. Differ. Geom. **17**, 357 (1982)
29. R.E. Gompf, A.I. Stipsicz, *An Introduction to 4-Manifolds and Kirby Calculus* (American Mathematical Society, Rhode Island, 1999)
30. Ž. Bižaca, An explicit family of exotic Casson handles. Proc. AMS **123**(4) (1995)
31. T. Kato, Spectral analysis on tree-like spaces from gauge theoretic point of view. Contemp. Math. **347** (2004)
32. T. Asselmeyer-Maluga, J. Król, Topological quantum D-branes and wild embeddings from exotic smooth R^4. Int. J. Mod. Phys. A **26**(20), 3421 (2011). arXiv:1105.1557

33. T. Asselmeyer-Maluga, J. Król, Quantum geometry and wild embeddings as quantum states. Int. J. Geom. Meth. Mod. Phys. **10**, 10 (2013)
34. A. Robinson, *Non-Standard Analysis*, Studies in Logic and the Foundations of Mathematics (North-Holland Publishing Co., Amsterdam, 1966)
35. K.J. Hintikka, Reductions in the theory of types, in *Two Papers on Symbolic Logic*, Acta Philosophica Fennica, No. 8 (Helsinki, 1955)
36. R. Montague, in *Reductions of Higher-order Logic*, ed. by J.W. Addison, L. Henkin, A. Tarski. The Theory of Models (North-Holland Publishing Co., Amsterdam, 1965), pp. 251-264
37. I. Moerdijk, G.E. Reyes, *Models for Smooth Infinitesimal Analysis* (Springer, New York, 1991)
38. I. Moerdijk, A model for intuitionistic non-standard arithmetic. Ann. Pure Appl. Logic **73**, 37 (1995)
39. R.C. Walker, *The Stone-Cech Compactification, Ergebnisse der Mathematik und ihrer Grenzgebiete, 83* (Springer, Berlin, 1974)
40. S. Shelah, *Proper Forcing*, Lec. N. Math. **940** (Springer, Berlin-New York, 1982)
41. J. Król, A model for spacetime: the role of interpretation in some Grothendieck topoi. Found. Phys. **36**(7), 1070 (2006)
42. C. Bergbauer, D. Kreimer, The Hopf algebra of rooted trees in Epstein-Glaser renormalization. Annales Henri Poincaré **6**, 343 (2005). arXiv:hep-th/0403207
43. T. Asselmeyer-Maluga, J. Król, Inflation and topological phase transition driven by exotic smoothness. Adv. High Energy Phys. Article ID 867460, special issue, Experimental Tests of Quantum Gravity and Exotic Quantum Field Theory Effects (QGEQ) (2014). http://dx.doi.org/10.1155/2014/867460
44. J. Król, (Quantum) gravity effects via exotic R^4. Ann. Phys. (Berlin) **19**(3–5), 355 (2010)
45. T. Asselmeyer-Maluga, J. Król, Abelian gerbes, generalized geometries and foliations of small exotic R^4 (2009). arXiv:0904.1276v5
46. S. Feferman, A theory of variable types, in *Proceedings of 5th Lattin American Symposium on Mathematical Logic* ed. by X. Caicedo, N.C.A. da Costa, R. Chuaqui (1985)

Chapter 14
Exotic Smoothness on Spheres

Duane Randall

Abstract In his article [13], "Differential Topology Forty-six Years Later" published in the Notices of the AMS in 2011, John Milnor posed the following problem. Is the finite abelian group of oriented diffeomorphism classes of closed smooth homotopy spheres of dimension n nontrivial for all dimensions $n > 6$ with n different from 12 and 61? He includes a table enumerating these groups of closed smooth homotopy spheres for all $n < 64$, n different from 4. Nontriviality of the group of distinct exotic smoothness structures on the n-dimensional sphere provides counterexamples to the differential Poincare hypothesis in dimension n. We note in this abstract that this problem posed by John Milnor has a nearly complete solution, principally due to constructions of infinite 2-primary families of nontrivial elements in the stable homotopy of spheres by numerous topologists and also by the recent spectacular work of Hill et al. [7] on the non-existence of Kervaire invariant one elements in all dimensions $2^j - 2$ with $j > 7$. We obtain the main theorem: The n-dimensional sphere, S^n, admits exotic smoothness structures of 2-primary order for all dimensions $n > 61$ such that n is not congruent to 4 modulo 8 and also n not equal to 125 or 126. Moreover, for any integer n congruent to 4 modulo 8 of the form $n = 4p(p-1) - 4$ for some odd prime p, S^n admits exotic differentiable structures of p-primary order.

14.1 Notes from T. Asselmeyer-Maluga (Editor)

With these notes, we will present an introduction to understand the importance of the result in the abstract.

In 1904, Poincaré proposed the following famous conjecture:

Poincaré conjecture: Let M be a closed 3-manifold. If M is simply connected, then M is homeomorphic to the 3-sphere

This conjecture was proved by Perelman [18] in the more wider context of Thurston's geometrization conjecture [15, 16]. This conjecture can be generalized to higher dimensions:

D. Randall (✉)
Loyola University, 6363 St. Charles Avenue, Box 92, New Orleans 70118, USA
e-mail: randall@loyno.edu

© Springer International Publishing Switzerland 2016
T. Asselmeyer-Maluga (ed.), *At the Frontier of Spacetime*,
Fundamental Theories of Physics 183, DOI 10.1007/978-3-319-31299-6_14

Question 1: Let M be a closed n-manifold. Suppose M is homotopy equivalent to S^n. Is M homeomorphic to S^n?

The answer turns out to be yes for all dimensions. For $n = 4$, it was proved by Freedman [5] in 1982. For $n \geq 5$, it was proved by Smale [20] in 1962, using the theory of h-cobordisms, and by Newman [17] in 1966 and by Connell [3] in 1967. Smale has to assume the existence of a smooth structure, while Newman and Connell proved it without this condition. Finally one obtains:

Theorem 14.1 *Any closed n-manifold that is homotopy equivalent to S^n is homeomorphic to S^n.*

One can also generalize this question into the smooth category, i.e.

Question 2: Let M be a closed n-manifold. Suppose M is homeomorphic to S^n. Is M diffeomorphic to S^n?

For $n = 3$, the answer is yes. It is due to Moise [14] that every closed 3-manifold has a unique smooth structure. In particular, the 3-sphere has a unique smooth structure. For $n = 4$, this question is wildly open. For higher dimensions, Milnor [12] constructed an exotic smooth structure on S^7. Furthermore, Kervaire and Milnor [8] showed that the answer is not true in general for $n \geq 5$. The construction can be simply summarized: Every n-sphere is given by a decomposition

$$S^n = D^n \cup_f D^n$$

into two n-disks D^n glued together by a diffeomorphism

$$f : \partial D^n = S^{n-1} \to \partial D^n = S^{n-1}$$

and we obtain by Theorem 14.1 that every topological n-sphere can be decomposed in this form. Using the h-cobordism theorem for $n > 6$, f depends on the connecting components of the diffeomorphism group i.e. the elements of $\pi_0(Diff(S^{n-1}))$ called isotopy classes. If f is connected to the identity component (the class $[f] = 0 \in \pi_0(Diff(S^{n-1}))$) then S^n admits the so-called standard smoothness structure (as induced by the embedding in $\mathbb{R}^{n+1}$). Therefore the answer to Question 2 is not true in general, leading to two natural questions:

Question 3: How many exotic structures are there on S^n?

Question 4: For which n's does there exist a unique smooth structure on S^n?

Kervaire and Milnor reduced the question above to a computation of the stable homotopy groups of spheres. In fact, Kervaire and Milnor constructed a group Θ_n, which is the group of h-cobordism classes of homotopy n-spheres. The group Θ_n classifies the differential structures on S^n for $n \geq 5$. This group Θ_n has a subgroup Θ_n^{bp}, which consists of homotopy spheres that bound parallelizable manifolds. For completeness we have to introduce the nth stable homotopy group of the sphere π_n. Let $S^{n+k} \to S^k$ be a map where the corresponding homotopy class is an element of $\pi_{n+k}(S^k)$. According to the stability (Freudenthal suspension theorem [6]), the group $\pi_{n+k}(S^k)$ for $k > n + 1$ depends only on n, the nth stable homotopy group of the sphere π_n. Furthermore there is a famous homomorphism between the homotopy groups of the group $SO(k)$ and the homotopy groups of the sphere called J-homomorphism

$$J : \pi_n(SO(k)) \rightarrow \pi_{n+k}(S^k)$$

defining in the stable case a map

$$J : \pi_n(SO) \rightarrow \pi_n$$

where SO is the stable special orthogonal group. Then the relation between Θ_n and π_n can be summarized by the following theorem.

Theorem 14.2 (Kervaire and Milnor [8]) *Suppose that $n \geq 5$.*

1. The subgroup Θ_n^{bp} is cyclic, and has the following order:

$$|\Theta_n^{bp}| = \begin{cases} 1, & \text{if } n \text{ is even,} \\ 1 \text{ or } 2, & \text{if } n \equiv 1 \ (mod\ 4), \\ 2^{2n-2}(2^{2n-2}-1)B(n), & \text{if } n \equiv 3 \ (mod\ 4). \end{cases}$$

Here $B(n)$ is the numerator of $4B_{2n}/n$ and B_{2n} is the Bernoulli number.
2. For $n \not\equiv 2 \ (mod\ 4)$, there is an exact sequence

$$0 \longrightarrow \Theta_n^{bp} \longrightarrow \Theta_n \longrightarrow \pi_n/J \longrightarrow 0.$$

Here π_n/J is the cokernel of the J-homomorphism.
3. For $n \equiv 2 \ (mod\ 4)$, there is an exact sequence

$$0 \longrightarrow \Theta_n^{bp} \longrightarrow \Theta_n \longrightarrow \pi_n/J \overset{\Phi}{\longrightarrow} \mathbb{Z}/2 \longrightarrow \Theta_{n-1}^{bp} \longrightarrow 0.$$

Here the map Φ is the Kervaire invariant.

By the argumentation above, this result implies that the number of exotic n-spheres is finite for $n \geq 5$ to answer Question 3. Then the classification of the exotic spheres given by Table 14.1 up to dimension 18. But what about the higher-dimensional examples. The result in the abstract gives a partial answer the result above, one has to go deeper into this theorem. In the first part of Theorem 14.2, the case $n \equiv 3 \ (mod\ 4)$ depends on the computation of the order of the image of the J-homomorphism. The case $n \equiv 1 \ (mod\ 4)$ depends on the Kervaire invariant in dimension $n + 1$. The computation of the image of the J-homomorphism at $4k - 1$ stems is a special case of the Adams conjecture. The proof was completed by Mahowald [10], and the full Adams conjecture was proved by Quillen [19], Sullivan [21], and by Becker and Gottlieb [1].

Table 14.1 number of homotopy spheres with respect to the dimension n, for $n > 4$ equivalent to the number of exotic spheres

n	1	2	3	4	5	6	7	8	9
Order of Θ_n	1	1	1	?	1	1	28	2	8
n	10	11	12	13	14	15	16	17	18
Order of Θ_n	6	992	1	3	2	16256	2	16	16

For Question 4, it is clear from Theorem 14.2 that, for $n = 4k + 3$ with $k \geq 1$, the smooth structure on the n-sphere is never unique. For $n = 4k + 1$ with $k \geq 1$, the answer depends on the existence of the Kervaire invariant one elements. In 2009, Hill et al. [7] showed that the only dimensions in which the Kervaire invariant one elements exist are 2, 6, 14, 30, 62 and possibly 126. That is, in other dimensions, the Kervaire invariant map

$$\pi_n / J \xrightarrow{\ \Phi\ } \mathbb{Z}/2$$

in part (3) of Theorem 14.2 is always zero and the group Θ_{n-1}^{bp} is $\mathbb{Z}/2$. Therefore, the only odd dimensional spheres that could have a unique smooth structure are S^1, S^3, S^5, S^{13}, S^{29}, S^{61} and S^{125}. Further, the cases S^{13} and S^{29} can be ruled out by May's [11] 3-primary computation of the stable homotopy groups of spheres. For dimension 61, it was shown in [22] that the sphere S^{61} has a unique smooth structure whereas the sphere S^{125} may not have a unique smooth structure. Here we have to note that these results have to be considered with care. Finally we decided to use it here because the results look promising. Finally, one obtains:

Theorem 14.3 *The only odd dimensional spheres with a unique smooth structure are S^1, S^3, S^5 and S^{61}.*

For even dimensions, since the subgroup Θ_n^{bp} is always zero, we need to understand the cokernel of the J-homomorphism to get the result in the abstract. In [13], Milnor states that up to dimension 64: the n-sphere has a unique smooth structure only for $n = 1, 2, 3, 5, 6, 12, 61$. This observation is based on the computation of 2-primary stable homotopy groups of spheres up to the 64 stem by Kochman and Mahowald [9] from 1995. Recently, Isaksen [4] discovered several errors in Kochman and Mahowald's computations, and he was able to give rigorous proofs of computations through the 59 stem. One major correction is that, instead of having order 4, π_{56} is of order 2 and is generated by a class in the image of J. Therefore one has:

Theorem (Isaksen) The sphere S^{56} has a unique smooth structure.

For $5 \leq n \leq 61$, S^n has a unique smooth structure only for $n = 5, 6, 12, 56$ and 61. Recent work of Behrens et al. [2] shows that the next sphere with a unique smooth structure, if exists, is in dimension at least 126.

The odd-dimensional result was covered by the arguments above. For the even case, Duane Randall obtained an existence result for even-dimensional exotic spheres of dimension higher than 61.

Main Theorem: The n-dimensional sphere, S^n, admits exotic smoothness structures of 2-primary order for all dimensions $n > 61$ such that n is not congruent to 4 modulo 8 and also n not equal to 125 or 126. Moreover, for any integer n congruent to 4 modulo 8 of the form $n = 4p(p - 1) - 4$ for some odd prime p, S^n admits exotic differentiable structures of p-primary order.

This result rules out the unknown dimension 126. Then the first relevant dimension which is not covered by the main theorem is 132, because it has dimension congruent to 4 modulo 8 but $32 \neq p(p - 1)$ for any odd prime.

Acknowledgments Finally, we wish to express our sincere gratitude to Professor Carl Brans for his collaboration and friendship at Loyola University. We also thank Torsten Asselmeyer-Maluga for his work on this Festschrift and his very interesting work with Carl on exotic smoothness in physics.

References

1. J.C. Becker, D.H. Gottlieb, The transfer map and fiber bundles. Topology **14**, 1–12 (1975)
2. M. Behrens, M. Hill, M.J. Hopkins, M. Mahowald, Exotic spheres detected by topological modular forms. Preprint
3. E.H. Connell, A topological H-cobordism theorem for $n \geq 5$. Ill. J. Math. **11**, 300–309 (1967)
4. C. Daniel, Isaksen. Stable Stems. arXiv:1407.8418
5. M.H. Freedman, The topology of four-dimensional manifolds. J. Differ. Geom. **17**, 357–453 (1982)
6. H. Freudenthal, Über die Klassen der Sphärenabbildungen. Comput. Math. **5**, 299–314 (1937)
7. M.A. Hill, M.J. Hopkins, C. Douglas, Ravenel. On the non-existence of elements of Kervaire invariant one. arXiv:0908.3724
8. M.A. Kervaire, J.W. Milnor, Groups of homotopy spheres: I. Ann. Math. (Princeton University Press) **77**(3), 504–537 (1963)
9. S.O. Kochman, M.E. Mahowald, On the computation of stable stems. Contemp. Math. **181**, 299–316 (1993)
10. M. Mahowald. The order of the image of the J-homomorphism, in *Proceedings of Advanced Study Inst. on Algebraic Topology*, Aarhus, vol. II, (Inst., Aarhus Univ, Mat, 1970), p. 376–384
11. J.P. May, The cohomology of restricted Lie algebras and of Hopf algebras; application to the Steenrod algebra. Thesis. (The Department of Mathematics, Princeton University, 1964)
12. J.W. Milnor, On manifolds homeomorphic to the 7-sphere. Ann. Math. **64**(2), 399–405 (1956)
13. J.W. Milnor, Differential topology forty-six years later. Not. Am. Math. Soc. **58**(6), 804–809 (2011)
14. E.E. Moise, Affine structures in 3-manifolds. V. The triangulation theorem and Hauptvermutung. Ann. Math. **56**, 96–114 (1952)
15. J.W. Morgan, G. Tian, *Ricci Flow and the Poincaré Conjecture*. Clay Mathematics Institute Monograph Vol. 3 (AMS and Clay Math. Institute 2007). http://www.claymath.org/library/monographs/cmim03c.pdf
16. J.W. Morgan, F.T.-H. Fong, *Ricci Flow and Geometrization of 3-Manifolds*. University Lecture Vol. 53 (AMS, 2010)
17. M.H.A. Newman, The engulfing theorem for topological manifolds. Ann. Math. **84**, 555–571 (1966)

18. G. Perelman, The entropy formula for the Ricci flow and its geometric applications. arXiv:math.DG/0211159
19. D. Quillen, The Adams conjecture. Topology **10**, 67–80 (1971)
20. S. Smale, Generalized Poincarés conjecture in dimensions greater than four. Ann. Math. **74**, 391–406 (1961)
21. D.P. Sullivan, Genetics of homotopy theory and the Adams conjecture. Ann. Math. **100**, 1–79 (1974)
22. G. Wang, Z. Xu, On the uniqueness of the smooth structure of the 61-sphere. arXiv:1601.02184

Chapter 15
Smooth Quantum Gravity: Exotic Smoothness and Quantum Gravity

Torsten Asselmeyer-Maluga

Abstract Over the last two decades, many unexpected relations between exotic smoothness, e.g. exotic $\mathbb{R}^4$, and quantum field theory were found. Some of these relations are rooted in a relation to superstring theory and quantum gravity. Therefore one would expect that exotic smoothness is directly related to the quantization of general relativity. In this article we will support this conjecture and develop a new approach to quantum gravity called *smooth quantum gravity* by using smooth 4-manifolds with an exotic smoothness structure. In particular we discuss the appearance of a wildly embedded 3-manifold which we identify with a quantum state. Furthermore, we analyze this quantum state by using foliation theory and relate it to an element in an operator algebra. Then we describe a set of geometric, non-commutative operators, the skein algebra, which can be used to determine the geometry of a 3-manifold. This operator algebra can be understood as a deformation quantization of the classical Poisson algebra of observables given by holonomies. The structure of this operator algebra induces an action by using the quantized calculus of Connes. The scaling behavior of this action is analyzed to obtain the classical theory of General Relativity (GRT) for large scales. This approach has some obvious properties: there are non-linear gravitons, a connection to lattice gauge field theory and a dimensional reduction from 4D to 2D. Some cosmological consequences like the appearance of an inflationary phase are also discussed. At the end we will get the simple picture that the change from the standard $\mathbb{R}^4$ to the exotic R^4 is a quantization of geometry.

15.1 Introduction

On the 25-th of November in 1915, Einstein presented his field equations, the basic equations of General Relativity, to the Prussian Academy of Sciences in Berlin. This equation had a tremendous impact on physics, in particular on cosmology. The essence of the theory was expressed by Wheeler by the words: *Spacetime tells matter how to move; matter tells spacetime how to curve.* Einsteins theory remained

T. Asselmeyer-Maluga (✉)
German Aerospace Center, Rosa-Luxemburg-Str. 2, D-10178 Berlin, Germany
e-mail: torsten.asselmeyer-maluga@dlr.de

© Springer International Publishing Switzerland 2016
T. Asselmeyer-Maluga (ed.), *At the Frontier of Spacetime*,
Fundamental Theories of Physics 183, DOI 10.1007/978-3-319-31299-6_15

unchanged for about 40 years. Then one started to investigate theories fulfilling Mach's principle leading to a variable gravitational constant. Brans-Dicke theory was the first realization of an extended Einstein theory with variable gravitational constant (Jordans proposal was not widely known). All experiments are, however, in good agreement with Einstein's theory and currently there is no demand to change it.

General relativity (GR) has changed our understanding of space-time. In parallel, the appearance of quantum field theory (QFT) has modified our view of particles, fields and the measurement process. The usual approach for the unification of QFT and GR to a quantum gravity, starts with a proposal to quantize GR and its underlying structure, space-time. There is a unique opinion in the community about the relation between geometry and quantum theory: The geometry as used in GR is classical and should emerge from a quantum gravity in the limit (Planck's constant tends to zero). Most theories went a step further and try to get a space-time from quantum theory. Then, the model of a smooth manifold is not suitable to describe quantum gravity, but there is no sign for a discrete space-time structure or higher dimensions in current experiments [50]. Therefore, we conjecture that the model of spacetime as a smooth 4-manifold can be used also in a quantum gravity regime, but then one has the problem to represent QFT by geometric methods (submanifolds for particles or fields etc.) as well to quantize GR. In particular, one must give meaning to the quantum state by geometric methods. Then one is able to construct the quantum theory without quantization. Here we implicitly assumed that the quantum state is real, i.e. the quantum state or the wave function has a real counterpart and is not a collection of future possibilities representing some observables. Experiments [28, 75, 83] supported this view. Then the wave function is not merely representing our limited knowledge of a system but it is in direct correspondence to reality! Then one has to go the reverse way: one has to show that the quantum state is produced by the quantization of a classical state. It is, however, not enough to have a geometric approach to quantum gravity (or the quantum field theory in general). What are the quantum fluctuations? What is the measurement process? What is decoherence and entanglement? In principle, all these questions have to be addressed too.

Here, the exotic smoothness structure of 4-manifolds can help finding a way. A lot of work was done in the last decades to fulfill this goal. It starts with the work of Brans and Randall [32] and of Brans alone [29–31] where the special situation in exotic 4-manifolds (in particular the exotic $\mathbb{R}^4$) was explained. One main result of this time was the *Brans conjecture*: exotic smoothness can serve as an additional source of gravity. I will not present the whole history where I refer to Carl's article. Here I will list only some key results which will be used in the following

- Exotic smoothness is an extra source of gravity (Brans conjecture is true), see Asselmeyer [5] for compact manifolds and Słdkowski [86, 87] for the exotic $\mathbb{R}^4$. Therefore an exotic $\mathbb{R}^4$ is always curved and cannot be flat!
- The exotic $\mathbb{R}^4$ cannot be a globally hyperbolic space (see [40] for instance), i.e. represented by $M \times \mathbb{R}$ for some 3-manifold. Instead it admits complicated folia- tions [17]. Using non-commutative geometry, we are able to study these foliations (the leaf space) and get relations to QFT. For instance, the von Neumann algebra

of a codimension-one foliation of an exotic $\mathbb{R}^4$ must contain a factor of type III_1 used in local algebraic QFT to describe the vacuum [11, 13, 19].

- The end of $\mathbb{R}^4$ (the part extending to infinity) is $S^3 \times \mathbb{R}$. If $\mathbb{R}^4$ is exotic then $S^3 \times \mathbb{R}$ admits also an exotic smoothness structure. Clearly, there is always a topologically embedded 3-sphere but there is no smoothly embedded one. Let us assume the well-known hyperbolic metric of the spacetime $S^3 \times \mathbb{R}$ using the trivial foliation into leafs $S^3 \times \{t\}$ for all $t \in \mathbb{R}$. Now we demand that $S^3 \times \mathbb{R}$ carries an exotic smoothness structure at the same time. Then we will get only topologically embedded 3-spheres, the leafs $S^3 \times \{t\}$. These topologically embedded 3-spheres are also known as wild 3-spheres. In [14], we presented a relation to quantum D-branes. Finally we proved in [16] that the deformation quantization of a tame embedding (the usual embedding) is a wild embedding.[1] Furthermore we obtained a geometric interpretation of quantum states: wild embedded submanifolds are quantum states. Importantly, this construction depends essentially on the continuum, because wild embedded submanifolds admit always infinite triangulations.
- For a special class of compact 4-manifolds we showed in [20] that exotic smoothness can generate fermions and gauge fields using the so-called knot surgery of Fintushel and Stern [51]. In the paper [10] we presented an approach using the exotic $\mathbb{R}^4$ where the matter can be generated (like in QFT).
- The path integral in quantum gravity is dominated by the exotic smoothness contribution (see [6, 49, 80] or by using string theory [12]).

The paper is organized as follows. In the following three sections we will explain exotic 4-manifolds and motivate the whole approach by using the path integral for the Einstein-Hilbert action. Here we will also present how to couple the matter and gauge fields to this theory. For a 4-manifold, there are two main invariants the Euler and Pontrjagin class which determine the main topological invariant of a 4-manifold, the intersection form. In Sect. 15.5, we will obtain the Einstein-Hilbert and Holst action by using these two classes. At the first view, this section is a little bit isolated from the previous and subsequent sections but we will use this result later during the study of the scaling. In the main Sect. 15.6, we will construct the foliation of an exotic $\mathbb{R}^4$ of codimension (equivalent to a Lorentz structure). Following Connes, [41] the leaf space is an operator algebra constructed from the geometrical information of the foliation (holonomy groupoid). This operator algebra is a factor III von Neumann algebra and we will use the Tomita-Takesaki modular theory to uncover the structure of the foliation. It is not the first time that this factor was used for quantum gravity and we refer to the paper [22] for a nice application. States in this operator algebra are represented by equivalence classes of knotted curves (element of the Kauffman bracket skein module). The reconstruction of the spatial space from the states gives a wild embedded 3-sphere as geometrical representation of the state. Surprisingly, it fits with the properties of the exotic $\mathbb{R}^4$. If one introduces a global foliation of the exotic $\mathbb{R}^4$ by a global time then one obtains a foliation into wild embedded 3-spheres. In contrast, if one uses a local but complicate foliation then

---

[1] A wild embedding is a topological embedding $I : N \to M$ so that the image $I(N) \subset M$ is an infinite polyhedron or the triangulation needs always infinitely many simplices.

this wild object can be omitted and one obtains a state given by a finite collection of knotted curves. Interestingly, the operator algebra can be understood as observable algebra given by a deformation quantization (Turaev-Drinfeld quantization [97, 98]) of the classical observable algebra (Poisson algebra of holonomies a la Goldman [61]). In Sect. 15.7, we will use the splitting of the operator algebra (15.10) given by Tomita-Takesaki modular theory to introduce the dynamics (see Connes and Rovelli [43] with similar ideas). Finally we will obtain a quantum action (15.15) in the quantized calculus of Connes [42]. Then the scaling behavior is studied in the next section. For large scales, the action can be interpreted as a non-linear sigma model. The renormalization group (RG) flow analysis [56] gives the Einstein equations for large scales. The short-scale analysis is much more involved, yielding for small fluctuations the Einstein-Hilbert action and a non-minimally coupled scalar field. In particular, we will obtain a $(2 + \epsilon)$-dimensional fractal structure. In Sect. 15.9 we will present some direct consequences of this approach: the nonlinear graviton [79], a relation to lattice gauge field theory with a discussion of discreteness and the appearance of dimensional reduction from 4D to 2D. In Sect. 15.10 we will discuss the answer to a fundamental question: where does the quantum fluctuations come from? The main result of this section can be written as: *The set of canonical pairs (as measurable variables in the theory) forms a fractal subset of the space of all holonomies. Then we can only determine the initial condition up to discrete value (given by the canonical pair) and the chaotic behavior of the foliation (i.e. the Anosov flow) makes the limit not predictable.* This interesting result is followed by a section where we will discuss the collapse of the wavefunction by the gravitational interaction by calculating the minimal decoherence time. Furthermore we will discuss entanglement and the measurement process. In Sect. 15.12 we will list our work in cosmology which uses partly the results of this paper. In the last Sect. 15.13, we will discuss some consequences and open questions. Some mathematical prerequisites are presented in three appendices.

This article is dedicated to my only teacher, Carl H. Brans for 20 years of collaboration and friendship. He is the founder of this research area. We had and will have many interesting discussions. Carl always asked the right question and put the finger on many open points. During the 7 years of writing our book, we had a very fruitful collaboration and I learned so much to complete even this work. Carl, I hope for many discussions with you in the future. I'm very glad to count on your advice. Happy Birthday!

15.2 What Is Exotic Smoothness?

Why am I going to concentrate on a concept like exotic smoothness? Einstein used the equivalence principle as a key principle in the development of general relativity. Every gravitational field can be locally eliminated by acceleration. Then, the spacetime is locally modeled as subsets of the flat $\mathbb{R}^4$ or the equivalence principle enforces us to use the concept of a manifold for spacetime. Together with the smoothness of the dynamics (usage of differential equations), we obtain a smooth 4-manifold as model for the spacetime in agreement with the current experimental situation. A manifold consists of charts and transition functions forming an atlas which covers the manifold completely. *The smooth atlas is called the smoothness structure of the manifold.* It was an open problem for a long time whether every topological manifold admits a unique smooth atlas. In 1957, Milnor found the first counterexample: the construction of a 7-sphere with at least 8 different smoothness structures. Later it was shown that all manifolds of dimension larger than 4 admit only a finite number of distinct smoothness structures. The real breakthrough for 4-manifolds came in the 80s where one constructed infinitely many different smoothness structures for many compact 4-manifolds (countably infinite) and for many non-compact 4-manifolds (uncountably infinite) including the $\mathbb{R}^4$. In all dimensions smaller than four, there is only one smoothness structure (up to diffeomorphisms), the standard structure. The standard $\mathbb{R}^4$ is simply characterized by the unique property to split smoothly like $\mathbb{R}^4 = \mathbb{R} \times \mathbb{R} \times \mathbb{R} \times \mathbb{R}$. All other distinct smoothness structures are called *exotic smoothness* structures. These structures are different, nonequivalent, smooth descriptions of the same topological manifold, a different atlas of charts. In case of the exotic $\mathbb{R}^4$, the difference is tremendous: the standard $\mathbb{R}^4$ needs one chart (and every other description can be reduced to it) whereas every known exotic $\mathbb{R}^4$ admits infinitely many charts (which cannot be reduced to a simpler description). So, the spacetime exhibits a much larger complexity by using an exotic smoothness structure, but why is dimension 4 so special? There is a good description in [55] and I will give a short account now. At first we have to discuss the question: how do I build an atlas for a smooth manifold? The answer is given by considering the construction of diffeomorphisms. Every diffeomorphism is locally given by the solution of $\dot{\mathbf{x}} = -\nabla f(\mathbf{x})$ for a real function f over the manifold. The fixed points of this equation are the critical points of f. In case of isolated critical points, one can reproduce the structure of the manifold (this is called Morse theory). Every critical point leads to the attachment of a handle, a submanifold like $D^{n-k} \times D^k$, i.e. the k-handle (where D^k is the k-disk). In many cases, the corresponding structure of the manifold, the handle body, can be very complicated but there are rules (handle sliding) to simplify them. In all dimensions except dimension 4. Therefore, two handle bodies can be described by the same 4-manifold topologically but differ in the smooth description.

15.3 The Main Example: Exotic $\mathbb{R}^4$

One of the most surprising aspects of exotic smoothness is the existence of exotic $\mathbb{R}^4$'s. In all other dimensions [88], the Euclidean space $\mathbb{R}^n$ with $n \neq 4$ admits a unique smoothness structure, up to diffeomorphisms. Beginning with the first examples [66], Taubes [93] and Freedman/DeMichelis [46] constructed countably many large and small exotic $\mathbb{R}^4$'s, respectively. A small exotic $\mathbb{R}^4$ embeds smoothly in the 4-sphere whereas a large exotic $\mathbb{R}^4$ cannot be embedded in that way. For the following we need some simple definitions: the connected sum # and the boundary connected sum ♮ of manifolds. Let M, N be two n-manifolds with boundaries ∂M, ∂N. The *connected sum* $M \# N$ is the procedure of cutting out a disk D^n from the interior $int(M) \backslash D^n$ and $int(N) \backslash D^n$ with the boundaries $S^{n-1} \sqcup \partial M$ and $S^{n-1} \sqcup \partial N$, respectively, and gluing them together along the common boundary component S^{n-1}. The boundary $\partial(M \# N) = \partial M \sqcup \partial N$ is the disjoint sum of the boundaries ∂M, ∂N. The *boundary connected sum* $M \natural N$ is the procedure of cutting out a disk D^{n-1} from the boundary $\partial M \backslash D^{n-1}$ and $\partial N \backslash D^{n-1}$ and gluing them together along S^{n-2} of the boundary. Then the boundary of this sum $M \natural N$ is the connected sum $\partial(M \natural N) = \partial M \# \partial N$ of the boundaries ∂M, ∂N.

15.3.1 Large Exotic $\mathbb{R}^4$

Large exotic $\mathbb{R}^4$ can be constructed using the failure to arbitrarily split a compact, simply-connected 4-manifold. For every topological 4-manifold one knows how to split this manifold *topologically* into simpler pieces using the work of Freedman [53]. Donaldson [47], however, that some of these 4-manifolds do not exist as smooth 4-manifolds. This contradiction between the continuous and the smooth case produces the first examples of exotic $\mathbb{R}^4$. Below we discuss one of these examples.

One starts with a compact, simply-connected 4-manifold X classified by the intersection form [53]

$$Q_X : H_2(X, \mathbb{Z}) \times H_2(X, \mathbb{Z}) \to \mathbb{Z}$$

a quadratic form over the second integer homology group. In the first construction of a large exotic $\mathbb{R}^4$, one starts with the K3 surface as 4-manifold having the intersection form

$$Q_{K3} = E_8 \oplus E_8 \oplus (\oplus_3 \begin{pmatrix} 0 & 1 \\ 1 & 0 \end{pmatrix}) := 2E_8 \oplus 3H \tag{15.1}$$

with the the matrix E_8:

$$E_8 = \begin{bmatrix} 2 & 1 & 0 & 0 & 0 & 0 & 0 & 0 \\ 1 & 2 & 1 & 0 & 0 & 0 & 0 & 0 \\ 0 & 1 & 2 & 1 & 0 & 0 & 0 & 0 \\ 0 & 0 & 1 & 2 & 1 & 0 & 0 & 0 \\ 0 & 0 & 0 & 1 & 2 & 1 & 0 & 1 \\ 0 & 0 & 0 & 0 & 1 & 2 & 1 & 0 \\ 0 & 0 & 0 & 0 & 0 & 1 & 2 & 0 \\ 0 & 0 & 0 & 0 & 1 & 0 & 0 & 2 \end{bmatrix}.$$

The work of Donaldson [47] shows that a closed, **smooth**, simply-connected, compact 4-manifold $X_{E_8 \oplus E_8}$ with intersection form $E_8 \oplus E_8$ does not exist. Freedman [53] showed, however, that there is a topological splitting

$$K3 = X_{E_8 \oplus E_8} \# \left(\#_3(S^2 \times S^2) \right) \tag{15.2}$$

with the m-times connected sum $\#_m$ (see above) which fails to be smooth. This splitting means that we glue together the two manifolds $\#_3(S^2 \times S^2) \backslash D^4$ and $X_{E_8 \oplus E_8} \backslash D^4$ along the common boundary $S^3 = \partial D^4$ (D^4 is the 4-disk or 4-ball). Now we define the interior $X = \#_3(S^2 \times S^2) \backslash Int \, D^4$. The splitting (15.2) gives a way to represent the $3H$ part of the intersection form (15.1) by using X but that fails smoothly. So, choosing a topological splitting

$$\begin{aligned} K3 &= X_{E_8 \oplus E_8} \# \left(\#_3(S^2 \times S^2) \right) \\ &= \left(X_{E_8 \oplus E_8} \backslash D^4 \right) \cup \left(S^3 \times [0, 1] \right) \cup \left(\#_3(S^2 \times S^2) \backslash D^4 \right) \end{aligned}$$

gives a $S^3 \times [0, 1]$ inside the $K3$. The interior of $S^3 \times [0, 1]$ defines a manifold $S^3 \times [0, 1)$ glued to a (topological) 4-disk $D^4 \subset \#_3(S^2 \times S^2) \backslash D^4$ along the common boundary, i.e. $W = D^4 \cup S^3 \times [0, 1)$ topologically. W is homeomorphic to $\mathbb{R}^4$ but the non-existence of the smooth splitting implies that it is an exotic $\mathbb{R}^4$ and there is no smooth embedded S^3 (otherwise the topological splitting is smooth). This failure for a smooth embedding implies also that such exotic $\mathbb{R}^4$'s do not embed in the 4-sphere, i.e. it is a large exotic $\mathbb{R}^4$. The details of the construction can be found in our book [8] (Sect. 8.4).

Gompf [64] introduced an important tool for finding new exotic $\mathbb{R}^4$ from others, the end-sum $\natural_e$. Let R, R' be two topological $\mathbb{R}^4$'s. The end-sum $R \natural_e R'$ is defined as follows: Let $\gamma : [0, \infty) \to R$ and $\gamma' : [0, \infty) \to R'$ be smooth properly embedded rays with tubular neighborhoods $\nu \subset R$ and $\nu' \subset R'$, respectively. For convenience, identify the two semi-infinite intervals with $[0, 1/2)$, and $(1/2, 1]$ leading to diffeomorphisms, $\phi : \nu \to [0, 1/2) \times \mathbb{R}^3$ and $\phi' : \nu' \to (1/2, 1] \times \mathbb{R}^3$. Then define

$$R \natural_e R' = R \cup_\phi I \times \mathbb{R}^3 \cup_{\phi'} R'$$

as the end sum of R and R'. With a little checking, it is easy to see that this construction leads to $R \natural_e R'$ as another topological $\mathbb{R}^4$. However, if R, R' are themselves exotic, then so will $R \natural_e R'$ and in fact, it will be a "new" exotic manifold, since it will not be diffeomorphic to either R or R'. Gompf used this technique to construct a class of exotic $\mathbb{R}^4$'s none of which can be embedded smoothly in the standard $\mathbb{R}^4$.

By an extension of Donaldson theory for a special class of open 4-manifolds, so-called end-periodic 4-manifolds, Taubes [93] gives a continuous family of exotic $\mathbb{R}^4$ which was extended by Gompf to a continuous 2-parameter family $R_{s,t}$.

15.3.2 Small Exotic $\mathbb{R}^4$

Small exotic $\mathbb{R}^4$'s are again the result of anomalous smoothness in 4-dimensional topology but of a different kind than for large exotic $\mathbb{R}^4$'s. In 4-manifold topology

[53], a homotopy-equivalence between two compact, closed, simply-connected 4-manifolds implies a homeomorphism between them (a so-called h cobordism), but Donaldson [48] provided the first smooth counterexample, i.e. both manifolds are generally not diffeomorphic to each other. The failure can be localized in some contractible submanifold (Akbulut cork) so that an open neighborhood of this sub-manifold is a small exotic $\mathbb{R}^4$. The whole procedure implies that this exotic $\mathbb{R}^4$ can be embedded in the 4-sphere S^4. Below we discuss the details for one of these examples.

In 1975 Casson (Lecture 3 in [39]) described a smooth 5-dimensional h-cobordism between compact 4-manifolds and showed that they "differ" by two proper homotopy $\mathbb{R}^4$'s (see below). Freedman knew, as an application of his proper h-cobordism theorem, that the proper homotopy $\mathbb{R}^4$'s were $\mathbb{R}^4$. After hearing about Donaldson's work in March 1983, Freedman realized that there should be exotic $\mathbb{R}^4$'s and, to find one, he produced the second part of the construction below involving the smooth embedding of the proper homotopy $\mathbb{R}^4$'s in S^4. Unfortunately, it was necessary to have a compact counterexample to the smooth h-cobordism conjecture, and Donaldson did not provide this until 1985 [48]. The idea of the construction is simply given by the fact that every such smooth h-cobordism between non-diffeomorphic 4-manifolds can be written as a product cobordism except for a compact contractible sub-h-cobordism V, the Akbulut cork. An open subset $U \subset V$ homeomorphic to $[0, 1] \times \mathbb{R}^4$ is the corresponding sub-h-cobordism between two exotic $\mathbb{R}^4$'s. These exotic $\mathbb{R}^4$'s are called ribbon $\mathbb{R}^4$'s. They have the important property of being diffeomorphic to open subsets of the standard $\mathbb{R}^4$. That stands in contrast to the previous defined examples of Kirby, Gompf and Taubes.

To be more precise, consider a pair (X_+, X_-) of homeomorphic, smooth, closed, simply-connected 4-manifolds. The transformation from X_- to X_+ visualized by a h-cobordism can be described by the following construction.

Let W be a smooth h-cobordism between closed, simply connected 4-manifolds X_- and X_+. Then there is an open subset $U \subset W$ homeomorphic to $[0, 1] \times \mathbb{R}^4$ with a compact subset $K \subset U$ such that the pair $(W \setminus K, U \setminus K)$ is diffeomorphic to a product $[0, 1] \times (X_- \setminus K, U \cap X_- \setminus K)$. The subsets $R_\pm = U \cap X_\pm$ (homeomorphic to $\mathbb{R}^4$) are diffeomorphic to open subsets of $\mathbb{R}^4$. If X_- and X_+ are not diffeomorphic, then there is no smooth 4-ball in $R_\pm$ containing the compact set $Y_\pm = K \cap R_\pm$, so both $R_\pm$ are exotic $\mathbb{R}^4$'s.

Thus, remove a certain contractible, smooth, compact 4-manifold $Y_- \subset X_-$ (called an Akbulut cork) from X_-, and re-glue it by an involution of ∂Y_-, i.e. a diffeomorphism $\tau : \partial Y_- \to \partial Y_-$ with $\tau \circ \tau = Id$ and $\tau(p) \neq \pm p$ for all $p \in \partial Y_-$. This argument was modified above so that it works for a contractible *open* subset $R_- \subset X_-$ with similar properties, such that R_- will be an exotic $\mathbb{R}^4$ if X_+ is not diffeomorphic to X_-. Furthermore R_- lies in a compact set, i.e. a 4-sphere or R_- is a small exotic $\mathbb{R}^4$. In the next subsection we will see how this results in the construction of handle bodies of exotic $\mathbb{R}^4$. In [46] Freedman and DeMichelis constructed also a continuous family of small exotic $\mathbb{R}^4$.

15.3.3 Main Property of (Small) Exotic $\mathbb{R}^4$

One of the characterizing properties of an exotic $\mathbb{R}^4$ (all known examples) is the existence of a compact subset $K \subset R^4$ which cannot be surrounded by any smoothly embedded 3-sphere (and homology 3-sphere bounding a contractible, smooth 4-manifold). Let $\mathbf{R}^4$ be the standard $\mathbb{R}^4$ (i.e. $\mathbf{R}^4 = \mathbb{R}^3 \times \mathbb{R}$ smoothly) and let R^4 be a small exotic $\mathbb{R}^4$ with compact subset $K \subset R^4$ which cannot be surrounded by a smoothly embedded 3-sphere. Then every completion $\overline{N(K)}$ of an open neighborhood $N(K) \subset R^4$ is not bounded by a 3-sphere $S^3 \neq \partial \overline{N(K)}$. However, R^4 is a small exotic $\mathbb{R}^4$ and there is a smooth embedding $E : R^4 \to \mathbf{R}^4$ in the standard $\mathbb{R}^4$. Then the completion of the image $\overline{E(R^4)}$ has the boundary $S^3 = \partial \overline{E(R^4)}$ as subset of $\mathbf{R}^4$. So, we have the strange situation that an open subset of the standard $\mathbf{R}^4$ represents a small exotic R^4. In case of the large exotic $\mathbb{R}^4$, the situation is much more complicated. A large exotic $\mathbb{R}^4$ does not embed in any smooth 4-manifold which is simpler than the manifold used for the construction of this exotic $\mathbb{R}^4$. Above we considered the example of a large exotic $\mathbb{R}^4$ constructed from a K3 surface. Therefore this large exotic $\mathbb{R}^4$ embeds in the K3 surface but not in simpler 4-manifolds like $\mathbb{C}P^2$.

15.3.4 Handle Decomposition of the Small Exotic $\mathbb{R}^4$ and Casson Handles

As of now, we only know of exotic $\mathbb{R}^4$'s represented by an infinite number of coordinate patches. This naturally makes it difficult to provide an explicit description of a metric. However, in [9], a suggestion to overcome this limitation is provided by the consideration of periodic explicitly described coordinate patches making use of more complex pieces, so-called handles, and even more complex gluing maps. Then one also gets infinite structures of handles but with a clear picture: the coordinate patches have a periodic structure.

Handles Every 4-manifold can be decomposed using standard pieces such as $D^k \times D^{4-k}$, the so-called k-handle attached along $\partial D^k \times D^{4-k}$ to the 0-handle $D^0 \times D^4 = D^4$. In the following we need two possible cases: the 1-handle $D^1 \times D^3$ and the 2-handle $D^2 \times D^2$. These handles are attached along their boundary components $S^0 \times D^3$ or $S^1 \times D^2$ to the boundary S^3 of the 0-handle D^4 (see [68] for the details). The attachment of a 2-handle is defined by a map $S^1 \times D^2 \to S^3$, the embedding of a circle S^1 into the 3-sphere S^3, i.e. a knot. This knot into S^3 can be thickened (to get a knotted solid torus). The important fact for our purposes is the freedom to twist this knotted solid torus (so-called Dehn twist). The (integer) number of these twists (with respect to the orientation) is called the framing number or the framing. Thus the gluing of the 2-handle on D^4 can be represented by a knot or link together with an integer framing. The simplest example is the unknot with framing ± 1 representing the complex projective space $\mathbb{C}P^2$ or with reversed orientation $\overline{\mathbb{C}P}^2$, respectively. The 1-handle will be glued by the map of $S^0 \times D^3 \to S^3$ represented by two disjoint solid 2-spheres D^3. Akbulut [2] introduced another description. He observed that a 1-handle is something like a cut-out 2-handle with a fixed framing. We remark that all details can be found in [68]. Now we are ready to discuss the handle body decomposition of an exotic $\mathbb{R}^4$ by Bizaca and Gompf [24].

Handle decomposition of small exotic $\mathbb{R}^4$ First it is very important to notice that the exotic $\mathbb{R}^4$ is the **interior** of the handle body described below (since the handle body has a non-null boundary and is compact). The construction of the handle body can be divided into two parts. The first part is a submanifold consisting of a pair of a 1- and a 2-handle. This pair can be canceled topologically by using a Casson handle and we obtain the topological 4-disk D^4 with $\mathbb{R}^4$ as interior. This submanifold is a smooth 4-manifold with a boundary that can be covered by a finite number of charts. The smoothness structure of the exotic $\mathbb{R}^4$, however, depends mainly on the infinite Casson handle.

Casson handle Now consider the Casson handle and its construction in more detail. Briefly, a Casson handle CH is the result of attempts to embed a disk D^2 into a 4-manifold. In most cases this attempt fails and Casson [39] looked for a substitute, which is now called a Casson handle. Freedman [53] showed that every Casson handle CH is homeomorphic to the open 2-handle $D^2 \times \mathbb{R}^2$ but in nearly all cases it is not diffeomorphic to the standard handle [63, 65]. The Casson handle is built by iteration, starting from an immersed disk in some 4-manifold M, i.e. an injective smooth map $D^2 \to M$. Every immersion $D^2 \to M$ is an embedding except on a countable set of points, the double points. One can kill one double point by immersing another disk into that point. These disks form the first stage of the Casson handle. By iteration one can produce the other stages. Finally consider not the immersed disk but rather a tubular neighborhood $D^2 \times D^2$ of the immersed disk including each stage. The union of all neighborhoods of all stages is the Casson handle CH. So, there are two input data involved with the construction of a CH: the number of double points in each stage and their orientation $\pm$. Thus we can visualize the Casson handle CH by a tree: the root is the immersion $D^2 \to M$ with k double points, the first stage forms the next level of the tree with k vertices connected with the root by edges etc. The edges are evaluated using the orientation $\pm$. Every Casson handle can be represented by such an infinite tree. The Casson handle $CH(R_+)$ having an immersed disk with one (positively oriented) self-intersection (or double point) is the simplest Casson handle represented by the simplest tree T_+ having one vertex in each level connected by one edge with evaluation $+$.

15.3.5 Small Exotic $\mathbb{R}^4$ as a Sequence of 3-Manifolds

One of the characterizing properties of an exotic $\mathbb{R}^4$ (all known examples) is the existence of a compact subset $K \subset R^4$ which cannot be surrounded by any smoothly embedded 3-sphere (and homology 3-sphere bounding a contractible, smooth 4-manifold). Let $\mathbf{R}^4$ be the standard $\mathbb{R}^4$ (i.e. $\mathbf{R}^4 = \mathbb{R}^3 \times \mathbb{R}$ smoothly) and let R^4 be a small exotic $\mathbb{R}^4$ with compact subset $K \subset R^4$ which cannot be surrounded by a

Fig. 15.1 Link picture for
the compact subset K

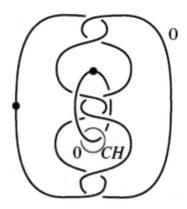

smoothly embedded 3-sphere. Then every completion $\overline{N(K)}$ of an open neighborhood $N(K) \subset R^4$ is not bounded by a 3-sphere $S^3 \neq \partial N(K)$, but R^4 is a small exotic $\mathbb{R}^4$ and there is a smooth embedding $E : R^4 \to \mathbf{R}^4$ in the standard $\mathbb{R}^4$. Then the completion of the image $\overline{E(R^4)}$ has the boundary $S^3 = \partial \overline{E(R^4)}$ as subset of $\mathbf{R}^4$. So, we have the strange situation that an open subset of the standard $\mathbf{R}^4$ represents a small exotic R^4.

Now we will describe R^4. Historically it was constructed by using a counterexample of the smooth h-cobordism theorem [24, 48]. Then the compact subset K is given by a non-canceling 1-/2-handle pair. The attachment of a Casson handle CH cancels this pair topologically. Then one obtains the 4-disk D^4 with interior $\mathbf{R}^4$, but this cancellation of the 1/2-handle pair cannot be done smoothly and one obtains a small exotic R^4 which is schematically given by $R^4 = K \cup CH$. Remember R^4 is a small exotic $\mathbb{R}^4$, i.e. R^4 is embedded into the standard $\mathbf{R}^4$ by definition. The completion $\bar{R}^4$ of $R^4 \subset \mathbf{R}^4$ has a boundary given by the 3-manifold Y_r. There is also the possibility to construct Y_r directly as the limit $n \to \infty$ of a sequence $\{Y_n\}$ of 3-manifolds. To construct this sequence of 3-manifolds [59], one can use the Kirby calculus, i.e. one represents the compact subset K by 1- and 2-handles pictured by a link say L_K where the 1-handles are represented by a dot (so that surgery along this link gives K) [68]. Then one attaches a Casson handle to this link [24]. As an example see Fig. 15.1.

The Casson handle is given by a sequence of Whitehead links (where the unknotted component has a dot) which are linked according to the tree (see the right figure of Fig. 15.2 for the building block and the left figure for the simplest Casson handle given by the unbranched tree).

For the construction of a 3-manifold which surrounds the compact K, one considers n-stages of the Casson handle and transforms the diagram to a real link (the dotted components are changed to usual components with framing 0). By handle manipulations one obtains a knot so that the n-th (untwisted) Whitehead double of this knot represents the desired 3-manifold (by using surgery). Then our example in Fig. 15.1 will result in the n-th untwisted Whitehead double of the pretzel knot $(-3, 3, -3)$, Fig. 15.3 (see [59] for the handle manipulations).

Fig. 15.2 Building block of every Casson handle (*right*) and the simplest Casson handle (*left*)

Fig. 15.3 Pretzel knot
$(-3, 3, -3)$ or the knot 9_{46}
in Rolfson notation
producing the 3-manifold Y_1
by 0-framed Dehn surgery

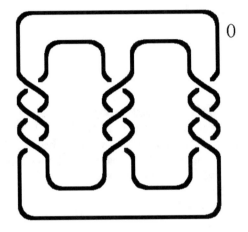

Then this sequence of 3-manifolds

$$Y_1 \rightarrow Y_2 \rightarrow \cdots \rightarrow Y_\infty = Y_r$$

characterizes the exotic smoothness structure of R^4. The limit of this sequence $n \rightarrow \infty$ gives a wild embedded 3-manifold Y_r whose physical relevance will be explained later.

15.4 Motivation: Path Integral Contribution by Exotic Smoothness

Here, we will motivate the appearance of exotic smoothness by discussing the path integral for the Einstein-Hilbert action. For simplicity, we consider general relativity without matter (using the notation of topological QFT). Space-time is a smooth oriented 4-manifold M which is non-compact and without boundary. From the formal point of view (no divergences of the metric) one is able to define a boundary ∂M at infinity. The classical theory is the study of the existence and uniqueness of (smooth) metric tensors g on M satisfying the Einstein equations subject to suitable boundary conditions. In the first order Hilbert–Palatini formulation, one specifies an

$SO(1, 3)$-connection A together with a cotetrad field e rather than a metric tensor. Fixing $A|_{\partial M}$ at the boundary, one can derive first order field equations in the interior (now called *bulk*) which are equivalent to the Einstein equations provided that the cotetrad is non-degenerate. The theory is invariant under space-time diffeomorphisms $M \to M$. In the particular case of the space-time $M = S^3 \times \mathbb{R}$ (topologically), we have to consider smooth 4-manifolds $M_{i,f}$ as parts of M whose boundary $\partial M_{i,f} = \Sigma_i \sqcup \Sigma_f$ is the disjoint union of two smooth 3-manifolds Σ_i and Σ_f to which we associate Hilbert spaces $\mathcal{H}_j$ of 3-geometries, $j = i, f$. These contain suitable wave functionals of connections $A|_{\Sigma_j}$. We denote the connection eigenstates by $|A|_{\Sigma_j}\rangle$. The path integral,

$$\langle A|_{\Sigma_f} |T_M| A|_{\Sigma_i} \rangle = \int_{A|\partial M_{i,f}} DA\, De\, \exp\left(\frac{i}{\hbar} S_{EH}[e, A, M_{i,f}]\right) \tag{15.3}$$

is the sum over all connections A matching $A|_{\partial M_{i,f}}$, and over all e. It yields the matrix elements of a linear map $T_M : \mathcal{H}_i \to \mathcal{H}_f$ between states of 3-geometry. Our basic gravitational variables will be cotetrad e^I_a and connection A^{IJ}_a on space-time M with the index a to present it as 1-forms and the indices I, J for an internal vector space V (used for the representation of the symmetry group). Cotetrads e are 'square-roots' of metrics and the transition from metrics to tetrads is motivated by the fact that tetrads are essential if one is to introduce spinorial matter. e^I_a is an isomorphism between the tangent space $T_p(M)$ at any point p and a fixed internal vector space V equipped with a metric η_{IJ} so that $g_{ab} = e^I_a e^J_b \eta_{IJ}$. Here we used the action

$$S_{EH}[e, A, M_{i,f}, \partial M_{i,f}] = \int_{M_{i,f}} \epsilon_{IJKL}(e^I \wedge e^J \wedge (dA + A \wedge A)^{KL}) + \int_{\partial M_{i,f}} \epsilon_{IJKL}(e^I \wedge e^J \wedge A^{KL}) \tag{15.4}$$

in the notation of [3, 4]. The boundary term $\epsilon_{IJKL}(e^I \wedge e^J \wedge A^{KL})$ is equal to twice the trace over the extrinsic curvature (or the mean curvature). For fixed boundary data, (15.3) is a diffeomorphism invariant in the bulk. If $\Sigma_i = \Sigma_f$ are diffeomorphic, we can identify $\Sigma = \Sigma_i = \Sigma_f$ and $\mathcal{H} = \mathcal{H}_i = \mathcal{H}_f$ i.e. we close the manifold $M_{i,f}$ by identifying the two boundaries to get the closed 4-manifold M'. Provided that the trace over $\mathcal{H}$ can be defined, the partition function,

$$Z(M') = tr_{\mathcal{H}} T_M = \int DA\, De\, \exp\left(\frac{i}{\hbar} S_{EH}[e, A, M, \partial M]\right) \tag{15.5}$$

where the integral is now unrestricted, is a dimensionless number which depends only on the diffeomorphism class of the smooth manifold M'. In case of the manifold $M_{i,f}$, the path integral (as transition amplitude) $\langle A|_{\Sigma_f} |T_M| A|_{\Sigma_i} \rangle$ is the diffeomorphism class of the smooth manifold relative to the boundary. The diffeomorphism class of the boundary, however, is unique and the value of the path integral depends on the topology of the boundary as well on the diffeomorphism class of the interior of $M_{i,f}$. Therefore we will shortly write

$$\langle \Sigma_f | T_M | \Sigma_i \rangle = \langle A |_{\Sigma_f} | T_M | A |_{\Sigma_i} \rangle$$

and consider the sum of manifolds like $M_{i,h} = M_{i,f} \cup_{\Sigma_f} M_{f,h}$ with the amplitudes

$$\langle \Sigma_h | T_M | \Sigma_i \rangle = \sum_{A | \Sigma_f} \langle \Sigma_h | T_M | \Sigma_f \rangle \langle \Sigma_f | T_M | \Sigma_i \rangle \tag{15.6}$$

where we sum (or integrate) over the connections and frames on Σ_h (see [69]). Then the boundary term

$$S_\partial[\Sigma_f] = \int_{\Sigma_f} \epsilon_{IJKL}(e^I \wedge e^J \wedge A^{KL}) = \int_{\Sigma_f} H\sqrt{h}d^3x \tag{15.7}$$

is needed where H is the mean curvature of Σ_f corresponding to the metric h at Σ_f (as restriction of the 4-metric). In the path integral (15.3), one integrates over the frames and connections. The possibility of singular frames was discussed at some places (see [103, 104]). The cotetrad field $e^I = e^I_a dx^a$ changes w.r.t. the smooth map $f : M \to M$ by $e^I_a(x)\,dx^a \mapsto e^I_a(x')\,dx'^a = e^I_a(f(x))(\partial_b f^a(x))dx^b$. The transformation matrix $(\partial_b f^a(x))$ has maximal rank 4 for every regular value of the smooth map, but at the critical points x_c of f, some derivatives vanish and one has a smaller rank at the point x_c, called a singular point. Then there is no inverse frame (or tetrad field) at this point. Usually singular frames are of this nature and one can decompose every singular frame into a product of a regular frame and a (singular) transformation induced by a smooth map. How can one interpret these singularities? At this point one needs some differential topology. A homeomorphism can be arbitrarily and accurately approximated by smooth mappings (see [70], Theorem 2.6), i.e. in a neighborhood of a homeomorphism one always finds a smooth map. Secondly, there is a special class of smooth maps, the stable maps. Here, two smooth maps are stable equivalent if both maps agree after a diffeomorphism of the corresponding manifolds [62]. Here we are interested into smooth mappings from 4-manifolds into 4-manifolds. By a deep result of Mather [76], stable mappings for this dimension are dense in all smooth mappings of 4-manifolds. In [8], we analyzed this situation: the approximation of a homeomorphism by a stable map. If this smooth map has no singularities then we can perturb them to a diffeomorphism. For a singular map, however, we showed that it induces a change of the smoothness structure. Then, a singular frame corresponds to a regular frame in a different smoothness structure. The path integral changed the domain of integration:

$$\int_{\text{regular+singular frames}} De \quad \to \quad \int_{\text{smoothness structures}} De$$

We remark that this change is unique for dimension four. No other dimension has this plethora of smoothness structures which can be used to express the singular frames.

The inclusion of exotic smoothness changed the description of trivial spaces like $\mathbb{R}^4$ completely. Instead of a single chart, we have now an infinite sequence of charts or an infinite sequence of 4-dimensional submanifolds. We will describe it more completely later. Each submanifold is bounded by a 3-manifold (different from a 3-sphere) and we obtain a sequence of 3-manifolds $Y_0 \to Y_1 \to Y_2 \to \cdots$ characterizing the smoothness structure. The sequence of 3-manifolds divides the path integral into a product

$$\langle Y_0 | T_M | Y_1 \rangle \langle Y_1 | T_M | Y_2 \rangle \langle Y_2 | T_M | Y_3 \rangle \cdots$$

and we have to think about the boundary term (15.7). In [10, 20] we analyzed this term: the boundary Y_n seen as embedding into the spacetime M can be described locally as spinor ψ and one obtains for the boundary term

$$\int_{Y_n} H \sqrt{h} d^3 x = \int_{Y_n} \bar{\psi} D \psi \sqrt{h} d^3 x \tag{15.8}$$

the Dirac action with the Dirac operator D and $|\psi|^2 = const.$ (see [57] for the construction of ψ). In particular we obtained the eigenvalue equation $D\psi = H\psi$, i.e. the mean curvature is the eigenvalue of the Dirac operator which has compact spectrum (from the compactness of Y_n) or we obtained discrete levels of geometry. This result enforced us to identify the 3-manifolds (or the parts) with the matter content. Furthermore the path integral of the boundary can be carried out by an integration along ψ (see [18]). With some effort [10, 20], one can extend this boundary term to a tubular neighborhood $Y_n \times [0, 1]$ of the boundary Y_n. However, the relation (15.8) is only true for simple (i.e. irreducible) 3-manifolds, i.e. for complements of a knot admitting hyperbolic structure. For more complex 3-manifolds, we have the following simple scheme: the knot complements are connected by torus bundles (locally written as $T^2 \times [0, 1]$). Therefore we also have to describe these bundles by using the boundary term. In [20] we described this situation by using the geometrical properties of these bundles and we will give a short account of these ideas in Sect. 15.9.1. Simply expressed, in this bundle one has a flow of constant curvature along the tube. The constant curvature connections are given by varying the Chern-Simons functional. Now following Floer [52], the 4-dimensional version of this flow equation is the instanton equation (or the self-dual equation) leading to the correct Yang-Mills functional (Chern-Simons gives the Pontrjagin class and the instanton equation makes it to the Yang-Mills functional). More importantly, the three possible types of torus bundles fit very good into the current scheme of three gauge field interactions (see [20] (Sect. 8)).

15.5   The Action Induced by Topology

Now we have the following picture: fermions as hyperbolic knot complements and
gauge bosons as torus bundles. Both components together are forming an irreducible
3-manifold which is connecting to the remaining space by a S^2-boundary (see the
prime decomposition in Appendix B). This connection via $S^2 \times [0, 1]$ (S^2-bundle) is
the only connection between matter and space. Here, there is only one interpretation:
this S^2-bundle must be interpreted as gravity. In this section we will support this
conjecture and construct the corresponding action. At first we will fix the model, i.e.
let Σ_M and Σ_S be the 3-manifolds for matter and space, respectively. The connected
sum # of both components represents the whole spatial component

$$\Sigma = \Sigma_M \# \Sigma_S = \Sigma_M \cup_{S^2} \left(S^2 \times [0, 1] \right) \cup_{S^2} \Sigma_S = \Sigma_M \# S^3 \# \Sigma_S$$

of the spacetime. The decomposition above showed the geometry of the S^2-bundle (in
the sense of Thurston, see Appendix B) to be the spherical geometry with isometry
group $SO(3)$. The idea of the following construction can be simply expressed: the
2-sphere S^2 explores locally the curvature of the space where the curvature is given
by the inverse volume $\frac{1}{vol(S^2)}$ of the 2-sphere S^2. The 2-sphere can be written as a
homogenous space $S^2 = SO(3)/SO(2)$ also known as Hopf bundle. As mentioned
above, the geometry of the bundle $S^2 \times [0, 1]$ (interpreted as an equator region of
S^3) is the spherical geometry with isometry group $SO(3)$. So, as a local model we
have an embedding of a 3-manifold (as the spatial component for a fixed time) into
the spacetime with local Lorentz symmetry (represented by $SO(3, 1)$). From the
mathematical point of view, it is a reductive Cartan geometry [101, 102] over the
homogenous space $SO(3, 1)/SO(3)$, the 3-dimensional hyperbolic space. For the
moment, let us extend this symmetry to the spacetime M itself. A Cartan connection
A decomposes as a $so(3)$-valued connection ω ($so(3)$ denotes the Lie algebra of
$SO(3)$) and a coframe field e (with values in $so(3, 1)/so(3)$) as

$$A = \omega + \frac{1}{\ell} e$$

by using the scale ℓ (in agreement with the physical units) and with curvature

$$F = dA + A \wedge A$$
$$= (d\omega + \omega \wedge \omega) + \frac{1}{\ell^2} e \wedge e = R + \frac{1}{\ell^2} e \wedge e$$

Then for the spacetime (as 4-manifold), we interpret the Cartan connection A as
the connection of the frame bundle (with respect to the Lorentz structure). Now we
have to think about what characterizes the S^2-bundle in a 4-manifold, i.e. a surface
bundle over a surface (at least locally). It is known that a surface bundle over a
surface is topologically described by the Euler class as well as the Pontrjagin class
(via the Hirzebruch signature theorem). Therefore we choose the sum of the Euler
and Pontrjagin class for the frame bundle as action

$$S = \int_M \left(\epsilon^{ABCD} F_{AB} \wedge F_{CD} + \gamma F \wedge F \right)$$

where the Pontrjagin class is weighted by a parameter γ. Using the rules above, we obtain

$$S = \frac{1}{\ell^2} \int_M \left(2\epsilon^{ABCD} e_A \wedge e_B \wedge R_{CD} + 2\gamma e \wedge e \wedge R + \frac{(1+\gamma)}{\ell^2} e \wedge e \wedge e \wedge e \right)$$
$$+ \int_M \left(\epsilon^{ABCD} R_{AB} \wedge R_{CD} + \gamma R \wedge R \right),$$

the Einstein-Hilbert action with cosmological constant and the Holst action with Immirizo parameter as well the Euler and Pontrjagin class for the reduced bundles. In this model, the curvature is changed locally by adding a S^2-bundle. Then the scale ℓ^2 has to agree with the volume of the S^2. In the action we have the coupling constant $\frac{1}{\ell^2}$ which has to agree with $1/L_P^2$ (L_P Planck length) to get in contact with Einsteins theory, i.e. we must set

$$\ell = L_P$$

The agreement with the Einstein-Hilbert action showed that this approach can describe gravity but it does not describe the global geometry. Later we can show, however, that it must be the de Sitter space $SO(4, 1)/SO(3, 1)$ globally.

15.6 Wild Embeddings: Geometric Expression for the Quantum State

In this section we will support our main hypothesis that an exotic $\mathbb{R}^4$ has automatically a quantum geometry, but as noted in the introduction, we must implicitly assume that the quantum-geometrical state is realized in the exotic $\mathbb{R}^4$. Interestingly, it follows from the physically motivated existence of a Lorentz metric which is induced by a codimension-one foliation. Therefore we will construct the foliation and the corresponding leaf space as the space of observables (using ideas of Connes). This leaf space is a non-commutative C^*-algebra with observable algebra a factor III_1 von Neumann algebra. A state in this algebra can be interpreted as a wild embedding which is also motivated by the exotic smoothness structure. The classical state is the tame embedding. Then, the deformation quantization of this tame embedding is the wild embedding (see [16]). In principle, the wild embedding determines the C^*-algebra completely. This algebra is generated by holonomies along connections of constant curvature. It is known from mathematics that this algebra (forming a so-called character variety [44]) determines the geometrical structure of the 3-manifold (along the way of Thurston [84, 95]). The main structure in this approach is the fundamental group, i.e. the group of closed, non-contractible curves in a manifold. The quantization of this group (as an expression of the classical geometry) gives the so-called skein algebra of knots in this manifold. We will relate this skein algebra to the leaf space above. On the way to show this relation, we will obtain the generator of the translation from one 3-manifold into another 3-manifold, i.e. the time together with the Hamiltonian.

15.6.1 Exotic $\mathbb{R}^4$ and Its Foliation

In Sect. 15.4, we described the sequence of 3-manifolds

$$Y_1 \rightarrow Y_2 \rightarrow \cdots \rightarrow Y_\infty = Y_r$$

characterizing the exotic smoothness structure of R^4. Then 0-framed surgery along this pretzel knot produces Y_1 whereas the n-th untwisted Whitehead double will give Y_n. For large n, the structure of the Casson handle is contained in the topology of Y_n and in the limit $n \rightarrow \infty$ we obtain Y_r (which is now a wild embedding $Y_r \subset \mathbf{R}^4$ in the standard $\mathbf{R}^4$ given by the embedding of the small exotic R^4, see above). What do we know about the structure of Y_n or Y_r in general? The compact subset K is a 4-manifold constructed by a pair of one 1-handle and one 2-handle which topologically cancel. The boundary of K is a compact 3-manifold having the first Betti number $b_1 = 1$. This information is also contained in Y_r. By the work of Freedman [53], every Casson handle is topologically $D^2 \times \mathbb{R}^2$ (relative to the attaching region) and therefore Y_r must be the boundary of D^4 (the Casson handle trivializes K to be D^4), i.e. Y_r *is a wild embedded 3-sphere* S^3. Then we obtain two different descriptions of R^4:

1. as a sequence of 3-manifolds Y_n (all having the first Betti number $b_1 = 1$) as boundaries of the neighborhood of K with increasing size and
2. as a global hyperbolic space of $R^4 \backslash K$ written as $S^3_\infty \times \mathbb{R}$ where S^3_∞ is a wild embedded 3-sphere (which looks differently for different $t \in \mathbb{R}$).

The first description gives a non-trivial but smooth foliation but there is no global spatial space. In contrast to this highly non-trivial foliation, the second description gives a global foliated spacetime containing a global spatial component, the wild embedded 3-sphere. In the first description we have a complex, relational description with no global time-like slices. Here, there only is a local coordinate system (with its own eigenzeit). This relational view has the big advantage that the simplest parts are also simple submanifolds (only finite surfaces with boundary). In contrast, the second description introduces a global foliation into equal time slices. Then the complexity is contained into the spatial component which is now a wild embedding (i.e. a space with an infinite number of polygons). This second approach will be described in the next subsection. So, lets concentrate on the first approach. Every 3-manifold Y_n admits a codimension-one ($PSL(2, \mathbb{R})$-invariant) foliation (see [17] for the details). By the description of the exotic R^4 using the sequence of 3-manifolds

$$Y_1 \rightarrow Y_2 \rightarrow Y_3 \rightarrow \cdots$$

we also get a foliation of the exotic R^4. The foliation on Y_n is defined by a $PSL(2, \mathbb{R})$-invariant one-form ω which is integrable $d\omega \wedge \omega = 0$ and defines another one-form η by $d\omega = -\eta \wedge \omega$. Then the integral

$$GV(Y_n) = \int_{Y_n} \eta \wedge d\eta$$

is known as Godbillon-Vey number $GV(Y_n)$ with the class $gv = \eta \wedge d\eta$. From the physics point of view, it is the abelian Chern-Simons functional. The Godbillon-Vey class characterizes the codimension-one foliation for the 3-manifold Y_n (see the Appendix B for more details). The foliation is very complicated. In [82] the local structure was analyzed. Let κ, τ be the curvature and torsion of a normal curve, respectively. Furthermore, let T, N, Z be the frame formed by this vector field dual to the one-forms ω, η, ξ and let l_T be the second fundamental form of leaf. Then the Godbillon-Vey class is locally given by

$$\eta \wedge d\eta = \kappa^2 \left(\tau + l_T(N, Z)\right) \omega \wedge \eta \wedge \xi$$

where $\tau \neq 0$ for $PSL(2, \mathbb{R})$ invariant foliations i.e. $[Z, N] = Z$, $[N, T] = T$ and $[Z, T] = N$. Recall that a foliation (M, F) of a manifold M is an integrable subbundle $F \subset TM$ of the tangent bundle TM. The leaves L of the foliation (M, F) are the maximal connected submanifolds $L \subset M$ with $T_x L = F_x \ \forall x \in L$. We denote with M/F the set of leaves or the leaf space. Now one can associate to the leaf space M/F a C^*-algebra $C(M, F)$ by using the smooth holonomy groupoid G of the foliation (see Connes [41]). According to Connes [42], one assigns to each leaf $\ell \in X$ the canonical Hilbert space of square-integrable half-densities $L^2(\ell)$. This assignment, i.e. a measurable map, is called a random operator forming a von Neumann $W(M, F)$. A deep theorem of Hurder and Katok [72] for foliations with non-zero Godbillon-Vey invariant states that this foliation has to contain a factor III von Neumann algebra. As shown in [13], the von Neumann algebra for the foliation of Y_n and for the exotic R^4 is a factor III_1-algebra. For the construction of this algebra, one needs the concept of a holonomy groupoid. Foliations are determined by the holonomies of closed curves in a leaf and the transport of this closed curve together with the holonomy from the given leaf to another leaf. Now one may ask why one considers only closed curves. Let PM the space of all paths in a manifold then this space admits a fibration over the space of closed paths ΩM (also called loop space) with fiber the constant paths (therefore homeomorphic to M), see [26]. Then, a curve is determined up to deformation (i.e. homotopy) by a closed path. Consider now a closed curve γ in a leaf ℓ and let act a diffeomorphism on ℓ. Then the curve γ is modified as well to γ' but γ and γ' are related by a (smooth) homotopy. Therefore to guarantee diffeomorphism invariance in this approach, one has to consider all closed curves up to homotopy. This structure can be made into a group (using concatenation

of paths as group operation) called the fundamental group $\pi_1(\ell)$ of the leaf. Above
we spoke about holonomy but a holonomy needs a connection of some bundle which
we did not introduce until now. But Connes [41] described a way to circumvent this
difficulty: Given a leaf ℓ of (M, F) and two points $x, y \in \ell$ of this leaf, any simple
path γ from x to y on the leaf ℓ uniquely determines a germ $h(\gamma)$ of a diffeomorphism
from a transverse neighborhood of x to a transverse neighborhood of y. The germ of
diffeomorphism $h(\gamma)$ only depends upon the homotopy class of γ in the fundamental
group of the leaf ℓ, and is called the holonomy of the path γ. All fundamental groups
of all leafs form the fundamental groupoid. The holonomy groupoid of a leaf ℓ is
the quotient of its fundamental groupoid by the equivalence relation which identifies
two paths γ and γ' from x to y (both in ℓ) iff $h(\gamma) = h(\gamma')$. Then the von Neumann
algebra of the foliation is the convolution algebra of the holonomy groupoid which
will be constructed later for the wild embedding.

15.6.1.1   Intermezzo: Factor *III* and Tomita-Takesaki Modular Theory

Remember a von Neumann algebra is an involutive subalgebra M of the algebra of
operators on a Hilbert space H that has the property of being the commutant of its
commutant: $(M')' = M$. This property is equivalent to saying that M is an involutive
algebra of operators that is closed under weak limits. A von Neumann algebra M is
said to be hyperfinite if it is generated by an increasing sequence of finite-dimensional
subalgebras. Furthermore we call M a factor if its center is equal to $\mathbb{C}$. It is a deep result
of Murray and von Neumann that every factor M can be decomposed into 3 types
of factors $M = M_I \oplus M_{II} \oplus M_{III}$. The factor I case divides into the two classes I_n
and I_∞ with the hyperfinite factors $I_n = M_n(\mathbb{C})$ the complex square matrices and
$I_\infty = \mathcal{L}(H)$ the algebra of all operators on an infinite-dimensional Hilbert space H.
The hyperfinite II factors are given by $II_1 = Cliff_\mathbb{C}(E)$, the Clifford algebra of an
infinite-dimensional Euclidean space E, and $II_\infty = II_1 \otimes I_\infty$. The case III remained
mysterious for a long time. Now we know that there are three cases parametrized
by a real number $\lambda \in [0, 1]$: $III_0 = R_W$ the Krieger factor induced by an ergodic
flow W, $III_\lambda = R_\lambda$ the Powers factor for $\lambda \in (0, 1)$ and $III_1 = R_\infty = R_{\lambda_1} \otimes R_{\lambda_2}$
the Araki-Woods factor for all λ_1, λ_2 with $\lambda_1/\lambda_2 \notin \mathbb{Q}$. We remark that all factor III
cases are induced by infinite tensor products of the other factors. One example of
such an infinite tensor space is the Fock space in quantum field theory.

   The modular theory of von Neumann algebras (see also [25]) has been discovered
by Tomita [96] in 1967 and put on solid grounds by Takesaki [91] around 1970. It is a
very deep theory that, to every von Neumann algebra $\mathcal{M} \subset \mathcal{B}(\mathcal{H})$ acting on a Hilbert
space $\mathcal{H}$, and to every vector $\xi \in \mathcal{H}$ that is cyclic, i.e. $\overline{(\mathcal{M}\xi)} = \mathcal{H}$, and separating,
i.e. for $A \in \mathcal{M}$, $A\xi = 0 \rightarrow A = 0$, associates:

- a one-parameter unitary group $t \mapsto \Delta^{it} \in \mathcal{B}(\mathcal{H})$
- and a conjugate-linear isometry $J : \mathcal{H} \rightarrow \mathcal{H}$ such that:

$$\Delta^{it} \mathcal{M} \Delta^{-it} = \mathcal{M}, \quad \forall t \in \mathbb{R}, \quad \text{and} \quad J \mathcal{M} J = \mathcal{M}',$$

where the commutant $\mathcal{M}'$ of $\mathcal{M}$ is defined by $\mathcal{M}' := \{A' \in \mathcal{B}(\mathcal{H}) \, | \, [A', A]_- = 0, \forall A \in \mathcal{B}(\mathcal{H})\}$.

More generally, given a von Neumann algebra $\mathcal{M}$ and a faithful normal state [2] (more generally for a faithful normal semi-finite weight) ω on the algebra $\mathcal{M}$, the modular theory allows to create a one-parameter group of $*$-automorphisms of the algebra $\mathcal{M}$,

$$\sigma^\omega : t \mapsto \sigma^\omega_t \in \mathrm{Aut}(\mathcal{M}), \quad \text{with} \quad t \in \mathbb{R},$$

such that:

- in the Gel'fand–Naĭmark–Segal representation π_ω induced by the weight ω, on the Hilbert space $\mathcal{H}_\omega$, the modular automorphism group σ^ω is implemented by a unitary one-parameter group $t \mapsto \Delta^{it}_\omega \in \mathcal{B}(\mathcal{H}_\omega)$ i.e. we have $\pi_\omega(\sigma^\omega_t(x)) = \Delta^{it}_\omega \pi_\omega(x) \Delta^{-it}_\omega$, for all $x \in \mathcal{M}$ and for all $t \in \mathbb{R}$;
- there is a conjugate-linear isometry $J_\omega : \mathcal{H}_\omega \to \mathcal{H}_\omega$, whose adjoint action implements a modular anti-isomorphism $\gamma_\omega : \pi_\omega(\mathcal{M}) \to \pi_\omega(\mathcal{M})'$, between $\pi_\omega(\mathcal{M})$ and its commutant $\pi_\omega(\mathcal{M})'$, i.e. for all $x \in \mathcal{M}$, we have $\gamma_\omega(\pi_\omega(x)) = J_\omega \pi_\omega(x) J_\omega$.

The operators J_ω and Δ_ω are called respectively the modular conjugation operator and the modular operator induced by the state (weight) ω. We will call "modular generator" the self-adjoint generator of the unitary one-parameter group $t \mapsto \Delta^{it}_\omega$ as defined by Stone's theorem i.e. the operator

$$K_\omega := \log \Delta_\omega, \quad \text{so that} \quad \Delta^{it}_\omega = e^{i K_\omega t}. \tag{15.9}$$

The modular automorphism group σ^ω associated to ω is the unique one-parameter automorphism group that satisfies the Kubo–Martin–Schwinger (KMS-condition) with respect to the state (or more generally a normal semi-finite faithful weight) ω, at inverse temperature $\beta = -1$, i.e.

$$\omega(\sigma^\omega_t(x)) = \omega(x), \quad \forall x \in \mathcal{M}$$

and for all $x, y \in \mathcal{M}$.

Using Tomita-Takesaki-theory, one has a continuous decomposition (as crossed product) of any factor *III* algebra M into a factor II_∞ algebra N together with a one-parameter group[3] $(\theta_\lambda)_{\lambda \in \mathbb{R}^*_+}$ of automorphisms $\theta_\lambda \in Aut(N)$ of N, i.e. one obtains

$$M = N \rtimes_\theta \mathbb{R}^*_+. \tag{15.10}$$

---

[2] ω is faithful if $\omega(x) = 0 \to x = 0$; it is normal if for every increasing bounded net of positive elements $x_\lambda \to x$, we have $\omega(x_\lambda) \to \omega(x)$.

[3] The group $\mathbb{R}^*_+$ is the group of positive real numbers with multiplication as group operation also known as Pontrjagin dual.

That means, there is a foliation induced from the foliation producing this II_∞ factor. Connes [42] (in Sect. I.4, p. 57ff) constructed the foliation F' canonically associated to the foliation F of factor III_1 above having the factor II_∞ as von Neumann algebra. In our case it is the horocycle flow: Let P the polygon on the hyperbolic space $\mathbb{H}^2$ determining the foliation above. P is equipped with the hyperbolic metric $2|dz|/(1 - |z|^2)$ together with the collection $T_1 P$ of unit tangent vectors to P. A horocycle in P is a circle contained in P which touches ∂P at one point, but from the classification of factors, we know that II_∞ is also splitted into

$$II_\infty = II_1 \otimes I_\infty$$

so that every factor III is determined by the factor II_1. The factor I_∞ are the compact operators in the Hilbert space. With an important observation we will close this intermezzo. The factor II_∞ admits an action of the group $\mathbb{R}_+^*$ by automorphisms so that the crossed product (15.10) is the factor III_1. The corresponding invariant, the flow of weights mod (M), was determined by Connes [42] to be the Godbillon-Vey invariant. Therefore *the modular generator above is given by the Godbillon-Vey invariant, i.e. this invariant is the Hamiltonian of the theory.*

15.6.1.2   Construction of a State

Then the C^*-algebra $C_r^*(M, F)$ of the foliation (M, F) is the C^*-algebra $C_r^*(G)$ of the smooth holonomy groupoid G. For completeness we will present the explicit construction (see [42] Sect. II.8). The basic elements of $C_r^*(M, F)$ are smooth half-densities with compact supports on G, $f \in C_c^\infty(G, \Omega^{1/2})$, where $\Omega_\gamma^{1/2}$ for $\gamma \in G$ is the one-dimensional complex vector space $\Omega_x^{1/2} \otimes \Omega_y^{1/2}$, where $s(\gamma) = x, t(\gamma) = y$, and $\Omega_x^{1/2}$ is the one-dimensional complex vector space of maps from the exterior power $\Lambda^k F_x, k = \dim F$, to $\mathbb{C}$ such that

$$\rho(\lambda \nu) = |\lambda|^{1/2} \rho(\nu) \qquad \forall \nu \in \Lambda^k F_x, \lambda \in \mathbb{R}.$$

For $f, g \in C_c^\infty(G, \Omega^{1/2})$, the convolution product $f * g$ is given by the equality

$$(f * g)(\gamma) = \int_{\gamma_1 \circ \gamma_2 = \gamma} f(\gamma_1) g(\gamma_2)$$

Then we define via $f^*(\gamma) = \overline{f(\gamma^{-1})}$ a $*$-operation making $C_c^\infty(G, \Omega^{1/2})$ into a $*$-algebra. For each leaf L of (M, F) one has a natural representation of $C_c^\infty(G, \Omega^{1/2})$ on the L^2 space of the holonomy covering $\tilde{L}$ of L. Fixing a base point $x \in L$, one identifies $\tilde{L}$ with $G_x = \{\gamma \in G, s(\gamma) = x\}$ and defines the representation

$$(\pi_x(f)\xi)(\gamma) = \int\limits_{\gamma_1 \circ \gamma_2 = \gamma} f(\gamma_1)\xi(\gamma_2) \quad \forall \xi \in L^2(G_x).$$

The completion of $C_c^\infty(G, \Omega^{1/2})$ with respect to the norm

$$\|f\| = \sup_{x \in M} \|\pi_x(f)\|$$

makes it into a C^*-algebra $C_r^*(M, F)$. Among all elements of the C^*-algebra, there are distinguished elements, idempotent operators or projectors having a geometric interpretation in the foliation. For later use, we will construct them explicitly (we follow [42] Sect. II.8. β closely). Let $N \subset M$ be a compact submanifold which is everywhere transverse to the foliation (thus $\dim(N) = \mathrm{codim}(F)$). A small tubular neighborhood N' of N in M defines an induced foliation F' of N' over N with fibers $\mathbb{R}^k$, $k = \dim F$. The corresponding C^*-algebra $C_r^*(N', F')$ is isomorphic to $C(N) \otimes \mathcal{K}$ with $\mathcal{K}$ the C^*-algebra of compact operators. In particular it contains an idempotent $e = e^2 = e^*, e = 1_N \otimes f \in C(N) \otimes \mathcal{K}$, where f is a minimal projection in $\mathcal{K}$. The inclusion $C_r^*(N', F') \subset C_r^*(M, F)$ induces an idempotent in $C_r^*(M, F)$ which is given by a closed curve in M transversal to the foliation.

In case of the foliation above (of the 3-manifolds Y_n), one has the foliation of the polygon P in $\mathbb{H}^2$ and a circle S^1 attached to every leaf of this foliation. Therefore we have the leafs $S^1 \times [0, 1]$ and the S^1 is the closed curve transversal to the foliation. Then every leaf defines (using the isomorphism $\pi_1(S^1 \times [0, 1]) = \pi_1(S^1) = \mathbb{Z}$) an idempotent represented by the fiber S^1 forming a base for the GNS representation of the C^*-algebra. Now we are able to construct a state in this algebra.

A state is a linear functional $\omega : C_r^*(M, F) \to \mathbb{C}$ so that $\omega(x \cdot x^*) \geq 0$ and $\omega(\mathbb{I}_{C_r^*(M,F)}) = 1$. Elements of $C_r^*(M, F)$ are half-densities with a support along some closed curve (as part of the holonomy groupoid). In a first step, one can use the GNS-representation of the C^*-algebra $C_r^*(M, F)$ by a map $C_r^*(M, F) \to \mathcal{B}(H)$ into the bounded operators of a Hilbert space. By the theorem of Fréchet-Riesz, every linear functional can be represented by the scalar product of the Hilbert space for some vector. To determine the linear functionals, we have to investigate the geometry of the foliation. The foliation was constructed to be $PSL(2, \mathbb{R})$-invariant, i.e. fixing the upper half space $\mathbb{H}^2$. Then we considered the unit tangent vectors of the tangent bundle over $\mathbb{H}^2$ defining the $\widetilde{SL}(2, \mathbb{R})$-geometry. But more is true. Every part of the 3-manifold Y_n is a knot/link complement with hyperbolic structure with isometry group $PSL(2, \mathbb{C})$ where the other geometric structures like $\widetilde{SL}(2, \mathbb{R})$ and $PSL(2, \mathbb{R})$ embed. Here we remark the known fact that every $PSL(2, \mathbb{C})$-geometry lifts uniquely to $SL(2, \mathbb{C})$ (the double cover). Therefore, to model the holonomy, we have to choose a flat $SL(2, \mathbb{C})$-connection and write it as the well-known integral of the connection 1-form along a closed curve. The linear functional is the trace of this integral (seen as matrix using a representation of $SL(2, \mathbb{C})$) known as Wilson loop. One can use the well-known identity

$$Tr(A) \cdot Tr(B) = Tr(AB) + Tr(AB^{-1})$$

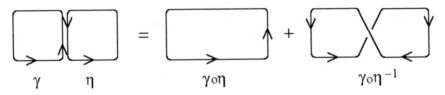

Fig. 15.4 Mandelstam identity as skein relation (see Fig. 15.7 and Sect. 15.6.3.2)

for $SL(2, \mathbb{C})$ which goes over to the Wilson loops. Let $W_\gamma[A]$ be the Wilson loop of a connection A along the closed curve γ. Then the relation of the Wilson loops

$$W_\gamma[A] \cdot W_\eta[A] = W_{\gamma \circ \eta}[A] + W_{\gamma \circ \eta^{-1}}[A]$$

for two intersecting curves γ and η is known as the Mandelstam identity for intersecting loops, see Fig. 15.4 for a visualization.

This relation is also known from another area: knot theory. There, it is the Kauffman bracket skein relation used to define the Kauffman knot polynomial. Therefore we obtain a state in the C^*-algebra by a closed curve in the leaf which extends to a knot (an embedded, closed curve) in a submanifold of the 3-manifold defined up to the skein relation. Finally:

State ω over leaf ℓ $\longleftrightarrow$ element of Kauffman skein module for $\ell \times [0, 1]$

We will later explain this correspondence as a deformation quantization. We will close this subsection by some remarks. Every representation $\pi_1(M) \to SL(2, \mathbb{C})$ defines (up to conjugacy) a flat connection. At the same time it defines also a hyperbolic structure on Y_n (for $M = Y_n$). By the argumentation above, the quantized version of this geometry (as defined by the C^*-algebra of the foliation) is given by the skein space (see Sect. 15.6.3.2 for the definition of the skein space).

15.6.2 The Wild Embedded 3-Sphere = Quantum (Geometric) State

Our previous work implied that the transition from the standard $\mathbf{R}^4$ to a small exotic R^4 has much to do with Quantum Gravity (QG). Therefore one would expect that a submanifold in the standard $\mathbf{R}^4$ with an appropriated geometry represents a classical state. Before we construct this state, there is a lot to say about the wild embedded 3-sphere as a quantum state.

Exotic $\mathbb{R}^4$

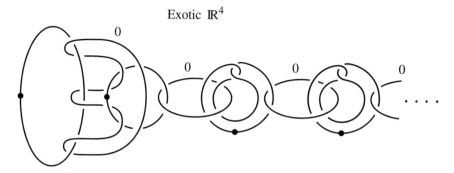

Fig. 15.5 Handle picture of the small exotic $\mathbb{R}^4$, the components with the dot are 1-handles and without the dot are 2-handles

15.6.2.1 The Wild Embedded 3-Sphere

To describe this wild 3-sphere, we will construct the sequence of Y_n by using the example of [23, 24] which was already partly explained in Sect. 15.3.5. At first we remark that the interior of the handle body in Fig. 15.5 is the R^4.

The Casson handle for this R^4 is given by the simplest tree $\mathcal{T}_+$, one positive self-intersection for each level. The compact 4-manifold inside of R^4 can be seen in Fig. 15.1 as a handle body. The 3-manifold Y_n surrounding this compact submanifold K is given by surgery (0-framed) along the link in Fig. 15.5 with a Casson handle of n-levels. In [59], this case is explicitly discussed. Y_n is given by 0-framed surgery along the n-th untwisted Whitehead double of the pretzel $(-3, 3 - 3)$ knot (see Fig. 15.3). Obviously, there is a sequence of inclusions

$$\cdots \subset Y_{n-1} \subset Y_n \subset Y_{n+1} \subset \cdots \rightarrow Y_{\mathcal{T}_+}$$

with the 3-manifold $Y_{\mathcal{T}_+}$ as limit. Let $\mathcal{K}_+$ be the corresponding (wild) knot, i.e. the ∞-th untwisted Whitehead double of the pretzel knot $(-3, 3, -3)$ (or the knot 9_{46} in Rolfson notation). The surgery description of $Y_{\mathcal{T}_+}$ induces the decomposition

$$Y_{\mathcal{T}_+} = C(\mathcal{K}_+) \cup \left(D^2 \times S^1\right) \qquad C(\mathcal{K}_+) = S^3 \backslash \left(\mathcal{K}_+ \times D^2\right) \qquad (15.11)$$

where $C(\mathcal{K}_+)$ is the knot complement of $\mathcal{K}_+$. In [33], the splitting of knot complements was described. Let $K_{9_{46}}$ be the pretzel knot $(-3, 3, -3)$ and let L_{Wh} be the Whitehead link (with two components). Then the complement $C(K_{9_{46}})$ has one torus boundary whereas the complement $C(L_{Wh})$ has two torus boundaries. Now according to [33], one obtains the splitting

$$C(\mathcal{K}_+) = C(L_{Wh}) \cup_{T^2} \cdots \cup_{T^2} C(L_{Wh}) \cup_{T^2} C(K_{9_{46}})$$

and we will describe each part separately (see Fig. 15.6).

Fig. 15.6 Schematic picture for the splitting of the knot complement $C(\mathcal{K}_+)$ (*above*) and in the more general case $C(\mathcal{K}_T)$ (*below*)

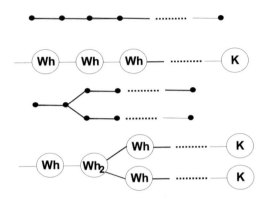

At first the knot $K_{9_{46}}$ is a hyperbolic knot, i.e. the interior of the 3-manifold $C(K_{9_{46}})$ admits a hyperbolic metric. By the work of Gabai [60], $C(K_{9_{46}})$admits a codimension-one foliation. The Whitehead link is a hyperbolic link but we need more: the Whitehead link is a fibered link of genus 1. That is, there is a fibration of the link complement $\pi : C(L_{Wh}) \rightarrow S^1$ over the circle so that $\pi^{-1}(p)$ is a surface of genus 1 (Seifert surface) for all $p \in S^1$. Now we will also describe the changes for a general tree. At first we will modify the Whitehead link: we duplicate the linked circle, i.e. there are as many circles as branching in the tree to get the link Wh_n with $n + 1$ components. Then the complement of Wh_n has also $n + 1$ torus boundaries and it also fibers over S^1. With the help of Wh_n we can build every tree T. Now the 3-manifold Y_T is given by 0-framed surgery along the ∞-th untwisted ramified (usage of Wh_n) Whitehead double of a knot k, denoted by the link $\mathcal{K}_T$. The tree T has one root, then Y_T is given by

$$Y_T = C(\mathcal{K}_T) \cup \left(D^2 \times S^1\right)$$

and the complement $C(\mathcal{K}_T)$ splits like the tree into complements of Wh_n and many copies of $C(k)$ (see Fig. 15.6). Using a deep result of Freedman [53], we obtain:

Y_T *is a wild embedded 3-sphere* S^3_∞.

15.6.2.2 Reconstruction of the Spatial Space by Using a State

Our result about the existence of a codimension-one foliation for Y_T can be simply expressed: foliations are characterized by the holonomy properties of the leafs. This principle is also the corner stone for the usage of non-commutative geometry as description of the leaf space. In the previous subsection, we already characterized the state as an element of the Kauffman skein module. Here we are interested in a reconstruction of the underlying space but now assuming a global foliation so that we will obtain the whole spatial space.

Starting point is the state constructed in the Sect. 15.6.1.2. Here, we got a relation between the state ω as linear functional over the algebra and the Kauffman skein module. Using this relation, we consider a leaf $\ell = S^1 \times [0, 1]$ and the 3-dimensional extension as solid torus $S^1 \times D^2$. The Kauffman skein module $K(S^1 \times D^2)$ is polynomial algebra with one generator (the loop around S^1). Now we consider one 3-manifold Y_n with the corresponding foliation. Using the splitting above, the Kauffman skein module $K(Y_n)$ is determined by the skein module for the parts, i.e. by the knot complements. Therefore we have to consider the skein module for hyperbolic 3-manifolds. Hyperbolic 3-manifolds contain special surfaces, called essential or incompressible surfaces, see Appendix C. It is known [36] that the skein module of 3-manifolds containing essential surfaces is not finitely generated. Therefore, the state itself is not finitely generated. If we use the leaf $S^1 \times D^2$ as a local model for one generator then we will obtain an infinitely complicated 3-manifold made from pieces $S^1 \times D^2$ so that the corresponding generators are not related to each other. An example of this structure is the Whitehead manifold having a non-finitely generated Kauffman skein module [1]. In general we will obtain a wild embedded 3-manifold by using this simple pieces. By the argumentation in the previous subsection we know that this wild embedded 3-manifold is the wild embedded 3-sphere Y_T. Finally we obtain:

$$\text{State } \omega \longleftrightarrow \text{ wild embedded 3-sphere } Y_T$$

the state ω is realized by some wild embedded 3-sphere.

15.6.2.3 Construction of the Operator Algebra

Following [16] we will construct a C^*-algebra from the wild embedded 3-sphere. Let $I : S^3 \to \mathbb{R}^4$ be a wild embedding of codimension-one so that $I(S^3) = S^3_\infty = Y_T$. Now we consider the complement $\mathbb{R}^4 \setminus I(S^3)$ which is non-trivial, i.e. $\pi_1(\mathbb{R}^4 \setminus I(S^3)) = \pi \neq 1$. Now we define the C^*-algebra $C^*(\mathcal{G}, \pi)$ associated to the complement $\mathcal{G} = \mathbb{R}^4 \setminus I(S^3)$ with group $\pi = \pi_1(\mathcal{G})$. If π is non-trivial then this group is not finitely generated. From an abstract point of view, we have a decomposition of $\mathcal{G}$ by an infinite union

$$\mathcal{G} = \bigcup_{i=0}^{\infty} C_i$$

of 'level sets' C_i. Then every element $\gamma \in \pi$ lies (up to homotopy) in a finite union of levels.

The basic elements of the C^*-algebra $C^*(\mathcal{G}, \pi)$ are smooth half-densities with compact supports on $\mathcal{G}$, $f \in C_c^\infty(\mathcal{G}, \Omega^{1/2})$, where $\Omega_\gamma^{1/2}$ for $\gamma \in \pi$ is the one-dimensional complex vector space of maps from the exterior power $\Lambda^k L$ (dim $L = k$), of the union of levels L representing γ, to $\mathbb{C}$ such that

$$\rho(\lambda \nu) = |\lambda|^{1/2} \rho(\nu) \qquad \forall \nu \in \Lambda^2 L, \lambda \in \mathbb{R}.$$

For $f, g \in C_c^\infty(\mathcal{G}, \Omega^{1/2})$, the convolution product $f * g$ is given by the equality

$$(f * g)(\gamma) = \int\limits_{\gamma_1 \circ \gamma_2 = \gamma} f(\gamma_1) g(\gamma_2)$$

with the group operation $\gamma_1 \circ \gamma_2$ in π. Then we define via $f^*(\gamma) = \overline{f(\gamma^{-1})}$ a $*$-operation making $C_c^\infty(\mathcal{G}, \Omega^{1/2})$ into a $*$-algebra. Each level set C_i consists of simple pieces denoted by T. For these pieces, one has a natural representation of $C_c^\infty(\mathcal{G}, \Omega^{1/2})$ on the L^2 space over T. Then one defines the representation

$$(\pi_x(f)\xi)(\gamma) = \int\limits_{\gamma_1 \circ \gamma_2 = \gamma} f(\gamma_1)\xi(\gamma_2) \qquad \forall \xi \in L^2(T), \forall x \in \gamma.$$

The completion of $C_c^\infty(\mathcal{G}, \Omega^{1/2})$ with respect to the norm

$$||f|| = \sup_{x \in \mathcal{G}} ||\pi_x(f)||$$

makes it into a C^*-algebra $C_c^\infty(\mathcal{G}, \pi)$. Finally we are able to define the C^*-algebra associated to the wild embedding. Using a result in [16], one can show that the corresponding von Neumann algebra is the factor III_1.

Among all elements of the C^*-algebra, there are distinguished elements, idempotent operators or projectors having a geometric interpretation. For later use, we will construct them explicitly (we follow [42] Sect. II.8. β closely). Let $Y_T \subset \mathbb{R}^4$ be the wild submanifold. A small tubular neighborhood N' of Y_T in $\mathbb{R}^4$ defines the corresponding C^*-algebra $C_c^\infty(N', \pi_1(\mathbb{R}^4 \backslash N'))$ is isomorphic to $C_c^\infty(\mathcal{G}, \pi_1(\mathbb{R}^4 \backslash I(S^3)) \otimes \mathcal{K}$ with $\mathcal{K}$ the C^*-algebra of compact operators. In particular it contains an idempotent $e = e^2 = e^*, e = 1_N \otimes f \in C_c^\infty(\mathcal{G}, \pi_1(\mathbb{R}^4 \backslash I(S^3))) \otimes \mathcal{K}$, where f is a minimal projection in $\mathcal{K}$. It induces an idempotent in $C_c^\infty(\mathcal{G}, \pi_1(\mathbb{R}^4 \backslash I(S^3)))$. By definition, this idempotent is given by a closed curve in the complement $\mathbb{R}^4 \backslash I(S^3)$. These projection operators form the basis in this algebra.

15.6.3 Reconstructing the Classical State

In this section we will describe a way from a (classical) Poisson algebra to a quantum algebra by using deformation quantization. Therefore we will obtain a positive answer to the question: Does the C^*-algebra of the foliation (as well of a wild (specific) embedding) comes from a (deformation) quantization? Of course, this question cannot be answered in all generality, but for our example we will show that the enveloping von Neumann algebra of foliation and of this wild embedding is the result of a deformation quantization using the classical Poisson algebra (of closed curves) of the tame embedding. This result shows two things: the wild embedding can be seen as a quantum state and the classical state is a tame embedding.

15.6.3.1 Intermezzo 1: The Observable Algebra and Its Poisson Structure

In this section we will describe the formal structure of a classical theory coming from the algebra of observables using the concept of a Poisson algebra. In quantum theory, an observable is represented by an hermitean operator having the spectral decomposition via projectors or idempotent operators. The coefficient of the projector is the eigenvalue of the observable or one possible result of a measurement. At least one of these projectors represents (via the GNS representation) a quasi-classical state. Thus, to construct the substitute of a classical observable algebra with Poisson algebra structure, we have to concentrate on the idempotents in the C^*-algebra. Now we will see that the set of closed curves on a surface has the structure of a Poisson algebra. Let us start with the definition of a Poisson algebra.

Let P be a commutative algebra with unit over $\mathbb{R}$ or $\mathbb{C}$. A *Poisson bracket* on P is a bilinearform $\{\,,\,\} : P \otimes P \to P$ fulfilling the following 3 conditions:

1. anti-symmetry $\{a, b\} = -\{b, a\}$
2. Jacobi identity $\{a, \{b, c\}\} + \{c, \{a, b\}\} + \{b, \{c, a\}\} = 0$
3. derivation $\{ab, c\} = a\{b, c\} + b\{a, c\}$.

Then a *Poisson algebra* is the algebra $(P, \{\,,\,\})$.

Now we consider a surface S together with a closed curve γ. Additionally we have a Lie group G given by the isometry group. The closed curve is one element of the fundamental group $\pi_1(S)$. From the theory of surfaces we know that $\pi_1(S)$ is a free abelian group. Denote by Z the free $\mathbb{K}$-module ($\mathbb{K}$ a ring with unit) with the basis $\pi_1(S)$, i.e. Z is a freely generated $\mathbb{K}$-module. Recall Goldman's definition of the Lie bracket in Z (see [61]). For a loop $\gamma : S^1 \to S$ we denote its class in $\pi_1(S)$ by $\langle \gamma \rangle$. Let α, β be two loops on S lying in general position. Denote the (finite) set $\alpha(S^1) \cap \beta(S^1)$ by $\alpha\#\beta$. For $q \in \alpha\#\beta$ denote by $\epsilon(q; \alpha, \beta) = \pm 1$ the intersection index of α and β in q. Denote by $\alpha_q \beta_q$ the product of the loops α, β based in q. Up to homotopy the loop $(\alpha_q \beta_q)(S^1)$ is obtained from $\alpha(S^1) \cup \beta(S^1)$ by the orientation preserving smoothing of the crossing in the point q. Set

$$[\langle \alpha \rangle, \langle \beta \rangle] = \sum_{q \in \alpha\#\beta} \epsilon(q; \alpha, \beta)(\alpha_q \beta_q). \tag{15.12}$$

According to Goldman [61] (Theorem 5.2), the bilinear pairing $[\,,\,] : Z \times Z \to Z$ given by (15.12) on the generators is well defined and makes Z a Lie algebra. The algebra $Sym(Z)$ of symmetric tensors is then a Poisson algebra (see Turaev [98]).

The whole approach seems natural for the construction of the Lie algebra Z but the introduction of the Poisson structure is an artificial act. From the physical point of view, the Poisson structure is not the essential part of classical mechanics. More important is the algebra of observables, i.e. functions over the configuration space forming the Poisson algebra. For the foliation discussed above, we already identified the observable algebra (the holonomy along closed curves) as well the corresponding group to be $SL(2, \mathbb{C})$. Therefore for the following, we will set $G = SL(2, \mathbb{C})$.

Now we introduce a principal G bundle on S, representing a geometry on the surface. This bundle is induced from a G bundle over $S \times [0, 1]$ having always a flat connection. Alternatively one can consider a homomorphism $\pi_1(S) \to G$ represented as holonomy functional

$$hol(\omega, \gamma) = \mathcal{P} \exp\left(\int_\gamma \omega\right) \in G \qquad (15.13)$$

with the path ordering operator $\mathcal{P}$ and ω as flat connection (i.e. inducing a flat curvature $\Omega = d\omega + \omega \wedge \omega = 0$). This functional is unique up to conjugation induced by a gauge transformation of the connection. Thus we have to consider the conjugation classes of maps

$$hol : \pi_1(S) \to G$$

forming the space $X(S, G)$ of gauge-invariant flat connections of principal G bundles over S. Now (see [85]) we can start with the construction of the Poisson structure on $X(S, G)$, based on the Cartan form as the unique bilinearform of a Lie algebra. As discussed above we will use the Lie group $G = SL(2, \mathbb{C})$ but the whole procedure works for every other group too. Now we consider the standard basis

$$X = \begin{pmatrix} 0 & 1 \\ 0 & 0 \end{pmatrix}, \qquad H = \begin{pmatrix} 1 & 0 \\ 0 & -1 \end{pmatrix}, \qquad Y = \begin{pmatrix} 0 & 0 \\ 1 & 0 \end{pmatrix} \qquad (15.14)$$

of the Lie algebra $sl(2, \mathbb{C})$ with $[X, Y] = H$, $[H, X] = 2X$, $[H, Y] = -2Y$. Furthermore there is the bilinearform $B : sl_2 \otimes sl_2 \to \mathbb{C}$ written in the standard basis as

$$\begin{pmatrix} 0 & 0 & -1 \\ 0 & -2 & 0 \\ -1 & 0 & 0 \end{pmatrix}$$

Now we consider the holomorphic function $f : SL(2, \mathbb{C}) \to \mathbb{C}$ and define the gradient $\delta_f(A)$ along f at the point A as $\delta_f(A) = Z$ with $B(Z, W) = df_A(W)$ and

$$df_A(W) = \frac{d}{dt} f(A \cdot \exp(t W))\bigg|_{t=0} .$$

The calculation of the gradient δ_{tr} for the trace tr along a matrix

$$A = \begin{pmatrix} a_{11} & a_{12} \\ a_{21} & a_{22} \end{pmatrix}$$

is given by

$$\delta_{tr}(A) = -a_{21}Y - a_{12}X - \frac{1}{2}(a_{11} - a_{22})H.$$

Given a representation $\rho \in X(S, SL(2, \mathbb{C}))$ of the fundamental group and an invariant function $f : SL(2, \mathbb{C}) \to \mathbb{R}$ extendable to $X(S, SL(2, \mathbb{C}))$. Then we consider two conjugacy classes $\gamma, \eta \in \pi_1(S)$ represented by two transversal intersecting loops P, Q and define the function $f_\gamma : X(S, SL(2, \mathbb{C})) \to \mathbb{C}$ by $f_\gamma(\rho) = f(\rho(\gamma))$. Let $x \in P \cap Q$ be the intersection point of the loops P, Q and c_x a path between the point x and the fixed base point in $\pi_1(S)$. Then we define $\gamma_x = c_x \gamma c_x^{-1}$ and $\eta_x = c_x \eta c_x^{-1}$. Finally we get the Poisson bracket

$$\{f_\gamma, f'_\eta\} = \sum_{x \in P \cap Q} sign(x) \, B(\delta_f(\rho(\gamma_x)), \delta_{f'}(\rho(\eta_x))),$$

where $sign(x)$ is the sign of the intersection point x. Thus,

The space $X(S, SL(2, \mathbb{C}))$ has a natural Poisson structure (induced by the bilinear form (15.12) on the group) and the Poisson algebra $(X(S, SL(2, \mathbb{C}), \{ , \})$ of complex functions over them is the algebra of observables.

15.6.3.2 Intermezzo 2: Drinfeld-Turaev Quantization

Now we introduce the ring $\mathbb{C}[[h]]$ of formal polynomials in h with values in $\mathbb{C}$. This ring has a topological structure, i.e. for a given power series $a \in \mathbb{C}[[h]]$ the set $a + h^n \mathbb{C}[[h]]$ forms a neighborhood. Now we define

A *Quantization* of a Poisson algebra P is a $\mathbb{C}[[h]]$ algebra P_h together with the $\mathbb{C}$-algebra isomorphism $\Theta : P_h/hP \to P$ so that

1. the module P_h is isomorphic to $V[[h]]$ for a $\mathbb{C}$ vector space V
2. let $a, b \in P$ and $a', b' \in P_h$ be $\Theta(a) = a', \Theta(b) = b'$ then

$$\Theta \left(\frac{a'b' - b'a'}{h} \right) = \{a, b\}.$$

One speaks of a deformation of the Poisson algebra by using a deformation parameter h to get a relation between the Poisson bracket and the commutator. Therefore we have the problem to find the deformation of the Poisson algebra $(X(S, SL(2, \mathbb{C})), \{ , \})$. The solution to this problem can be found via two steps:

Fig. 15.7 Crossings
L_∞, L_o, L_{oo}

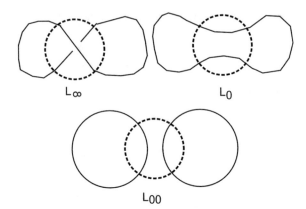

1. at first find another description of the Poisson algebra by a structure with one
   parameter at a special value and
2. secondly vary this parameter to get the deformation.

Fortunately both problems were already solved (see [97, 98]). The solution of the
first problem is expressed in the theorem:

   The skein module $K_{-1}(S \times [0, 1])$ *(i.e.* $t = -1$*) has the structure of an algebra
isomorphic to the Poisson algebra* $(X(S, SL(2, \mathbb{C})), \{, \})$. *(see also [34, 38])*

   Then we have also the solution of the second problem:

   The skein algebra $K_t(S \times [0, 1])$ *is the quantization of the Poisson algebra*
$(X(S, SL(2, \mathbb{C})), \{, \})$ *with the deformation parameter* $t = \exp(h/4)$. *(see also*
[38])

   To understand these solutions we have to introduce the skein module $K_t(M)$ of a
3-manifold M (see [81]). For that purpose we consider the set of links $\mathcal{L}(M)$ in M
up to isotopy and construct the vector space $\mathbb{C}\mathcal{L}(M)$ with basis $\mathcal{L}(M)$. Then one can
define $\mathbb{C}\mathcal{L}[[t]]$ as ring of formal polynomials having coefficients in $\mathbb{C}\mathcal{L}(M)$. Now
we consider the link diagram of a link, i.e. the projection of the link to the $\mathbb{R}^2$ having
the crossings in mind. Choosing a disk in $\mathbb{R}^2$ so that one crossing is inside this disk.
If the three links differ by the three crossings L_{oo}, L_o, L_∞ (see Fig. 15.7) inside of
the disk then these links are skein-related.

   Then in $\mathbb{C}\mathcal{L}[[t]]$ one writes the skein relation[4] $L_\infty - tL_o - t^{-1}L_{oo}$. Furthermore
let $L \sqcup O$ be the disjoint union of the link with a circle and one writes the fram-
ing relation $L \sqcup O + (t^2 + t^{-2})L$. Let $S(M)$ be the smallest submodule of $\mathbb{C}\mathcal{L}[[t]]$
containing both relations. Then we define the Kauffman bracket skein module by
$K_t(M) = \mathbb{C}\mathcal{L}[[t]]/S(M)$. We list the following general results about this module:

- The module $K_{-1}(M)$ for $t = -1$ is a commutative algebra.
- Let S be a surface, then $K_t(S \times [0, 1])$ carries the structure of an algebra.

The algebraic structure of $K_t(S \times [0, 1])$ can be simply seen by using the diffeomorphism between the sum $S \times [0, 1] \cup_S S \times [0, 1]$ along S and $S \times [0, 1]$. Then the

---

[4]The relation depends on the group $SL(2, \mathbb{C})$.

product ab of two elements $a, b \in K_t(S \times [0, 1])$ is a link in $S \times [0, 1] \cup_S S \times [0, 1]$ corresponding to a link in $S \times [0, 1]$ via the diffeomorphism. The algebra $K_t(S \times [0, 1])$ is in general non-commutative for $t \neq -1$. For the following we will omit the interval $[0, 1]$ and denote the skein algebra by $K_t(S)$.

In Sect. 15.6.1.2, we described the state as an element of the Kauffman skein module $K_t(\ell)$ of the leaf ℓ. Now we obtained also that the observable algebra is the Kauffman skein module again. How does this whole story fit into the description of the observable algebra for the foliation as factor III_1? In [58], it was shown that the Kauffman bracket skein module of a cylinder over the torus embeds as a subalgebra of the noncommutative torus. However, the noncommutative torus can be seen as the leaf space of the Kronecker foliation of the torus leading to the factor II_∞. Then by using (15.10), we obtain the factor III_1 back. We will use this relation in the next section to get the quantum action.

15.7 Action at the Quantum Level

Above, we used the foliation to get quantum states which agreed with the deformation quantization of a classical state. Central point in our argumentation is the construction of the C^*-algebra with the corresponding von Neumann algebra as observable algebra. This von Neumann algebra is a factor III_1. By using the Tomita-Takesaki modular theory, there is a relation to the factor II_∞ by using an action of the group $\mathbb{R}_+^*$ by automorphisms of a Lebesgue measure space leading to the decomposition of the factor III_1. This action is related to an invariant, the flow of weights mod(M). The main property of the factor III_1 is the constant flow of weights mod(M). Connes [41, 42] described the flow of weights as a bundle of densities over the leaf space, i.e. the $\mathbb{R}_+^*$ homogeneous space of nonzero maps. In case of the foliation considered above, this density is constant and we can naturally identify this density with the volume of the submanifold defining the foliation. By definition, this volume is given by the Godbillon-Vey invariant (see Eq. (15.34) in Appendix B, the circle in the fiber has unit size). This invariant can be seen as an element of $H^3(BG, \mathbb{R})$ with the holonomy groupoid G of the foliation. As shown by Connes [41, 42], the Godbillon-Vey class GV can be expressed as a cyclic cohomology class (the so-called flow of weights)

$$GV_{HC} \in HC^2(C_c^\infty(G))$$

of the C^*-algebra for the foliation. Then we define an expression

$$S = Tr_\omega (GV_{HC})$$

uniquely associated to the foliation (Tr_ω is the Dixmier trace). The expression S generates the action on the factor by

$$\Delta_\omega^{it} = exp(i\, S)$$

so that S is the action or the Hamiltonian multiplied by the time (see (15.9)). It is an operator which defines the dynamics by acting on the states. For explicit calculations we have to evaluate this operator. One way is the usage of the relation between the foliation and the wild embedding. This wild embedding is determined by the fundamental group π of its complement. In [16], we discussed the properties of this group π. It is a perfect group, i.e. every element is generated by a commutator. Then a representation of this group into some other group like $GL(\mathbb{C})$ (the limit of $GL(n, \mathbb{C})$ for $n \to \infty$) reduces to the representation of the maximal perfect subgroup. For that purpose we consider the representation of the group π into the group $E(\mathbb{C})$ of elementary matrices, which is the perfect subgroup of $GL(\mathbb{C})$. Then we obtain matrix-valued functions $X^\mu \in C_c^\infty(E(\mathbb{C}))$ as the image of the generators of π w.r.t. the representation $\pi \to E(\mathbb{C})$ labeled by the dimension $\mu = 1, \ldots, 4$ of the embedding space R^n. Via the representation $\iota : \pi \to E(\mathbb{C})$, we obtain a cyclic cocycle in $HC^2(C_c^\infty(E(\mathbb{C}))$ generated by a suitable Fredholm operator F. Here we use the standard choice $F = D|D|^{-1}$ with the Dirac operator D acting on the functions in $C_c^\infty(E(\mathbb{C}))$. Then the cocycle in $HC^2(C_c^\infty(E(\mathbb{C}))$ can be expressed by

$$\iota_* GV_{HC} = \eta_{\mu\nu}[F, X^\mu][F, X^\nu]$$

using a metric $\eta_{\mu\nu}$ in R^4 via the pull-back using the representation $\iota : \pi \to E(\mathbb{C})$. Finally we obtain the action

$$S = Tr_\omega([F, X^\mu][F, X_\mu]) = Tr_\omega([D, X^\mu][D, X_\mu]|D|^{-2}) \qquad (15.15)$$

which can be evaluated by using the heat-kernel of the Dirac operator D. The appearance of the heat kernel is a sign for a relation to quantum field theory where the heat kernel is a very convenient tool for studying one-loop divergences, anomalies and various asymptotics of the effective action.

Away from this operator expression for the Godbillon-Vey invariant, there are geometrical evaluations which are not defined on the leaf space but rather on the whole manifold. As mentioned above, this invariant admits values in the real numbers and we can evaluate them according to the type of the number: for integer values one obtains the Euler class and for rational numbers the Pontrjagin class (for the corresponding bundles). Therefore using the ideas of Sect. 15.5, we obtain the Einstein-Hilbert and the Holst action but also a correction given by irrational values of the Godbillon-Vey number.

15.8   The Scaling Behavior of the Action

A good test for the theory is the dependence of the action (15.15) on the scale. The theory has a strong geometrical flavor and therefore the scaling behavior can be understood by a geometrical construction using the exotic R^4. As explained above, the central point in the construction is the Casson handle. From the scaling point of view, the Casson handle contains disks of any size (with respect to the embedding $R^4 \hookrightarrow \mathbf{R}^4$). The long scales are given by the first levels of the Casson handle whereas the small scales are represented by the higher levels of the Casson handle.

15.8.1   Long-Scale Behavior (Einstein-Hilbert Action)

Let us consider the small exotic R^4. From the physics point of view, the large scale is given by the first levels of the Casson handle. In the construction of the foliation of R^4, the first levels describe a polygon in the hyperbolic space $\mathbb{H}^2$ with a finite and small number of vertices. The Godbillon-Vey number of this foliation is given by the volume of this polygon. In principle, it is also true for the inclusion of the higher levels (and also for the whole Casson handle) but every higher level gives only a very small contribution to the Godbillon-Vey number. Therefore, the first levels of the Casson handle can be simply characterized by the Godbillon-Vey number, i.e. by the size of the polygon in the scale r. Then the Godbillon-Vey number is given by $GV = r^2$. In [16] we analyzed this situation and found the relation

$$GV \overset{r \to \infty}{=} r^2 \int\limits_{D^2} \left(g_{\mu\nu} \partial_k \xi^\mu \partial^k \xi^\nu \right) d^2 x$$

to the Godbillon-Vey number. Here we integrate over the disk (equal to the polygon) which is used to define the foliation. This model is the non-linear sigma model (for the embedding of the disk into Y_n with metric g) depending on the scale r^2. The scaling behavior of this model was studied in [56] and one obtains the RG flow equation

$$\frac{\partial}{\partial r^2} g_{\mu\nu} = R_{\mu\nu} + \frac{1}{r^2} \left(R_{\mu\lambda\kappa\iota} R_\nu^{\lambda\kappa\iota} \right) + O(r^{-4}) \tag{15.16}$$

reducing to the Ricci flow equations for large scales ($r \to \infty$). The fixed point of this flow are geometries of constant curvature (used to prove the Thurston geometrization conjecture). Therefore in the classical limit of large scales, we obtain a geometry of the 3-manifold of constant curvature whereas for small scales one has to take into account higher curvature corrections. On the spacetime, one has also flow equations from one 3-manifold of constant curvature to another 3-manifold of constant curvature. This flow equation is equivalent to the (anti-)self-dual curvature (or instantons) by using the gradient flow of the Chern-Simons functional [52]. This approach has much in common with the non-linear graviton of Penrose [79]. We will explain these ideas in Sect. 15.9.1.

15.8.2 Short-Scale Behavior

For the short scale, we need the full power of the Casson handle. As a first step we can evaluate the action (15.15) so that the Dirac operator D acts on usual square-integrable functions, so that $[D, X^\mu] = dX^\mu$ is finite. Then the action (15.15) reduces to

$$S = Tr_\omega(\eta_{\mu\nu}(\partial_k X^\mu \partial^k X^\nu)|D|^{-2})$$

where $\mu, \nu = 1, \ldots, 4$ is the index for the coordinates on R^4 and $k = 1, 2$ represents the index of the disk (inside of the Casson handle). Now we will choose a small fluctuation ξ^k of a fixed embedding of the disk in the Casson handle given by $X^\mu = (x^k + \xi^\mu)\delta_k^\mu$ with $\partial_l x^k = \delta_l^k$. Then we obtain

$$\partial_k X^\mu \partial^k X^\nu = \delta_k^\mu \delta_k^\nu (1 + \partial_k \xi^\mu)(1 + \partial_k \xi^\nu)$$

and we use a standard argument to neglect the terms linear in $\partial\xi$: fluctuations have no preferred direction and therefore only the square contributes. Then we have

$$S = Tr_\omega(\eta_{\mu\nu}(\delta_k^\mu \delta_k^\nu + \partial_k \xi^\mu \partial^k \xi^\nu)|D|^{-2})$$

for the action. By using a result of [42] one obtains for the Dixmier trace

$$Tr_\omega(|D|^{-2}) = 2 \int_{D^2} *(\Phi_1)$$

with the first coefficient Φ_1 of the heat kernel expansion [21]

$$\Phi_1 = \frac{1}{6} R$$

and the action simplifies to

$$S = \int_{D^2} \left(\frac{2}{3}R + \partial_k \xi^\mu \partial^k \xi^\nu \frac{1}{3}R\right) dvol(D^2) \qquad (15.17)$$

for the main contributions where R is the scalar curvature of the embedded disk D^2. Again, but now for small fluctuations, we obtain the flow equation (15.16) but we have to consider the small case $r \to 0$. Then we have to take arbitrary curvature contributions into account. This short calculation showed that the short-scale behavior is given by a two-dimensional action. In the next section we will understand this behavior geometrically. For small fluctuations we obtained a disk but what happens for larger fluctuations? Then we have to take even the higher levels of the Casson handle into account. These higher levels form a complicated surface with a fractal structure (a generalization of the Cantor set). Then the action (15.17) has to be replaced by an integral over this fractal space. For the evaluation of the quantum action (15.15) one can use the ideas of noncommutative geometry as used for fractals and quasi-Fuchsian groups, see [42] (Sect. IV.3).

15.9 Some Properties of the Theory

In this section we will present some properties of the theory. For an impression, it is enough to present the main ideas. The details will be published separately.

15.9.1 The Graviton

By using the large scale behavior in Sect. 15.8.1, we have to consider Ricci-flat spaces and an easy calculation gives the well-known propagator in the linearized version, however, we are not interested in the linearized version. GRT is a highly non-linear theory and therefore one has to take this non-linearity into account. The Ricci-flatness of the spacetime goes over to the 3-manifold as the spatial component where it implies a 3-manifold of constant curvature (as fixed point of the Ricci flow). Then as shown by Witten [103–105], the 3-dimensional Einstein-Hilbert action

$$\int_N R_{(3)} \sqrt{h}\, d^3x = L \cdot CS(N, A)$$

is related to the Chern-Simons action $CS(N, A)$ with respect to the (Levi-Civita) connection A and the length L. By using the Stokes theorem we obtain

$$S_{EH}(N \times [0, 1]) = \int_{M_T} tr(F \wedge F),$$

i.e. the action for the 4-manifold $N \times [0, 1]$ (as local spacetime) with the curvature $F = DA$, i.e. the action is the (topological) Pontrjagin class of the 4-manifold. From the formal point of view, the curvature 2-form $F = DA$ is generated by a $SO(3, 1)$ connection A in the frame bundle, which can be lifted uniquely to a $SL(2, \mathbb{C})$-(Spin-) connection. According to the Ambrose-Singer theorem, the components of the curvature tensor are determined by the values of holonomy which is in general a subgroup of $SL(2, \mathbb{C})$. Thus we start with a suitable curvature 2-form $F = DA$ with values in the Lie algebra $\mathfrak{g}$ of the Lie group G as subgroup of the $SL(2, \mathbb{C})$. The variation of the Chern-Simons action gets flat connections $DA = 0$ as solutions. The flow of solutions $A(t)$ in $N \times [0, 1]$ (parametrized by the variable t, the 'time') between the flat connection $A(0)$ in $N \times \{0\}$ to the flat connection $A(1)$ in $N \times \{1\}$ will be given by the gradient flow equation (see for instance [52])

$$\frac{d}{dt} A(t) = \pm * F(A) = \pm * DA \tag{15.18}$$

where the coordinate t is normal to N. Therefore we are able to introduce a connection $\tilde{A}$ in $N \times [0, 1]$ so that the covariant derivative in t-direction agrees with $\partial/\partial t$. Then we have for the curvature $\tilde{F} = D\tilde{A}$, where the fourth component is given by $\tilde{F}_{4\mu} = d\tilde{A}_{\mu}/dt$. Thus we will get the instanton equation with (anti-) self-dual curvature

$$\tilde{F} = \pm * \tilde{F}.$$

It follows

$$S_{EH}([0, 1] \times N) = \int_{N \times [0,1]} tr(\tilde{F} \wedge \tilde{F}) = \pm \int_{N \times [0,1]} tr(\tilde{F} \wedge *\tilde{F}),$$

(i.e. the MacDowell–Mansouri action).

We remark the main point in this argumentation: we obtain a self-dual curvature as gradient flow between two 3-manifolds of constant curvature. Of course, (anti-)self-dual curvatures are also solutions of Einsteins equation (but the reverse is not true). Following Penrose [79], we call these solutions the nonlinear graviton.

15.9.2   Relations to the Quantum Groups

Above we constructed the observable algebra from the foliation leading to the Kauffman bracket skein module. In the subsection we will discuss the relation to lattice gauge field theory. Main source for this discussion is the work of Bullock, Frohman and Kania-Bartoszyńska [35–37, 58]. In this paper the authors realize that gauge fields come from the restricted dual of the Hopf algebra on which the theory is based. This leads to a coordinate free formulation. Then they comultiply connections in a way that implies the usual exchange relations for fields while preserving their evaluability. Their new foundations allow them to compute Wilson loops and many other operators using a simple extension of tangle functors. Then they analyzed the structure of the algebra of observables. In their viewpoint, the observables correspond to quantum groups seen as rings of invariants of n-tuples of matrices under conjugation. The connection with lattice gauge field theory is that each n-tuple of matrices corresponds to a connection on a lattice with one vertex and n-edges, with the gauge fields based on a classical group. The construction given in this paper leads to an algebra of "characters" of a surface group with respect to any ribbon Hopf algebra. The algebras are interesting from many points of view: They generalize objects studied in invariant theory; they should provide tools for investigating the structure of the mapping class groups of surfaces; and they should give a way of understanding quantum invariants of 3-manifolds. The algebra of observables based on the enveloped Lie algebra $U(\mathfrak{g})$ is proved to be the ring of G-characters of the fundamental group of the associated surface. Then, given the ring of G-characters of a surface group, they showed that the observables based on the corresponding Drinfeld-Jimbo algebra form a quantization with respect to the usual Poisson structure. Furthermore they proved for the classical groups that the algebra of observables is generated by Wilson loops. Finally, invoking a quantized Cayley-Hamilton identity, they obtain a new proof, that the $U_h(sl_2)$-characters of a surface are exactly the Kauffman bracket skein module of a cylinder over that surface. The power of lattice gauge field theory is that it places the representation theory of the underlying manifold and the quantum invariants in the same setting. Ultimately the asymptotic analysis of the quantum invariants of a 3-manifold in terms of the representations of its fundamental group should flow out of this setting. The identification of the representation theory of a quantum group with that of a compact Lie group leads to rigorous integral formulas for quantum invariants of 3-manifolds. This should in turn lead to a simple explication of the relationship between quantum invariants and more classical invariants of 3-manifolds.

This relation to lattice gauge field theory seems to imply an underlying discrete structure of the space and/or spacetime, but the approach in the paper uncovers the reason [35–37, 58]: the Kauffman bracket skein module is discrete structure containing only a finite amount of information. Therefore, any description has to be discrete as well including the approach via gauge fields. This idea can be extended to the 4-manifold. As explained above, every smooth 4-manifold can be effectively described by handles and one only needs a finite number to describe every compact 4-manifold. Then the handles can be simply triangulated by using simplices to end up with a piecewise-linear (or PL) structure. The surprising result of Cerf for manifolds of dimension smaller than 7 was simple: PL-structure (or triangulations) and smoothness structure are the same. This implies that every PL-structure can be smoothed to a smoothness structure and vice versa. Therefore *the discrete approach (via triangulations) and the smooth approach to defining a manifold are the same!* So, our spacetime admits a kind of duality: it contains discrete information in its handle structure but it is a continuous space at the same time. Both approaches are interchangeable. Therefore the underlying structure of the spacetime is discrete but the spacetime itself is a smooth 4-manifold. Or, the information contained in a smooth 4-manifold is finite.

15.9.3 Dimensional Reduction and Exotic Smooth Black Holes

In [9] we described an exotic black hole by constructing a smooth metric for the interior. Here we will present the main argument shortly.

In [31] the existence of an exotic Black hole (as exotic Kruskal space) using an exotic $\mathbb{R}^4$ was suggested. The idea was simply to consider the complement $\mathbb{R}^4 \backslash (D^3 \times \mathbb{R}) = S^2 \times \mathbb{R}^2$ where $\times$ was only understood topologically. In case of the exotic small

$\mathbb{R}^4$ given by a Casson handle, we can reproduce our construction of an exotic $S^2 \times \mathbb{R}^2$ by using a Casson handle. Therefore we will here concentrate on the representation of the exotic $S^2 \times \mathbb{R}^2$ by using the Casson handle CH to get

$$S^2 \times_\Theta \mathbb{R}^2 = D^2 \cup_{\partial CH} CH.$$

In [74] the analytical properties of the Casson handle were discussed. The main idea is the usage of the theory of end-periodic manifolds, i.e. an infinite periodic structure generated by W glued along a compact set K to get

$$S^2 \times_\Theta \mathbb{R}^2 = K \cup_N W \cup_N W \cup_N \cdots$$

the end-periodic manifold. The definition of an end-periodic manifold is very formal (see [93]) and we omit it here. All Casson handles generated by a balanced tree have the structure of end-periodic manifolds as shown in [74]. By using the theory of Taubes [93] one can construct a metric on $\cdots \cup_N W \cup_N W \cup_N \cdots$ by using the metric on W. Then a metric g in $S^2 \times_\Theta \mathbb{R}^2$ transforms to a periodic function $\hat{g}$ on the infinite periodic manifold

$$\tilde{Y} = \cdots \cup_N W_{-1} \cup_N W_0 \cup_N W_1 \cup_N \cdots$$

where W_i is the building block W at the i-th place. To reflect the number of the building block, we have to extend $\hat{g}$ to $Y \times \mathbb{C}^*$ by using a metric $\hat{g}_z$ holomorphic in $z \in \mathbb{C}^* = S^1$ with $Y = W/i$ where i identifies the two boundaries of W. From the formal point of view we have the *generalized Fourier-Laplace transform (or Fourier-Laplace transform for short)*

$$\hat{g}_z(.) = \sum_{n=0}^{\infty} a_n z^n \cdot \hat{g}(.) \tag{15.19}$$

where the coefficient a_n represents the building block W_n in $\tilde{Y}$. Without loss of generality we can choose the coordinates x in M so that the 0-th component x_0 is related to the integer $n = [x_0]$ via its integer part [ ]. Using the inverse transformation we can construct a smooth metric g in $\tilde{Y}$ at the n-th building block via

$$(\tilde{T}^n g)(x) = \frac{1}{2\pi i} \int_{|z|=s} z^{-n} \hat{g}_z(\pi(x)) \frac{dz}{z} \tag{15.20}$$

for $x \in \tilde{Y}$, $s \in (0, \infty)$, $n = [x_0]$ with the projection $\pi : \tilde{Y} \to Y$ (mathematically: $\tilde{Y}$ is the universal cover of Y like $\mathbb{R}$ is the universal cover of S^1). In the case of the Kruskal space we have the metric

$$ds^2 = \left(\frac{2M^3}{r}\right) \exp\left(-\frac{r}{2M}\right) \left(-dv^2 + du^2\right) + r^2 \left(d\theta^2 + \sin^2\theta d\phi^2\right) \quad (15.21)$$

in the usual units with a singularity at $r = 0$ used for the whole space $S^2 \times \mathbb{R}^2$. The coordinates (u, v) together with the relation

$$u^2 - v^2 = \left(\frac{r}{2M} - 1\right) \exp\left(-\frac{r}{2M}\right) \quad (15.22)$$

represent $\mathbb{R}^2$ and the angles (θ, ϕ) the 2-sphere S^2 parametrized by the radius r. Clearly this metric can be also used for each building block W having the topological structure $D^3 \times S^1 = W$ with two attaching regions topologically given by $D^2 \times S^1 = N$ forming the boundary $S^2 \times S^1 = \partial(D^3 \times S^1)$ (see the description of a Casson handle above). Remember that the Casson handle is topologically the subset $D^2 \times \mathbb{R}^2 \subset S^2 \times \mathbb{R}^2$. Now we consider the decomposition $W = D^2 \times (D^1 \times S^1)$ and the part $D^1 \times S^1$ will be later the $\mathbb{R}^2$ part of the Casson handle. The size of the D^2 is parametrized by r as above. Then we obtain the metric (15.21) for the building block W.

Our model of the black hole based on the implicit dependence of the two coordinates (u, v) on the parameter r, the radius of the 2-sphere. Therefore we choose for the coordinate $z \in \mathbb{C}^*$ the relation $z = \exp(ir)$ and obtain a metric $\hat{g}$ on $Y \times \mathbb{C}^*$. So, we make the assumptions:

1. The coordinate z is related to the radius by $z = \exp(ir)$.
2. Only the (u, v) part of the metric is periodic and we do not change the other component $r^2 \left(d\theta^2 + \sin^2\theta d\phi^2\right)$ of the metric.
3. The integer part $n = [v]$ of the coordinate v gives the number of the building block W_n in the Casson handle (seen as end-periodic manifold).
4. The metric on $S^2 \times_\theta \mathbb{R}^2$ is given by a Fourier transformation (15.20) of the (u, v) part of the metric in the building block W.

Some more comments are in order. The number $n = [v]$ is related to the coordinate v as substitute of "time". The metric g in $\tilde{Y}$ is smooth with respect to v and we obtain the number of the building block by $n = [v]$. To express this property we have to identify (u, v) with the coordinates of $D^1 \times S^1$. Then we obtain the metric on $S^2 \times_\theta \mathbb{R}^2$ by the generalized Fourier-Laplace transformation of the metric on $Y = W/i$ using the metric of the building block W and the coordinate z similar to (15.20)

$$g(v, u, \theta, \phi) = \int \exp(irv) \, \hat{g}_r(v, u, \theta, \phi) dr$$

Especially the singular part of the metric (i.e. the (u, v) part) on the building block W

$$(\hat{g}_r)_{00} = \left(\frac{2M^3}{r}\right) \exp\left(-\frac{r}{2M}\right) = (\hat{g}_r)_{11}$$

Fig. 15.8 Hyperbolic triangle with increasing curvature (from *left* to *right*), the tree is the limit and a dimensional reduction 2D to 1D

transforms to the Heaviside jump function

$$g_{00} = g_{11} = 2M^3 \Theta (v^2 - u^2 - 1)$$

using the relation (15.22), having no singularity. The metric vanishes, however, for large values of v in the interior of the black hole. This sketch of some arguments gives a hint that the transformation of the smoothness to exotic smoothness could possibly smooth out some singularity in the black hole case. This metric vanishes along two directions or *one obtains a dimensional reduction from 4D to 2D*. But, is there a geometrical reason for this reduction? A hyperbolic 3-manifold M admits a hyperbolic structure by fixing a homomorphism $\pi_1(M) \to SL(2, \mathbb{C})$ (up to conjugation). From the physics point of view, this homomorphism is given by the holonomy along a closed curve (as element in $\pi_1(M)$) for a flat connection. A sequence of these holonomies does not converge but it is possible to compactify the space of flat $SL(2, \mathbb{C})$ connections. This limit can be understood geometrically: the hyperbolic 3-manifold is triangulated by tetrahedrons. However, because of the hyperbolic geometry, the edge between two vertices is not the usual line but rather a geodesics in the hyperbolic geometry. The curvature of this geodesics depends on the hyperbolic structure. In the limit, all geodesics of the tetrahedron meet and one obtains a tree instead of tetrahedrons. Therefore in the limit of large curvature, one obtains a reduction from 3D (=tetrahedrons) to 1D (=tree). Figure 15.8 visualizes the transition from 2D(triangle) to 1D(tree).

15.10 Where Do the Quantum Fluctuations Come From?

In a purely geometrical theory, one has to answer this question. It cannot be shifted to assume the appearance of quantum fluctuations. Instead we have to understand the root of these quantum fluctuations. Starting point of our approach is the foliation of the exotic R^4 by using the Anosov flow. Main point in the argumentation above is the appearance of the hyperbolic geometry in 3- and 4-dimensional submanifolds. The foliation can, however, be interpreted differently: a foliation defines a dynamics at a manifold leading to a splitting into leafs (the integral curves of the dynamics). Therefore, a tiny variation in the initial conditions will lead to a strong variation of the corresponding integral curve. This chaotic behavior is a natural consequence of the exotic smoothness structure (leading to the non-trivial $PSL(2, \mathbb{R})$-foliation). For completeness we will describe this dynamics, called the Anosov flow. For that purpose we consider the standard basis

$$J = \frac{1}{2}\begin{pmatrix} 1 & 0 \\ 0 & -1 \end{pmatrix}, \ X = \begin{pmatrix} 0 & 1 \\ 0 & 0 \end{pmatrix}, \ Y = \begin{pmatrix} 0 & 0 \\ 1 & 0 \end{pmatrix} \tag{15.23}$$

of the Lie algebra $sl(2, \mathbb{R})$ with

$$[J, X] = X \qquad [J, Y] = -Y \qquad [X, Y] = 2J$$

leading to the exponential maps

$$g_t = \exp(tJ) = \begin{pmatrix} e^{t/2} & 0 \\ 0 & e^{-t/2} \end{pmatrix} \ h_t^* = \exp(tX) = \begin{pmatrix} 1 & t \\ 0 & 1 \end{pmatrix} \ h_t = \exp(tY) = \begin{pmatrix} 1 & 0 \\ t & 1 \end{pmatrix}$$

defining right-invariant flows on the unit tangent bundle $T_1 \mathbb{H} = PSL(2, \mathbb{R})$ of the hyperbolic space. The connection to the Anosov flow comes from the realization that g_t is the geodesic flow on $P = T_1 \mathbb{H}$. With Lie vector fields being (by definition) left invariant under the action of a group element, one has that these fields are left invariant under the specific elements g_t of the geodesic flow. This flow goes over to a surface $M = \mathbb{H}/\Gamma$ defined by a subgroup $\Gamma \subset PSL(2, \mathbb{R})$ with $Q = T_1 M$. Now the geodesic flow g_t acts on the exponential maps g_s, h_t^*, h_t so that the geodesic flow itself is invariant, $g_s g_t = g_t g_s = g_{s+t}$, but the other two shrink and expand: $g_s h_t^* = h_{t \cdot \exp(-s)}^* g_s$ and $g_s h_t = h_{t \cdot \exp(s)} g_s$. Then the bundle TQ splits into three subbundles

$$TQ = E^+ \oplus E^0 \oplus E^-$$

where one bundle E^+ expands, one bundle E^- contracts and one bundle E^0 is invariant w.r.t. geodesic flow. This property is crucial for the following discussion. Because of the expanding behavior of one subbundle, the Anosov flow is the generator of a chaotic dynamics. Therefore, two geodesics diverge exponentially in this foliation, but this behavior goes over to the holonomies characterizing the geometry. The transport of a holonomy along two diverging geodesics can lead to totally different holonomies. Currently this dynamics is deterministic, i.e. if we choose exactly the same initial condition then we will end at the state (seen as limit point). This situation changes if we are unable to choose the initial condition exactly (by choosing real numbers) but instead we can only choose a rational number where this rational number is the characterizing property of the state. Then all initial conditions (represented by all real numbers) in this class represent the same state but have totally different limit points of the corresponding dynamics. Now we will describe this dynamics.

Starting point is the observable algebra $X(S, SL(2, \mathbb{C}))$, i.e. the space of holonomies $\pi_1(S) \to SL(2, \mathbb{C})$ (i.e. homomorphisms) up to conjugation, see Sect. 15.6.3.1. The deformation quantization (see Sect. 15.6.3.2) is the Kauffman bracket skein module. Here we made use of the identity

$$tr(A) \cdot tr(B) = tr(AB) + tr(AB^{-1})$$

between two elements $A, B \in SL(2, \mathbb{C})$ (w.r.t. a representation). Using the group commutator $[A, B] = ABA^{-1}B^{-1}$ one also obtains

$$2 + tr([A, B]) = (tr(A))^2 + (tr(B))^2 + (tr(AB))^2 - tr(A)tr(B)tr(AB)$$

According to deformation procedure, pairs of elements $A, B \in SL(2, \mathbb{C})$ coming from closed curves via the holonomy and fulfilling

$$tr([A, B]) = \pm 2$$

can be a canonical pair w.r.t. the symplectic structure. The sign is purely convention and we choose $tr([A, B]) = -2$. Then the canonical pair has to fulfill the equation

$$(tr(A))^2 + (tr(B))^2 + (tr(AB))^2 - tr(A)tr(B)tr(AB) = 0$$

which can be written in a more familiar form

$$x^2 + y^2 + z^2 - 3xyz = 0 \tag{15.24}$$

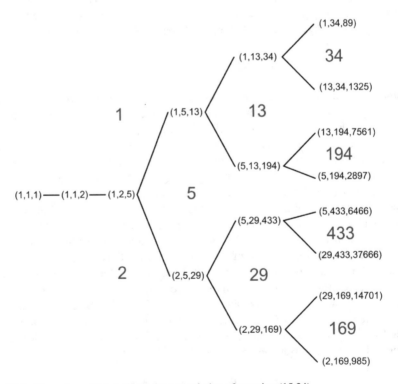

Fig. 15.9 Binary tree of Markoff numbers as solution of equation (15.24)

by using $3x = tr(A), 3y = tr(B), 3z = tr(AB)$. Because of the discreteness of $\pi_1(S)$, we have to look for rational solutions of this equation (Diophantine equation). The solutions of the equation are Markoff triples forming a binary tree (see Fig. 15.9).

The set of these elements A, B corresponding to discrete groups is known to be fractal in nature [27]. It is the large class of quasi-Fuchsian groups having a fractal curve (Julia set) as limit set. Then we have the desired behavior:

The set of canonical pairs (as measurable variables in the theory) forms a fractal subset of the space of all holonomies. Then we can only determine the initial condition up to discrete value (given by the canonical pair) and the chaotic behavior of the foliation (i.e. the Anosov flow) makes the limit not predictable.

At the end of this section, one remark about the role of the canonical pair. It is always possible to construct a classical continuous random field that has the same probability density as the quantum vacuum state. Furthermore it is known that a random field can be generated by a chaotic dynamics. There is, however, a large difference between the classical random field and a quantum field: there are pairs of not equally accurate measurable observables (mostly the canonical pairs) for quantum fields impossible for the classical random fields. With our approach, we showed the same behavior for the canonical pairs.

15.11 Decoherence, Entanglement and Measurement

Our geometrical approach should also lead to a description of the measurement process (including the collapse of the wave function). In Sect. 15.6, we constructed the geometrical expression for a quantum state given by a wild embedding (the wild S^3). The reduction of the quantum state (as linear combination) to an eigenstate (or the collapse of the wave function) is equivalent to a reduction of the wild embedding to a tame embedding. Therefore we need a mechanism to reduce the wild embedding to a tame one. The construction of the wild S^3 is strongly related to the Casson handle. The exoticness of the smooth structure of R^4 and the wildness of the S^3 depend both on the self-intersection of some disk. If we are able to remove these self-intersections then we will obtain the desired reduction. According to the discussion in Sect. 15.3.2, one needs a Casson handle for the cancellation. How many levels of the Casson handle are needed to cancel the self-intersection? This question was answered by Freedman [54]: one needs three levels (a three-level Casson tower)! At the same time, however, one produces more self-intersections in the higher levels. Therefore one needs a little bit more: a Casson tower where a complete Casson handle can be embedded. Then this Casson handle is able to cancel the self-intersection and we will obtain a tame embedding or a classical state. As shown by Freedman [53] and Gompf/Singh [67], one needs a 5-stage Casson tower so that a Casson handle with the same attaching circle can be embedded into this 5-stage tower. We obtain a process which is "the collapse of the wave function". What is the cause of this collapse? As explained above, we cannot choose a single disk to remove the self-intersections. Instead we have to choose a Casson tower where each stage is a boundary-connected sum of $S^1 \times D^3$, i.e. its boundary is the sum $S^1 \times S^2 \# S^1 \times S^2 \# \cdots$ where the number of components is equal to the number of self-intersections. So, every piece $S^1 \times S^2$ of the boundary is given by the identification of the two boundary components for $S^2 \times [0, 1]$. In Sect. 15.5, we identified this 3-manifold with the graviton or *the collapse of the wave function is caused by a gravitational interaction*. The corresponding process is known as decoherence. In the following we will calculate the minimal decoherence time for the gravitational interaction. The 5-stage Casson tower can be also understood as a cobordism between the 3-manifold

$$\Sigma_0 = S^1 \times S^2$$

(the S^1 defines the attaching circle) and a 3-manifold having the same homology. In case of the simplest Casson tower, it is given by five complements of the Whitehead link $C(Wh)$ closed by two solid tori, i.e.

$$\Sigma_1 = (D^2 \times S^1) \cup C(Wh) \cup C(Wh) \cup C(Wh) \cup C(Wh) \cup C(Wh) \cup (D^2 \times S^1)$$

and this manifold can be very complicated for more complex towers. Now we will add some geometry to calculate the decoherence time. As shown by Witten [103–105], the action

$$\int_{\Sigma_{0,1}} {}^3R\sqrt{h}\,d^3x = L \cdot CS(\Sigma_{0,1}) \tag{15.25}$$

for every 3-manifold (in particular for Σ_0 and Σ_1 denoted by $\Sigma_{0,1}$) is related to the Chern-Simons action $CS(\Sigma_{0,1})$. The scaling factor L is related to the volume by $L = \sqrt[3]{vol(\Sigma_{0,1})}$ and we obtain formally

$$L \cdot CS(\Sigma_{0,1}, A) = L^3 \cdot \frac{CS(\Sigma_{0,1})}{L^2} = \int_{\Sigma_{0,1}} \frac{CS(\Sigma_{0,1})}{L^2 \cdot vol(\Sigma_{0,1})}\sqrt{h}\,d^3x \tag{15.26}$$

by using

$$L^3 \cdot vol(\Sigma_{0,1}) = \int_{\Sigma_{0,1}} \sqrt{h}\,d^3x$$

with the (unit) volume $vol(\Sigma_{0,1})$. If $\Sigma_{0,1}$ is a hyperbolic 3-manifold then the (unit) volume is a topological invariant which cannot be normalized to 1. Together with

$$^3R = \frac{3k}{a^2}$$

one can compare the kernels of the integrals of (15.25) and (15.26) to get for a fixed time

$$\frac{3k}{a^2} = \frac{CS(\Sigma_{0,1})}{L^2 \cdot vol(\Sigma_{0,1})} .$$

This gives the scaling factor

$$\vartheta = \frac{a^2}{L^2} = \frac{3 \cdot vol(\Sigma_{0,1})}{CS(\Sigma_{0,1})} \tag{15.27}$$

where we set $k = 1$ in the following. The hyperbolic geometry of the cobordism is best expressed by the metric

$$ds^2 = dt^2 - a(t)^2 h_{ik} dx^i dx^k \tag{15.28}$$

also called the Friedmann-Robertson-Walker metric (FRW metric) with the scaling function $a(t)$ for the (spatial) 3-manifold. Mostow rigidity enforces us to choose

$$\left(\frac{\dot{a}}{a}\right)^2 = \frac{1}{L^2}$$

in the length scale L of the hyperbolic structure. In the following we will switch to quadratic expressions because we will determine the expectation value of the area. A second reason for the consideration of quadratic expressions is again the hyperbolic structure of $\mathbb{H}^2$. We needed this structure for the construction of the foliation which is given by a polygon in $\mathbb{H}^2$. This polygon defines a compact surface of genus $g > 1$. Then the foliation of the polygon induces a foliation of the small exotic R^4. The area of the polygon is mainly the Godbillon-Vey invariant of the foliation. It is known that foliations of surfaces are given by quadratic differentials of the form defined below. Here, there are deep connections to trees and $SL(2, \mathbb{C})$ flat connections, i.e. a tree defines a quadratic differential and vice versa [45, 71, 90, 106].

Using the previous equation, we obtain

$$da^2 = \frac{a^2}{L^2} dt^2 = \vartheta \, dt^2 \tag{15.29}$$

with respect to the scale ϑ. By using the tree of the Casson handle, we obtain a countable infinite sum of contributions for (15.29). Before we start we will clarify the geometry of the Casson handle. A Casson handle admits a hyperbolic geometry. Therefore the tree corresponding to the Casson handle must be interpreted as a metric tree with hyperbolic structure in $\mathbb{H}^2$ and metric $ds^2 = (dx^2 + dy^2)/y^2$. The embedding of the Casson handle in the cobordism is given by the rules

1. The direction of the increasing levels $n \to n+1$ is identified with dy^2 and dx^2 is the number of edges for a fixed level with scaling parameter ϑ.
2. The contribution of every level in the tree is determined by the previous level best expressed in the scaling parameter ϑ.
3. An immersed disk at level n needs at least one disk to resolve the self-intersection point. This disk forms the level $n+1$ but this disk is connected to the previous disk. So we obtain for $da^2|_{n+1}$ at level $n+1$

$$da^2|_{n+1} \sim \vartheta \cdot da^2|_n$$

up to a constant.

By using the metric $ds^2 = (dx^2 + dy^2)/y^2$ with the embedding ($y^2 \to n+1$, $dx^2 \to \vartheta$) we obtain for the change dx^2/y^2 along the x-direction (i.e. for a fixed y) $\frac{\vartheta}{n+1}$. This change determines the scaling from the level n to $n+1$, i.e.

$$da^2|_{n+1} = \frac{\vartheta}{n+1} \cdot da^2|_n = \frac{\vartheta^{n+1}}{(n+1)!} \cdot da^2|_0$$

and after the whole summation (as substitute for an integral in case of discrete values) we obtain for the relative scaling

$$a^2 = \sum_{n=0}^{\infty} (da^2|_n) = a_0^2 \cdot \sum_{n=0}^{\infty} \frac{1}{n!} \vartheta^n = a_0^2 \cdot \exp(\vartheta) = a_0^2 \cdot l_{scale} \tag{15.30}$$

with $da^2|_0 = a_0^2$. The Chern-Simons invariant for Σ_0 vanishes and we are left with

$$CS(\Sigma_1) = \frac{5}{8}$$

and the complements $C(Wh)$ are hyperbolic 3-manifolds with

$$vol(\Sigma_1) = 5 \cdot vol(C(Wh)) \approx 18.31931...$$

by using the software Snapea. Finally for the scaling we obtain

$$\vartheta \approx 87.932688...$$

and for the time we have to choose

$$T_{decoherence} = T_0 \cdot \exp\left(\frac{\vartheta}{2}\right)$$

using the well-known relation $a_0 = cT_0$ between length and time, i.e. we see one coordinate along the Casson handle as time axis. The time T_0 has to be identified

Fig. 15.10 Two disjoint circles get linked after the application of the skein relation for the area marked by the *small circles*

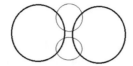

with the Planck time $T_0 \approx 10^{-43} s$ (see Sect. 15.5) so that

$$T_{decoherence} \approx 10^{-24} s$$

is the minimal decoherence time for the gravitational interaction.

Now we also discuss the entanglement which has to be also geometrically expressed. A quantum state is an element of the skein algebra $K_t(S)$ for $S \times [0, 1]$. For two disjoint surfaces $S_0 \sqcup S_1$ one has

$$K_t(S_0 \sqcup S_1) = K_t(S_0) \otimes K_t(S_1).$$

Now let us choose a knot $|0\rangle$ in $S_0 \times [0, 1]$ as element of $K_t(S_0)$ as well a knot $|1\rangle$ in $K_t(S_1)$. Then $|0\rangle \otimes |1\rangle$ is an element of $K_t(S_0 \sqcup S_1)$. Furthermore we can assume that the knots $|1\rangle$ and $|0\rangle$ can be also an element of $K_t(S_0)$ and $K_t(S_1)$, respectively. Then the element

$$|0\rangle \otimes |1\rangle + |1\rangle \otimes |0\rangle$$

exists but now as an element of $K_t(S)$ with $S_0 \sqcup S_1 \subset S$. Using the skein relations in $K_t(S)$, see Fig. 15.7, we obtain a linking between the corresponding knots, i.e. $|0\rangle$ and $|1\rangle$ forming a link. Figure 15.10 visualizes the transition from disjoint circles (=disjoint states) to linked circles (=entangled states). Then entanglement is reduced to a linking!

Next we have to think about the measurement which reduces the entangled state to one product state. Here we will only present some rough ideas for the description of the measurement process, but at first we have to define a measurement device. In this proposal, it is a union of Casson handles which can be used to unlink two linked components. At the level of skein algebras, the Casson handle is also given by elements of a skein algebra (given by closed, knotted curves at the levels). The particular structure of the Casson handle is not determined (see also Sect. 15.10). Now a given quantum state is linked to this Casson handle. The limit point of the Casson handle (i.e. the leafs of the tree) give the result of the unlinking. All limit points of the Casson handle have a fractal structure (a Cantor set) expressing our inability to know the outcome of the measurement. The tree structure of the Casson handle has also another effect: the limit points are exponentially separated from each other and can be seen as classical states. With these speculations, we will close this section.

15.12   Some Implications for Cosmology

In the last section we will collect some implications for a cosmological model. Let us assume the topology $S^3 \times \mathbb{R}$ for the spacetime but with an exotic smoothness structure $S^3 \times_\theta \mathbb{R}$. One can construct this spacetime from the exotic R^4 by $R^4 \setminus D^4 = S^3 \times_\theta \mathbb{R}$. From previous work, we know:

- Cosmological anomalies like dark matter and dark energy are (conjecturally) rooted in exotic smoothness [7].
- The initial state of the cosmos must be a wild 3-sphere representing a quantum state [16].
- Then there is an inflationary phase [18] driven by a decoherence which can be described by the Starobinsky model.
  In this model, we have a topological transition from a 3-manifold Σ_0 to another 3-manifold Σ_1. Both 3-manifolds are homology 3-spheres. Therefore let us describe this change (a so-called homology cobordism) between two homology 3-spheres Σ_0 and Σ_1. The situation can be described by a diagram

$$\begin{array}{ccc} \Sigma_1 & \xrightarrow{\Psi} & \mathbb{R} \\ \phi\downarrow & \circlearrowleft & \updownarrow id \\ \Sigma_0 & \xrightarrow{\psi} & \mathbb{R} \end{array} \qquad (15.31)$$

which commutes. The two functions ψ and Ψ are the Morse function of Σ_0 and Σ_1, respectively, with $\Psi = \psi \circ \phi$. The Morse function over $\Sigma_{0,1}$ is a function $\Sigma_{0,1} \to \mathbb{R}$ having only isolated, non-degenerated, critical points (i.e. with vanishing first derivatives at these points). A homology 3-sphere has two critical points (located at the two poles). The Morse function looks like $\pm ||x||^2$ at these critical points. The transition $y = \phi(x)$ represented by the (homology) cobordism $M(\Sigma_0, \Sigma_1)$ maps the Morse function $\psi(y) = ||y||^2$ on Σ_0 to the Morse function $\Psi(x) = ||\phi(x)||^2$ on Σ_1. The function $-||\phi||^2$ represents also the critical point of the cobordism $M(\Sigma_0, \Sigma_1)$. As we learned above, this cobordism has a hyperbolic geometry and we have to interpret the function $||\phi(x)||^2$ not as an Euclidean form but change it to the hyperbolic geometry so that

$$-||\phi||^2 = -\left(\phi_1^2 + \phi_2^2 + \phi_3^2\right) \to -e^{-2\phi_1}(1 + \phi_2^2 + \phi_3^2)$$

i.e. we have a preferred direction represented by a single scalar field $\phi_1 : \Sigma_1 \to \mathbb{R}$. Therefore, the transition $\Sigma_0 \to \Sigma_1$ is represented by a single scalar field $\phi_1 : \Sigma_1 \to \mathbb{R}$ and we identify this field as the moduli. Finally we interpret this Morse function in the interior of the cobordism $M(\Sigma_0, \Sigma_1)$ as the potential (shifted away from the point 0) of the scalar field ϕ with Lagrangian

$$L = R + (\partial_\mu \phi)^2 - \frac{\rho}{2}(1 - \exp(-\lambda\phi))^2$$

with two free constants ρ and λ. For the value $\lambda = \sqrt{2/3}$ and $\rho = 3M^2$ we obtain the Starobinski model [89] (by a conformal transformation using ϕ and a redefinition of the scalar field [100])

$$L = R + \frac{1}{6M^2} R^2 \qquad (15.32)$$

with the mass scale $M \ll M_P$ much smaller than the Planck mass. From our discussion above, the appearance of this model is not totally surprising. It favors a surface to be incompressible (which is compatible with the properties of hyperbolic manifolds).

- This inflationary phase is followed by another exponentially increasing phase leading to a hyperbolic 4-manifold with constant curvature which is rigid by Mostow rigidity [18]. Here, we obtained the global geometry of the spacetime: it is a de Sitter space $SO(4, 1)/SO(3, 1)$ with a cosmological constant which is the curvature of the spacetime.
- This constant curvature can be identified with the cosmological constant in good agreement with the Planck satellite results [18]. The cosmological constant is constant by Mostow rigidity (but now for the 4-manifold).
- The topology of the spatial component (seen as 3-manifold) is strongly restricted [15] by the smoothness of the spacetime.
- The inclusion of matter can be done naturally as direct consequence of exotic smoothness [10].
- The interior of black holes can be described by exotic smoothness where the singularity is smoothed out [9].

15.13   Conclusion and Open Questions

Smooth Quantum Gravity, the usage of exotic smoothness structures on 4-manifolds, are the attempt to obtain a consistent theory of quantum gravity without any further assumptions. For us, the change of the smoothness structure is the next step in extending General Relativity, where non-Euclidean geometry was used to describe gravity and all accelerations. Then, two different smoothness structures represent two different physical systems. In particular I think that the standard smoothness structure represents the case of a spacetime without matter and non-gravitational fields. In this paper we are going a more radical way to construct a quantum theory without quantization but by using purely geometrical ideas from mathematical topics like differential and geometric topology. The flow of ideas can be simply described by the following points:

- An exotic $\mathbb{R}^4$ is given by an infinite handlebody (so one needs infinitely many charts) and one finds also the description by an infinite sequence of 3-manifolds together with 4-dimensional cobordisms connecting them.

- Every 3-manifold admits a codimension-one foliation which goes over to the 4-dimensional cobordisms. The leaf space of this foliation is an operator algebra with a strong connection to algebraic quantum field theory.
- The states (as linear functionals in the algebra) depend on knotted curves and are elements of the Kauffman bracket skein algebra. The reconstruction of the spatial space gives a wild embedded 3-sphere which is therefore related to the state, or the quantum state can be identified with the wild embedding. The classical state is a tame (i.e. usual) embedding where the deformation quantization of a tame embedding is a wild embedding.
- The structure of the operator algebra can be analyzed by the Tomita-Takesaki modular theory. Then it is possible to construct the quantum action by using the quantized calculus of Connes.
- For large scales, one gets the Einstein-Hilbert action. Whereas for small scales, one obtains a dimensional reduced action.
- The foliation is given by a hyperbolic dynamics having a chaotic behavior. For our states, one gets an unpredictable behavior so that the dynamics can generate the quantum fluctuations.

This list shows the current state but there are many open points, where we list only the most important here:

- What is the Hamiltonian of the theory? In principle we constructed this operator but have a problem connecting to Loop quantum gravity.
- What are the states seen as knots? The states are knots but the skein and Mandelstam identities give a class of knots: the states are conjecturally the concordance class of knots.
- Is the state a solution of the Hamiltonian? Here we conjecture that the concordance class of the knot lies already in the kernel of the Hamiltonian (therefore it is a solution of the Hamiltonian constraint).

A lot is done but there are also many open problems.

Happy Birthday Carl!

Acknowledgments I have to thank Carl for 20 years of friendship and collaboration as well numerous discussions. Special thanks to Jerzy Król for our work and many discussions about fundamental problems in math and physics. Now I understand the importance of Model theory. Many thanks to Paul Schultz for reading and corrections.

Appendix A Casson Handles and Labeled Trees

Let us now consider the basic construction of the Casson handle CH. Let M be a smooth, compact, simply-connected 4-manifold and $f : D^2 \to M$ a (codimension-2) mapping. By using diffeomorphisms of D^2 and M, one can deform the mapping f to get an immersion (i.e. injective differential) generically with only double points (i.e. $\#|f^{-1}(f(x))| = 2$) as singularities [62]. But to incorporate the generic location of the disk, one is rather interesting in the mapping of a 2-handle $D^2 \times D^2$ induced by $f \times id : D^2 \times D^2 \to M$ from f. Then every double point (or self-intersection) of $f(D^2)$ leads to self-plumbings of the 2-handle $D^2 \times D^2$. A self-plumbing is an identification of $D_0^2 \times D^2$ with $D_1^2 \times D^2$ where $D_0^2, D_1^2 \subset D^2$ are disjoint sub-disks of the first factor disk. In complex coordinates the plumbing may be written as $(z, w) \mapsto (w, z)$ or $(z, w) \mapsto (\bar{w}, \bar{z})$ creating either a positive or negative (respectively) double point on the disk $D^2 \times 0$. Consider the pair $(D^2 \times D^2, \partial D^2 \times D^2)$ and produce finitely many self-plumbings away from the attaching region $\partial D^2 \times D^2$ to get a kinky handle $(k, \partial^- k)$ where $\partial^- k$ denotes the attaching region of the kinky handle. A kinky handle $(k, \partial^- k)$ is a one-stage tower $(T_1, \partial^- T_1)$ and an $(n + 1)$-stage tower $(T_{n+1}, \partial^- T_{n+1})$ is an n-stage tower union of kinky handles $\bigcup_{\ell=1}^n (T_\ell, \partial^- T_\ell)$ where two towers are attached along $\partial^- T_\ell$. Let T_n^- be (interior T_n) $\cup \partial^- T_n$ and the Casson handle

$$CH = \bigcup_{\ell=0} T_\ell^-$$

is the union of towers (with direct limit topology induced from the inclusions $T_n \hookrightarrow T_{n+1}$). A Casson handle is specified up to (orientation preserving) diffeomorphism (of pairs) by a labeled finitely-branching tree with base-point *, having all edge paths infinitely extendable away from *. Each edge should be given a label $+$ or $-$ and each vertex corresponds to a kinky handle; the self-plumbing number of that kinky handle equals the number of branches leaving the vertex. The sign on each branch corresponds to the sign of the associated self plumbing. The whole process generates a tree with infinite many levels. In principle, every tree with a finite number of branches per level realizes a corresponding Casson handle. The simplest non-trivial Casson handle is represented by the tree $Tree_+$: each level has one branching point with positive sign $+$. The reverse construction of a Casson handle CH_T by using a labeled tree T can be found in the appendix A. Let T_1 and T_2 be two trees with $T_1 \subset T_2$ (it is the subtree) then $CH_{T_2} \subset CH_{T_1}$. Given a labeled based tree Q, let us describe a subset U_Q of $D^2 \times D^2$. Now we will construct a $(U_Q, \partial D^2 \times D^2)$ which is diffeomorphic to the Casson handle associated to Q. In $D^2 \times D^2$ embed a ramified Whitehead link with one Whitehead link component for every edge labeled by $+$ leaving * and one mirror image Whitehead link component for every edge labeled by $-$(minus) leaving *. Corresponding to each first level node of Q we have already found a (normally framed) solid torus embedded in $D^2 \times \partial D^2$. In each of these solid tori embed a ramified Whitehead link, ramified according to the number of $+$ and $-$ labeled branches leaving that node. We can do that process for every level of Q. Let the disjoint union of the (closed) solid tori in the n-th family (one solid torus for each branch at level n in Q) be denoted by X_n. Q tells us how to construct an infinite chain of inclusions:

$$\cdots \subset X_{n+1} \subset X_n \subset X_{n-1} \subset \cdots \subset X_1 \subset D^2 \times \partial D^2$$

and we define the Whitehead decomposition $WhC_Q = \bigcap_{n=1}^\infty X_n$ of Q. WhC_Q is the Whitehead continuum [99] for the simplest unbranched tree. We define U_Q to be

$$U_Q = D^2 \times D^2 \setminus (D^2 \times \partial D^2 \cup \text{closure}(WhC_Q))$$

alternatively one can also write

$$U_Q = D^2 \times D^2 \setminus \text{cone}(WhC_Q) \qquad (15.33)$$

where cone() is the cone of a space

$$cone(A) = A \times [0, 1]/(x, 0) \sim (x', 0) \qquad \forall x, x' \in A$$

over the point $(0, 0) \in D^2 \times D^2$. As Freedman (see [53] Theorem 2.2) showed U_Q is diffeomorphic to the Casson handle CH_Q given by the tree Q.

Appendix B Thurston Foliation of a 3-Manifold

In [94] Thurston constructed a foliation of the 3-sphere S^3 which depends on a polygon P in the hyperbolic plane $\mathbb{H}^2$ so that two foliations are non-cobordant if the corresponding polygons have different areas. For later usage, we will present the main ideas of this construction only (see also the book [92] Chap. VIII for the details). Starting point is the hyperbolic plane $\mathbb{H}^2$ with a convex polygon $K \subset \mathbb{H}^2$ having k sides $s_1, \ldots, s_k$. Assuming the upper half plane model of $\mathbb{H}^2$ then the sides are circular arcs. The construction of the foliation depends mainly on the isometry group $PSL(2, \mathbb{R})$ of $\mathbb{H}^2$ realized as rational transformations (and this group can be lifted to $SL(2, \mathbb{R})$). The followings steps are needed in the construction:

1. The polygon K is doubled along one side, say s_1, to get a polygon K'. The sides are identified by (isometric) transformations $s_i \to s_i'$ (as elements of $SL(2, \mathbb{R})$).
2. Take ϵ-neighborhoods $U_\epsilon(p_i)$, $U_\epsilon(p_i')$ with $\epsilon > 0$ sufficient small and set

$$V^2 = \left(K \cup K'\right) \setminus \bigcup_{i=1}^{k} \left(U_\epsilon(p_i) \cup U_\epsilon(p_i')\right)$$

$$= S^2 \setminus \bigcup_{i=1}^{k} D_i^2$$

having the topology of $V^2 = S^2 \setminus \{k \text{ punctures}\}$ and we set $P = K \cup K'$.
3. Now consider the unit tangent bundle $U\mathbb{H}^2$, i.e. a S^1-bundle over $\mathbb{H}^2$ (or the tangent bundle where every vector has norm one). The restricted bundle over V^2 is trivial so that $UV^2 = V^2 \times S^1$. Let L, L' be circular arcs (geodesics) in $\mathbb{H}^2$ (invariant w.r.t. $SL(2, \mathbb{R})$) starting at a common point which define parallel tangent vectors w.r.t. the metrics of the upper half plane model. The foliation of V^2 is given by geodesics transverse to the boundary and we obtain a foliation of $V^2 \times S^1$ (as unit tangent bundle). This foliation is given by a $SL(2, \mathbb{R})$-invariant smooth

1-form ω (so that $\omega = const.$defines the leaves) which is integrable $d\omega \wedge \omega = 0$. ($SL(2, \mathbb{R})$-invariant Foliation $\mathcal{F}_{SL}$)
4. With the relation $D^2 = V^2 \cup D_1^2 \cup \cdots \cup D_{k-1}^2$, we obtain $D^2 \times S^1 = V^2 \times S^1 \cup$ $\left(D_1^2 \times S^1\right) \cup \cdots \cup \left(D_{k-1}^2 \times S^1\right)$ or the gluing of $k - 1$ solid tori to $V^2 \times S^1$ gives a solid tori. Every glued solid torus will be foliated by a Reeb foliation. Finally using $S^3 = (D^2 \times S^1) \cup (S^1 \times D^2)$ (the Heegard decomposition of the 3-sphere) again with a solid torus with Reeb foliation, we obtain a foliation on the 3-sphere.

The construction of this foliation $\mathcal{F}_{Thurston}$ (Thurston foliation) will be also work for any 3-manifold. Thurston [94] obtains for the Godbillon-Vey number

$$GV(V^2 \times S^1, \mathcal{F}_{SL}) = 4\pi \cdot vol(P) = 8\pi \cdot vol(K)$$

and

$$GV(S^3, \mathcal{F}_{Thurston}) = 4\pi \cdot Area(P) \qquad (15.34)$$

so that *any real number can be realized by a suitable foliation of this type.* Furthermore, two cobordant foliations have the same Godbillon-Vey number (but the reverse is in general wrong). Let $[1] \in H^3(S^3, \mathbb{R})$ be the dual of the fundamental class $[S^3]$ defined by the volume form, then the Godbillon-Vey class can be represented by

$$\Gamma_{\mathcal{F}_a} = 4\pi \cdot Area(P)[1] \qquad (15.35)$$

The Godbillon-Vey class is an element of the deRham cohomology $H^3(S^3, \mathbb{R})$. Now we will discuss the general case of a compact 3-manifold carrying a foliation of the same type like the 3-sphere above. The main idea of the construction is very simple and uses a general representation of all compact 3-manifolds by Dehn surgery. Here we will use an alternative representation of surgery by using the Dehn-Lickorish theorem ([81] Corollary 12.4 at p. 84). Let Σ be a compact 3-manifold without boundary. There is now a natural number $k \in \mathbb{N}$ so that any orientable 3-manifold can be obtained by cutting out k solid tori from the 3-sphere S^3 and then pasting them back in, but along different diffeomorphisms of their boundaries. Moreover, it can be assumed that all these solid tori in S^3 are unknotted. Then any 3-manifold Σ can be written as

$$\Sigma = \left(S^3 \backslash \left(\bigsqcup_{i=1}^{k} D_i^2 \times S^1\right)\right) \cup_{\phi_1} \left(D_1^2 \times S^1\right) \cup_{\phi_2} \cdots \cup_{\phi_k} \left(D_k^2 \times S^1\right)$$

where $\phi_i : \partial\left(S^3 \backslash \left(\bigsqcup_{i=1}^{k} D_i^2 \times S^1\right)\right) \to \partial D_i^2 \times S^1$ is the gluing map from each boundary component of $\left(S^3 \backslash \left(\bigsqcup_{i=1}^{k} D_i^2 \times S^1\right)\right)$ to the boundary of $\partial D_i^2 \times S^1$. This gluing map is a diffeomorphism of tori $T^2 \to T^2$ (where $T^2 = S^1 \times S^1$). The Dehn-Lickorish theorem describes all diffeomorphisms of a surface: Every diffeomorphism of a surface is the composition of Dehn twists and coordinate transformations (or small diffeomorphisms). The decomposition

$$S^3 = \left(V^2 \times S^1\right) \cup \left(D_1^2 \times S^1\right) \cup \cdots \cup \left(D_{k-1}^2 \times S^1\right) \cup \left(S^1 \times D_k^2\right) \quad (15.36)$$

of the 3-sphere can be used to get a decomposition of Σ by

$$\Sigma = \left(V^2 \times S^1\right) \cup_{\phi_1} \left(D_1^2 \times S^1\right) \cup_{\phi_2} \cdots \cup_{\phi_k} \left(D_k^2 \times S^1\right)$$

which will guide us to the construction of a foliation on Σ:

- Construct a foliation $\mathcal{F}_{\Sigma,SL}$ on $V^2 \times S^1$ using a polygon P (see above) and
- Glue in k Reeb foliations of the solid tori using the diffeomorphisms ϕ_i.

Finally we get a foliation $\mathcal{F}_{\Sigma,Thurston}$ on Σ. According to the rules above, we are able to calculate the Godbillon-Vey number

$$GV(\Sigma, \mathcal{F}_{\Sigma,Thurston}) = 4\pi \cdot vol(P)$$

Therefore for any foliation of S^3, we can construct a foliation on any compact 3-manifold Σ with the same Godbillon-Vey number. Both foliations $\mathcal{F}_{Thurston}$ and $\mathcal{F}_{\Sigma,Thurston}$ agree for the common submanifold $V^2 \times S^1$ or there is a foliated cobordism between $V^2 \times S^1 \subset \Sigma$ and $V^2 \times S^1 \subset S^3$. Of course, S^3 and Σ differ by the gluing of the solid tori but every solid torus carries a Reeb foliation which does not contribute to the Godbillon-Vey number.

Appendix C 3-Manifolds and Geometric Structures

A connected 3-manifold N is prime if it cannot be obtained as a connected sum of two manifolds $N_1 \# N_2$ neither of which is the 3-sphere S^3 (or, equivalently, neither of which is the homeomorphic to N). Examples are the 3-torus T^3 and $S^1 \times S^2$ but also the Poincare sphere. According to [77], any compact, oriented 3-manifold is the connected sum of a unique (up to homeomorphism) collection of prime 3-manifolds (prime decomposition). A subset of prime manifolds are the irreducible 3-manifolds. A connected 3-manifold is irreducible if every differentiable submanifold S homeomorphic to a sphere S^2 bounds a subset D (i.e. $\partial D = S$) which is homeomorphic to the closed ball D^3. The only prime but reducible 3-manifold is $S^1 \times S^2$. For the geometric properties (to meet Thurstons geometrization theorem) we need a finer decomposition induced by incompressible tori. A properly embedded connected surface $S \subset N$ is called 2-sided[5] if its normal bundle is trivial, and 1-sided if its normal bundle is nontrivial. A 2-sided connected surface S other than S^2 or D^2 is called incompressible if for each disk $D \subset N$ with $D \cap S = \partial D$ there is a disk $D' \subset S$ with $\partial D' = \partial D$. The boundary of a 3-manifold is an incompressible surface. Most importantly, the 3-sphere S^3, $S^2 \times S^1$ and the 3-manifolds S^3/Γ with $\Gamma \subset SO(4)$

---

[5]The 'sides' of S then correspond to the components of the complement of S in a tubular neighborhood $S \times [0, 1] \subset N$.

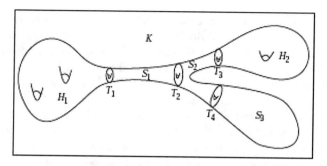

Fig. 15.11 Torus (JSJ-) decomposition, H_i hyperbolic manifold, S_i Graph-manifold, T_i Tori

a finite subgroup do not contain incompressible surfaces. The class of 3-manifolds S^3/Γ (the spherical 3-manifolds) include cases like the Poincare sphere ($\Gamma = I^*$ the binary icosaeder group) or lens spaces ($\Gamma = \mathbb{Z}_p$ the cyclic group). Let K_i be irreducible 3-manifolds containing incompressible surfaces then we can N split into pieces (along embedded S^2)

$$N = K_1 \# \cdots \# K_{n_1} \#_{n_2} S^1 \times S^2 \#_{n_3} S^3/\Gamma , \tag{15.37}$$

where $\#_n$ denotes the n-fold connected sum and $\Gamma \subset SO(4)$ is a finite subgroup. The decomposition of N is unique up to the order of the factors. The irreducible 3-manifolds $K_1, \ldots, K_{n_1}$ are able to contain incompressible tori and one can split K_i along the tori into simpler pieces $K = H \cup_{T^2} G$ [73] (called the JSJ decomposition). The two classes G and H are the graph manifold G and the hyperbolic 3-manifold H (see Fig. 15.11).

The hyperbolic 3-manifold H has a torus boundary $T^2 = \partial H$, i.e. H admits a hyperbolic structure in the interior only. In this paper we need the splitting of the link/knot complement. As shown in [33], the Whitehead double of a knot leads to JSJ decomposition of the complement into the knot complement and the complement of the Whitehead link (along one torus boundary of the Whitehead link complement).

One property of hyperbolic 3-manifolds is central: Mostow rigidity. As shown by Mostow [78], every hyperbolic n-manifold $n > 2$ with finite volume has this property: *Every diffeomorphism (especially every conformal transformation) of a hyperbolic n-manifold with finite volume is induced by an isometry.* Therefore one cannot scale a hyperbolic 3-manifold and the volume is a topological invariant. Together with the prime and JSJ decomposition

$$N = (H_1 \cup_{T^2} G_1) \# \cdots \# \left(H_{n_1} \cup_{T^2} G_{n_1}\right) \#_{n_2} S^1 \times S^2 \#_{n_3} S^3/\Gamma ,$$

we can discuss the geometric properties central to Thurstons geometrization theorem: *Every oriented closed prime 3-manifold can be cut along tori (JSJ decomposition), so that the interior of each of the resulting manifolds has a geometric structure with finite volume.* Now, we have to clarify the term geometric structure's. A model geometry is a simply connected smooth manifold X together with a transitive action of a Lie group G on X with compact stabilizers. A geometric structure on a manifold N is a diffeomorphism from N to X/Γ for some model geometry X, where Γ is a discrete subgroup of G acting freely on X. t is a surprising fact that there are also a finite number of three-dimensional model geometries, i.e. 8 geometries with the following models: spherical (S^3, $O_4(\mathbb{R})$), Euclidean ($\mathbb{E}^3$, $O_3(\mathbb{R}) \ltimes \mathbb{R}^3$), hyperbolic ($\mathbb{H}^3$, $O_{1,3}(\mathbb{R})^+$), mixed spherical-Euclidean ($S^2 \times \mathbb{R}$, $O_3(\mathbb{R}) \times \mathbb{R} \times \mathbb{Z}_2$), mixed hyperbolic-Euclidean ($\mathbb{H}^2 \times \mathbb{R}$, $O_{1,3}(\mathbb{R})^+ \times \mathbb{R} \times \mathbb{Z}_2$) and 3 exceptional cases called $\widetilde{SL}_2$ (twisted version of $\mathbb{H}^2 \times \mathbb{R}$), NIL (geometry of the Heisenberg group as twisted version of $\mathbb{E}^3$), SOL (split extension of $\mathbb{R}^2$ by $\mathbb{R}$, i.e. the Lie algebra of the group of isometries of 2-dimensional Minkowski space). We refer to [84] for the details.

References

1. H. Abchir, TQFT invariants at infinity for the Whitehead manifold, in *Knots in Hellas '98*, ed. by C. McA, V.F.R. Gordan, L. Jones, S. Lambropoulou Kauffman, J.H. Przytycki (World Scientific, Singapore, 1998), pp. 1–17
2. S. Akbulut, R. Kirby, Mazur manifolds. Mich. Math. J. **26**, 259–284 (1979)
3. A. Ashtekar, J. Engle, D. Sloan, Asymptotics and Hamiltonians in a first order formalism. Class. Quant. Grav. **25**, 095020 (2008), arXiv:0802.2527
4. A. Ashtekar, D. Sloan, Action and Hamiltonians in higher dimensional general relativity: first order framework. Class. Quant. Grav. **25**, 225025 (2008), arXiv:0808.2069
5. T. Asselmeyer, Generation of source terms in general relativity by differential structures. Class. Quant. Grav. **14**, 749–758 (1996)
6. T. Asselmeyer-Maluga, Exotic smoothness and quantum gravity. Class. Q. Grav. **27**, 165002 (2010), arXiv:1003.5506
7. T. Asselmeyer-Maluga, C.H. Brans, Cosmological anomalies and exotic smoothness structures. Gen. Rel. Grav. **34**, 1767–1771 (2002)
8. T. Asselmeyer-Maluga, C.H. Brans, *Exotic Smoothness and Physics* (World Scientific Publishing, Singapore, 2007)
9. T. Asselmeyer-Maluga, C.H. Brans, *Smoothly Exotic Black Holes*, Space Science, Exploration and Policies (NOVA publishers, 2012), pp. 139–156
10. T. Asselmeyer-Maluga, C.H. Brans, How to include fermions into general relativity by exotic smoothness. Gen. Relativ. Grav. **47**, 30 (2015), doi:10.1007/s10714-015-1872-x, arXiv:1502.02087
11. T. Asselmeyer-Maluga, J. Król, Exotic smooth $\mathbb{R}^4$, noncommutative algebras and quantization (2010), arXiv:1001.0882
12. T. Asselmeyer-Maluga, J. Król, Small exotic smooth R^4 and string theory, in *International Congress of Mathematicians ICM 2010 Short Communications Abstracts Book*, ed. by R. Bathia (Hindustan Book Agency, 2010), p. 400
13. T. Asselmeyer-Maluga, J. Król, Constructing a quantum field theory from spacetime (2011), arXiv:1107.3458
14. T. Asselmeyer-Maluga, J. Król, Topological quantum d-branes and wild embeddings from exotic smooth R^4. Int. J. Mod. Phys. A **26**, 3421–3437 (2011), arXiv:1105.1557
15. T. Asselmeyer-Maluga, J. Król, On topological restrictions of the spacetime in cosmology. Mod. Phys. Lett. A **27**, 1250135 (2012), arXiv:1206.4796
16. T. Asselmeyer-Maluga, J. Król, Quantum geometry and wild embeddings as quantum states. Int. J. Geom. Methods Modern Phys. **10**(10) (2013), will be published in Nov. 2013, arXiv:1211.3012

17. T. Asselmeyer-Maluga, J. Król, Abelian gerbes, generalized geometries and foliations of small exotic R^4, arXiv:0904.1276v5, subm. to Rev. Math. Phys. (2014)
18. T. Asselmeyer-Maluga, J. Król, Inflation and topological phase transition driven by exotic smoothness. Adv. HEP, Article ID 867460, 14p (2014), doi:10.1155/2014/867460
19. T. Asselmeyer-Maluga, R. Mader, Exotic R^4 and quantum field theory, in *7th International Conference on Quantum Theory and Symmetries (QTS7)*, ed. by C. Burdik et al. (IOP Publishing, Bristol, UK, 2012), p. 012011, arXiv:1112.4885, doi:10.1088/1742-6596/343/1/012011
20. T. Asselmeyer-Maluga, H. Rosé, On the geometrization of matter by exotic smoothness. Gen. Rel. Grav. **44**, 2825–2856 (2012), doi:10.1007/s10714-012-1419-3, arXiv:1006.2230
21. N. Berline, M. Vergne, E. Getzler, *Heat kernels and Dirac Operators* (Springer, New York, 1992)
22. P. Bertozzini, R. Conti, W. Lewkeeratiyutkul, Modular theory, non-commutative geometry and quantum gravity. SIGMA **6**, 47pp (2010), arXiv:1007.4094
23. Z. Bizaca, An explicit family of exotic Casson handles. Proc. AMS **123**, 1297–1302 (1995)
24. Ž. Bižaca, R Gompf, Elliptic surfaces and some simple exotic $\mathbb{R}^4$'s. J. Diff. Geom. **43**, 458–504 (1996)
25. H.J. Borchers, On revolutionizing quantum field theory with Tomita's modular theory. J. Math. Phys. **41**, 3604–3673 (2000)
26. R. Bott, L.W. Tu, *Differential Forms in Algebraic Topology*. Graduate Texts in Mathematics, vol. 82 (Springer, 1995)
27. B.H. Bowditch, Markoff triples and quasifuchsian groups. Proc. Lond. Math. Soc. **77**, 697–736 (1998)
28. C. Branciard, How Ψ-epistemic models fail at explaining the indistinguishability of quantum states. Phys. Rev. Lett. **113**, 020409 (2014), doi:10.1103/PhysRevLett.113.020409
29. C. Brans, Absolulte spacetime: the twentieth century ether. Gen. Rel. Grav. **31**, 597 (1999)
30. C.H. Brans, Exotic smoothness and physics. J. Math. Phys. **35**, 5494–5506 (1994)
31. C.H. Brans, Localized exotic smoothness. Class. Quant. Grav. **11**, 1785–1792 (1994)
32. C.H. Brans, D. Randall, Exotic differentiable structures and general relativity. Gen. Rel. Grav. **25**, 205 (1993)
33. R. Budney, JSJ-decompositions of knot and link complements in the 3-sphere. L'enseignement Mathématique **52**, 319–359 (2006), arXiv:math/0506523
34. D. Bullock, A finite set of generators for the Kauffman bracket skein algebra. Math. Z. **231**, 91–101 (1999)
35. D. Bullock, C. Frohman, J. Kania-Bartoszyńska, Topological interpretations of lattice gauge field theory. Comm. Math. Phys. **198**, 47–81 (1998), arXiv:q-alg/9710003
36. D. Bullock, C. Frohman, J. Kania-Bartoszyńska, Understanding the Kauffman bracket skein module. J. Knot Theory Ramif. **8**, 265–277 (1999), arXiv:q-alg/9604013
37. D. Bullock, C. Frohman, J. Kania-Bartoszyńska, The Kauffman bracket skein as an algebra of observables. Proc. AMS **130**, 2479–2485 (2002), arXiv:math/0010330
38. D. Bullock, J.H. Przytycki, Multiplicative structure of Kauffman bracket skein module quantization. Proc. AMS **128**, 923–931 (1999)
39. A. Casson, *Three Lectures on New Infinite Constructions in 4-Dimensional Manifolds*, vol. 62. Birkhäuser, Progress in Mathematics Edition (1986). Notes by Lucian Guillou, first published 1973
40. V. Chernov, S. Nemirovski, Smooth cosmic censorship. Comm. Math. Phys. **320**, 469–473 (2013), arXiv:1201.6070
41. A. Connes, A survey of foliations and operator algebras. Proc. Symp. Pure Math. **38**, 521–628 (1984). See www.alainconnes.org
42. A. Connes, *Non-commutative Geometry* (Academic Press, 1994)
43. A. Connes, C. Rovelli, Von neumann algebra automorphisms and time-thermodynamics relation in general covariant quantum theories. Class. Quant. Grav. **11**, 2899–2918 (1994), arXiv:gr-qc/9406019
44. M. Culler, P.B. Shalen, Varieties of group representations and splittings of 3-manifolds. Ann. Math. **117**, 109–146 (1983)

45. G. Daskalopoulos, S. Dostoglou, R. Wentworth, Character varieties and harmonic maps to r-trees. Math. Res. Lett. **5**, 523–533 (1998)
46. S. DeMichelis, M.H. Freedman, Uncountable many exotic R^4's in standard 4-space. J. Diff. Geom. **35**, 219–254 (1992)
47. S. Donaldson, An application of gauge theory to the topology of 4-manifolds. J. Diff. Geom. **18**, 279–315 (1983)
48. S. Donaldson, Irrationality and the h-cobordism conjecture. J. Diff. Geom. **26**, 141–168 (1987)
49. C. Duston, Exotic smoothness in 4 dimensions and semiclassical Euclidean quantum gravity. Int. J. Geom. Meth. Mod. Phys. **8**, 459–484 (2010), arXiv:0911.4068
50. Fermi GBM/LAT Collaborations, Testing Einstein's special relativity with Fermi's short hard gamma-ray burst GRB090510. Nature **462**, 331–334 (2009), arXiv:0908.1832
51. R. Fintushel, R. Stern, Knots, links, and 4-manifolds. Inv. Math, **134**, 363–400 (1998), arXiv:dg-ga/9612014
52. A. Floer, An instanton invariant for 3-manifolds. Comm. Math. Phys. **118**, 215–240 (1988)
53. M.H. Freedman, The topology of four-dimensional manifolds. J. Diff. Geom. **17**, 357–454 (1982)
54. M.H. Freedman, The disk problem for four-dimensional manifolds. Proc. Internat. Cong. Math. Warzawa **17**, 647–663 (1983)
55. M.H. Freedman, There is no room to spare in four-dimensional space. Not. Am. Math. Soc. **31**, 3–6 (1984)
56. D. Friedan, Nonlinear models in $2 + \epsilon$ dimensions. Phys. Rev. Lett. **45**, 1057–1060 (1980)
57. T. Friedrich, On the spinor representation of surfaces in euclidean 3-space. J. Geom. Phys. **28**, 143–157 (1998), arXiv:dg-ga/9712021v1
58. C. Frohman, R. Gelca, Skein modules and the noncommutative torus. Trans. AMS **352**, 4877–4888 (2000), arXiv:math/9806107
59. S. Ganzell, Ends of 4-manifolds. Top. Proc. **30**, 223–236 (2006), http://faculty.smcm.edu/sganzell/ends.pdf
60. D. Gabai, Foliations and the Topology of 3-Manifolds. J. Diff. Geom. **18**, 445–503 (1983)
61. W.M. Goldman, The symplectic nature of the fundamental groups of surfaces. Adv. Math. **54**, 200–225 (1984)
62. M. Golubitsky, V. Guillemin, *Stable Mappings and their Singularities*. Graduate Texts in Mathematics, vol. 14 (Springer, New York-Heidelberg-Berlin, 1973)
63. R. Gompf, Infinite families of casson handles and topological disks. Topology **23**, 395–400 (1984)
64. R. Gompf, An infinite set of exotic $\mathbb{R}^4$'s. J. Diff. Geom. **21**, 283–300 (1985)
65. R. Gompf, Periodic ends and knot concordance. Top. Appl. **32**, 141–148 (1989)
66. R.E. Gompf, Three exotic R^4's and other anomalies. J. Diff. Geom. **18**, 317–328 (1983)
67. R.E. Gompf, S. Singh, On Freedman's reimbedding theorems, in *Four-Manifold Theory*, vol. 35, ed. by C. Gordan, R. Kirby (AMS, Providence Rhode Island, 1984), pp. 277–310
68. R.E. Gompf, A.I. Stipsicz, *4-manifolds and Kirby Calculus* (American Mathematical Society, 1999)
69. S.W. Hawking, The path-integral approach to quantum gravity, in *General Relativity. An Einstein Centenary Survey*, ed. by I. Hawking (Cambridge University Press, Cambridge, 1979), pp. 746–789
70. M.W. Hirsch, *Differential Topology* (Springer, New York, 1976)
71. J. Hubbard, H. Masur, Quadratic differentials and foliations. Acta Math. **142**, 221–274 (1979)
72. S. Hurder, A. Katok, Secondary classes and trasnverse measure theory of a foliation. Bull. AMS **11**, 347–349 (1984). Announced results only
73. W. Jaco, P. Shalen, *Seifert Fibered Spaces in 3-manifolds*, Memoirs of the American Mathematical Society, vol. 21 (AMS, 1979)
74. T. Kato, ASD moduli space over four-manifolds with tree-like ends. Geom. Top. **8**, 779–830 (2004), arXiv:math.GT/0405443
75. M.S. Leifer, Is the quantum state real? an extended review of ψ-ontology theorems. Quanta **3**, 67–155 (2014), doi:10.12743/quanta.v3i1.22

76. J. Mather, Stability of C^∞ mappings. VI: the nice dimensions, in *Proccedings of the Liverpool Singularities Symposium*, pp. 207–253. Springer Lecture Notes in Mathematyics, vol. 192 (1971)
77. J. Milnor, A unique decomposition theorem for 3-manifolds. Am. J. Math. **84**, 1–7 (1962)
78. G.D. Mostow, Quasi-conformal mappings in *n*-space and the rigidity of hyperbolic space forms. Publ. Math. IHÉS **34**, 53–104 (1968)
79. R. Penrose, Nonlinear gravitons and curved twistor theory. Gen. Relativ. Grav. **7**, 31–52 (1976)
80. H. Pfeiffer, Quantum general relativity and the classification of smooth manifolds. Report number: DAMTP 2004-32 (2004)
81. V.V. Prasolov, A.B. Sossinisky, *Knots, Links, Braids and 3-Manifolds* (AMS, Providence, 1997)
82. B.L. Reinhart, J.W. Wood, A metric formula for the Godbillon-Vey invariant for foliations. Proc. AMS **38**, 427–430 (1973)
83. M. Ringbauer, B. Duffus, C. Branciard, E.G. Cavalcanti, A.G. White, A. Fedrizzi, Measurements on the reality of the wavefunction. Nat. Phys. **11**, 249–254 (2015), doi:10.1038/NPHYS3233
84. P. Scott, The geometries of 3-manifolds. Bull. Lond. Math. Soc. **15**, 401–487 (1983)
85. A.R. Skovborg, The Moduli Space of Flat Connections on a Surface Poisson Structures and Quantization. Ph.D. thesis, Universty Aarhus (2006)
86. J. Sładkowski, Strongly gravitating empty spaces. Preprint arXiv:gr-qc/9906037 (1999)
87. J. Sładkowski, Gravity on exotic $\mathbb{R}^4$ with few symmetries. Int. J. Mod. Phys. D **10**, 311–313 (2001)
88. J. Stallings, Piecewise-linear structure of euclidean space. Proc. Camb. Phil. Soc. **58**, 481 (1962)
89. A.A. Starobinski, A new type of isotropic cosmological models without singularity. Phys. Lett. **91B**, 99–102 (1980)
90. K. Strebel, *Quadratic Differentials*. A Series of Modern Surveys in Mathematics, vol. 5 (Springer, Berlin-Heidelberg, 1984)
91. M. Takesaki, *Tomita's Theory of Modular Hilbert Algebras and its Applications*, Lecture Notes in Mathematics, vol. 128 (Springer, Berlin, 1970)
92. I. Tamura, *Topology of Foliations: An Introduction*. Translations of Mathematical Monographs, vol. 97 (AMS, Providence, 1992)
93. C.H. Taubes, Gauge theory on asymptotically periodic 4-manifolds. J. Diff. Geom. **25**, 363–430 (1987)
94. W. Thurston, Noncobordant foliations of S^3. Bull. AMS **78**, 511–514 (1972)
95. W. Thurston, *Three-Dimensional Geometry and Topology*, 1st edn. (Princeton University Press, Princeton, 1997)
96. M. Tomita, On canonical forms of von neumann algebras, in *Fifth Functional Analysis Sympos (Tôhoku Univ., Sendai, 1967)*, pp. 101–102, Sendai (1967). Tôhoku Univ., Math. Inst
97. V. Turaev, Algebras of loops on surfaces, algebras of knots, and quantization. Adv. Ser. Math. Phys. **9**, 59–95 (1989)
98. V.G. Turaev, Skein quantization of poisson algebras of loops on surfaces. Ann. Sci. de l'ENS **24**, 635–704 (1991)
99. J.H.C. Whitehead, A certain open manifold whose group is unity. Quart. J. Math. Oxf. **6**, 268–279 (1935)
100. B. Whitt, Fourth order gravity as general relativity plus matter. Phys. Lett. **145B**, 176–178 (1984)
101. D.K. Wise, Symmetric space Cartan connections and gravity in three and four dimensions. SIGMA **5**, 080 (2009), arXiv:0904.1738
102. D.K. Wise, Macdowell-Mansouri gravity and Cartan geometry. Class. Quantum Grav. **27**, 155010 (2010), arXiv:gr-qc/0611154
103. E. Witten, 2+1 dimensional gravity as an exactly soluble system. Nucl. Phys. **B311**, 46–78 (1988/89)

104. E. Witten, Topology-changing amplitudes in 2+1 dimensional gravity. Nucl. Phys. B **323**, 113–140 (1989)
105. E. Witten, Quantization of Chern-Simons gauge theory with complex gauge group. Comm. Math. Phys. **137**, 29–66 (1991)
106. M. Wolf, Harmonic maps from surfaces to **R**-trees. Math. Z. **218**, 577–593 (1995)

Printed in the United States
By Bookmasters